Sustainable Viticulture: Soil Fertility, Plant Nutrition and Grape Quality

Sustainable Viticulture: Soil Fertility, Plant Nutrition and Grape Quality

Guest Editors

Miguel A. Olego
Roberto Lopez
Fernando Visconti Reluy

MDPI

Basel • Beijing • Wuhan • Barcelona • Belgrade • Novi Sad • Cluj • Manchester

Guest Editors
Miguel A. Olego
Research Institute of Vine and Wine
Universidad de León
León
Spain

Roberto Lopez
Research Institute of Vine and Wine
Universidad de León
León
Spain

Fernando Visconti Reluy
Centre for the Development of Sustainable Agriculture
Valencian Institute of Agricultural Research
Moncada
Spain

Editorial Office
MDPI AG
Grosspeteranlage 5
4052 Basel, Switzerland

This is a reprint of the Special Issue, published open access by the journal *Horticulturae* (ISSN 2311-7524), freely accessible at: https://www.mdpi.com/journal/horticulturae/special_issues/8GKWO935XF.

For citation purposes, cite each article independently as indicated on the article page online and as indicated below:

Lastname, A.A.; Lastname, B.B. Article Title. *Journal Name* **Year**, *Volume Number*, Page Range.

ISBN 978-3-7258-4671-9 (Hbk)
ISBN 978-3-7258-4672-6 (PDF)
https://doi.org/10.3390/books978-3-7258-4672-6

Cover image courtesy of Miguel A. Olego

Contents

horticulturae

MDPI

Editorial

Special Issue: 'Sustainable Viticulture: Soil Fertility, Plant Nutrition and Grape Quality'

Fernando Visconti [1], Roberto López [2] and Miguel Ángel Olego [2,*]

[1] Centre for the Development of Sustainable Agriculture, Valencian Institute of Agricultural Research, Crta. CV-315, km. 10.7, Moncada, 46113 València, Spain; visconti_fer@gva.es

[2] Research Institute of Vine and Wine, Universidad de León, Avenida de Portugal, 41, 24071 León, Spain; rlopg@unileon.es

* Correspondence: molem@unileon.es

1. Introduction

Sustainable viticulture is the practice of growing grapes indefinitely—for winemaking, juice production, or fresh or dry consumption—while upholding the highest standards of product quality and environmental stewardship [1–3]. Achieving this goal requires the adoption of integrated soil, water, nutrient, vine, and post-harvest management practices that enhance vineyard health, protect the environment, and promote consumer well-being, in line with the "One Health–One World" paradigm [4].

While the concept of sustainability in viticulture is readily defined, its practical implementation is considerably more complex. The multifaceted interactions among viticultural production factors, coupled with the wide variability of grape-growing conditions globally, present significant challenges for viticulturists [5]. Striking a balance between production and grape quality objectives, while optimizing the vineyard's edaphoclimatic potential, requires careful management to avoid compromising long-term vineyard viability and environmental sustainability [6–8], [Contribution 10].

Therefore, addressing the challenge of sustainability in viticulture requires a robust foundation of knowledge, and it is not surprising that science can serve as the critical source of insights capable of driving meaningful progress. This need has been clearly recognized by the agricultural research community, as evidenced by the remarkable rise in scientific publications tackling the subject of viticulture sustainability, from only 1 paper in 1999 to 214 in 2024 according to a SCOPUS search for the words viticult* and sustainab* in the title, the abstract and keywords of papers indexed therein.

It is within this context that this Special Issue was conceived. It was launched with the aim of gathering a representative collection of recent research advances in the pursuit of sustainable viticulture. As a result, it features ten new studies focused on the impacts of several practices dealing with soil, water, nutrients, vines, and grapes on the health of viticultural systems.

These investigations vary in their methodological approaches depending on the specific research questions addressed in each case, encompassing three of the most commonly used types in scientific research: primary experimental [Contribution 1, Contribution 4, Contributions 8–9], primary observational [Contributions 2–3, Contributions 5–7], and secondary research [Contribution 10]. Notably, primary modeling research is the only major category not represented.

Furthermore, the primary investigations in this Special Issue originate from a variety of viticultural regions scattered throughout the world within the 34–43° N and the 22–31° S latitude belts, featuring three continents, in which Europe is the most represented, with four

 https://doi.org/10.3390/horticulturae11060649

works developed in southern Italy [Contribution 1], northwestern Spain [Contributions 7–8], and northern–central Spain [Contribution 3]; followed by the Americas with three works developed in southwest USA [Contribution 2] and southern Brazil [Contribution 6, Contribution 9]; and finally Asia, with two works developed in northern–central China [Contribution 5] and southern–central Turkey [Contribution 4].

As a result of their global distribution, the investigations presented in this Special Issue reflect substantial diversity in environmental conditions, which in turn influence management strategies for grape cultivation. These include vineyard altitudes ranging from 50 to 1400 m and climate characteristics—according to the soil temperature and moisture regimes defined by Soil Survey Staff [9]—spanning from mesic to thermic temperature regimes and from perudic to aridic moisture regimes, encompassing all the intermediate udic, ustic, and xeric, leading to a range of water management strategies from rainfed to fully irrigated systems. Finally, the soil properties also vary widely, with pH values ranging from 4.6 to 8.1, clay varying from 10 to 35%, sand from 30 to 70%, and soil organic matter (SOM) from 0.3% to 9.0%.

In the next section, an overview of the research articles included in this Special Issue is provided. For clarity and ease of reference, the articles are grouped into three thematic categories: (1) Soil Management for Vineyard Health, (2) Soil Water, Nutrients, and Vine Management for Grape Yield and Quality, and (3) Post-Harvest Treatments for Grape Quality.

2. Overview of the Published Articles

2.1. Soil Management for Vineyard Health

The adoption of conservation and regenerative agricultural practices (CRAPs) is essential for improving soil health both on site and beyond. For perennial crops like grapevines, soil-centered minimum-to-no-tillage CRAPs can be broadly categorized into four main types. These are (i) mulching with organic materials, (ii) incorporating organic matter into the soil, (iii) inoculating the soil with beneficial organisms, and (iv) establishing cover crops [Contribution 10]. In this Special Issue, the impacts of these four soil-centered minimum-to-no-tillage CRAPs in viticulture are reviewed [Contribution 10]. In addition, this Special Issue also includes newly focused studies about the effects on soil health of two of these CRAPs, namely mulching [Contribution 1] and cover crop establishment [Contribution 8]. Additionally, novel investigations addressing the effects on vineyards' soil health of irrigating with untreated wastewater [Contribution 2] and of SOM in zinc-polluted vineyards [Contribution 7] are presented.

The synthesis of primary research by Visconti et al. [Contribution 10] reviews the factors that affect a vineyard's soil health in depth. This is a narrative review in which the authors address the concept of soil health from a holistic perspective and how it applies to the special case of viticulture. In fact, wine, the most precious produce of vineyards, boasts obtaining some of its most distinctive traits from the soil on which grapevines are cultivated. These traits are wrapped within the terroir; however, since viticulture also makes harsh impacts on soils, is the terroir not threatened? The authors try to address this inconvenient question by presenting the different aspects into which soil health splits, how conventional agricultural practices affect soil health, how it could be enhanced, how it is related to the terroir, and how soil health status can be assessed.

One significant threat to soil health in vineyards is copper (Cu) accumulation, resulting from its widespread use as the active ingredient in grapevine fungicides [10–12]. In this Special Issue, the primary experimental, climate-chamber-based study by Vázquez-Blanco et al. [Contribution 8] explored the potential of perennial ryegrass (*Lolium perenne* L.) to enhance soil health, including its likely capacity for Cu phytoremediation, across a broad range of soils representative of the renowned northwestern Spanish viticultural region. Perennial ryegrass is revealed to exhibit a notable ability to exclude Cu uptake, and therefore it is attributed with a low potential for effective phytoremediation. However, viewed from another perspective, the strong performance of perennial ryegrass across diverse soil types, ranging from acidic to mildly alkaline, and with varying levels of Cu pollution, suggests that it may be highly suitable as a cover crop, particularly where soil Cu is high. Specifically, its use could contribute to soil health improvement in vineyards, capitalizing on the multiple soil ameliorating effects of cover crops [13–15], while posing minimal risk of Cu mobilization to the other environmental compartments. Particularly significant in this regard is its role in protecting the biota from Cu toxicity.

Zinc (Zn) accumulation poses another heavy metal risk to vineyards' soil health, primarily due to its use as an alternative or complement to Cu in managing fungal diseases in viticulture [16]. In this Special Issue, the primary observational, survey-based study by Pérez-Rodríguez et al. [Contribution 7] investigates Zn availability in vineyard soils and how it may shift following afforestation resulting from vineyard abandonment. Interestingly, SOM, particularly its degree of transformation (humification), as indicated by decreasing carbon-to-nitrogen (C/N) ratios, emerges as a key factor influencing Zn bioavailability and mobility in the environment. This occurs because, as SOM mineralizes, it releases low-molecular-weight, water-soluble organic ligands that complex $Zn2+$, stabilizing this heavy metal in the soil solution and hence increasing its availability and mobility. This investigation, therefore, highlights a case in which an increase in SOM might negatively impact soil and environmental health, by enhancing the mobilization of potentially toxic elements [17,18].

In addition to cover crop development, organic mulching is another CRAP that can promote soil health recovery in vineyards, since it has been shown to be particularly effective in mitigating raindrop impact and in buffering thermal and moisture soil regimes, with eventual SOM-enhancing effects in agriculture [19,20]. This Special Issue features a primary experimental, field-based study by Blanco et al. [Contribution 1], which examines the impact of mulching with municipal solid waste (MSW) compost on the row line soil health of a 5-year-old vineyard situated on flat terrain in southern Italy. The cooling effect of MSW compost on soil temperature in this hot summer environment was clearly evident. However, the anticipated SOM-increasing impact from this thermal buffering was less apparent. It is hypothesized that the likely SOM boost may be offset by a "microbial inoculation effect" associated with the MSW compost, potentially stimulating SOM decomposition. Notwithstanding, further research is needed, as the potential contribution of the MSW compost itself, as well as that of vine residues, to the fresh SOM pool, and consequently to overall SOM enhancement, was not assessed in this investigation.

In addition to solid organic materials like composts, untreated wastewater may serve as an effective source of both organic matter and nutrients to improve vineyard SOM and soil nutrients' concentrations in available form [21]. In this Special Issue, Mpanga et al. [Contribution 2] explore this possibility through a primary observational, field-based study conducted in a 20-year-old vineyard in central–northern Arizona (USA). The vineyard is irrigated with two types of water, ditch and fishpond, at an approximate rate of 3.3 mm per week. Although both water sources are similar in physicochemical properties, the fishpond water appears to contribute organic matter to the soil due to its algae content. This characteristic may explain how soils irrigated with fishpond water outperform those receiving ditch water in terms of SOM, phosphorus, and soil respiration, factors that are ultimately associated with a remarkably enhanced grape yield. However, further research is required, given the inherent limitations of observational studies in establishing causal relationships [22,23], and also given the associated health hazards that untreated wastewaters may bear [21].

2.2. Soil Water, Nutrients, and Vine Management for Grape Yield and Quality

Soil management is not the only factor influencing vineyard health and terroir—soil water and canopy management also play crucial roles [Contribution 10], above all under water scarcity conditions. Therefore, a comprehensive approach that integrates all three aspects is essential for developing sustainable viticultural systems. Specifically, the management of the physical and chemical fertility of the soil significantly regulates vine vigor, berry development, and ultimately, grape yield and quality [24,25]. Additionally, the regulation of grapevine fertility through pruning has a substantial impact on these same traits [26–28]. In this Special Issue, three newly focused studies explore the effects of soil physical fertility attributes and pruning practices on grapevine performance.

First, the study by Martínez-Vidaurre et al. [Contribution 3] examined how water stress—assessed through the model-simulated ratio of available soil water (ASW) to soil water-holding capacity (SWHC)—influences vine nutritional status, vigor, yield, and grape and wine composition. This primary observational field-based study was conducted in four rain-fed *Vitis vinifera* cv. 'Tempranillo' vineyards, aged 20 to 35 years, located in Spain's Rioja wine region, whose vines were grown over five seasons. The results showed that soils with a higher SWHC increased the ASW-to-SWHC ratio, thus reducing grapevine water stress, particularly during the critical fruit set to ripening stage, which in turn enhanced nutrient status, vine vigor, and grape yield, but led to lower total soluble solids (TSSs) in the grapes. Specifically, the period between veraison and ripening was identified as particularly sensitive to water stress, as increases in the ASW-to-SWHC ratio during this stage led to higher berry weight and acidity, alongside reductions in the anthocyanin concentration and total polyphenols. Interestingly, the effects of water stress on grape composition were reflected in the wine for anthocyanins and polyphenols, but not for TSSs (i.e., alcohol concentration) or acidity.

Consistent with the findings of Martínez-Vidaurre et al. [Contribution 3], Uriarte et al. [29] have observed in a mega-analysis that lower water stress—typically associated with higher irrigation rates—leads to overall increased yield and vigor, but to reduced TSSs and concentrations of anthocyanins and phenolics in grapes. Regarding the impact of water stress timing, Theocharis et al. [30] observed that when it happens prior to veraison, it can enhance anthocyanin accumulation in red grape varieties.

Therefore, the findings of Martínez-Vidaurre et al. [Contribution 3] reinforce the broader consensus on the effects of water stress on grapevine development and quality, and that the effects on grape composition—and consequently on wine quality—are modulated by the phenological stage at which the stress occurs. Furthermore, the study by Martínez-Vidaurre et al. [Contribution 3] underscores the importance of developing integrated water management strategies that consider the SWHC, which depends on the product of the soil effective depth (SED), one minus the coarse fragment fraction ($1—w_{CF}$), and the difference between the soil water content at field capacity and at wilting point ($\theta_{FC}–\theta_{WP}$). Therefore, this study suggests that rising SED, $1—w_{CF}$ or $\theta_{FC}–\theta_{WP}$ —such as by regeneratively avoiding soil compaction, removing gravels, stones, and cobbles from the subsurface soil layers, and increasing SOM —will be essential for sustaining grape yield and quality amid increasingly variable and arid climatic conditions, thereby supporting the long-term sustainability of viticulture, particularly in the context of climate change.

The ASW depends, in addition to the SWHC, on climate wetness. However, edapho-climatic conditions may impact grapevines not only through their effects on the ASW, but through the rest of the soil properties that traditionally define fertility and, specifically, chemical fertility and, furthermore, through all climate features. In this issue, Andrade et al. [Contribution 6] investigated how grapevine, soil, and climate characteristics influence yield variation by means of several statistical methodologies with the aim of reducing the uncertainty of grape yield predictions during pre-harvest negotiations, thus supporting more informed decision-making throughout the viticultural supply chain. This primary observational field-based study was conducted on rain-fed grapevines aged 2 to 40 years, encompassing 27 cultivars grown for 14 seasons in a commercial vineyard located in the Campanha Gaúcha emerging wine region in southern Brazil. Among the most influential factors driving yield variation were plant age, plot location, start-of-dormancy temperatures, and the available soil concentrations of key nutrients (Zn, Cu, potassium (K), and manganese (Mn)). Interestingly, SOM did not emerge as an influential factor, which may be due to the overall low average SOM content (0.75%), in conjunction with its narrow range of variation in the study area (0.5–1.5%). Regarding the macronutrients, the absence of N data in the study dataset unfortunately prevented the evaluation of its effect, while the likely sufficient supply of P in the vineyard, due to the site's commercial management, may have masked any potential impact.

Besides Andrade et al. [Contribution 6], the impact of soil characteristics on grapevine performance is further explored in this Special Issue by Li et al. [Contribution 5]. Their primary observational field study was conducted across six *Vitis vinifera* cv. 'Cabernet Franc' vineyards located in the rapidly developing wine-producing region on the Eastern Foothills of the Helan Mountains in China. In contrast to the study by Andrade et al. [Contribution 6], the work by Li et al. [Contribution 5] expands the analysis beyond overall yield to include yield components and grape quality traits, encompassing both technological and phenolic ripeness characteristics. Yet, consistent with Andrade et al. [Contribution 6], Li et al. [Contribution 5] found that SOM —with an average of 1.2% and a range of 0.3–3.7%—was not associated with grape yield, despite being strongly associated with the soil concentrations of macronutrients—nitrate nitrogen ($N\text{-}NO_3^-$) and P—as well as with available micronutrients such as iron (Fe) and manganese (Mn). However, SOM emerged as particularly linked to grape quality.

Interestingly, grape technological ripeness traits were separated into two groups: TSS was moderately associated with Mn and Zn, while acidity-related traits were more strongly linked to SOM and its associated nutrients in the study area (N, P, Fe, and Mn), along with Zn. Phenolic ripeness traits also exhibited a split: tannins and anthocyanins were moderately associated with SOM and its related nutrients (N, P, Fe, and Mn), whereas flavonol concentration was also associated with SOM and the same nutrients, but more strongly, and additionally, to soil pH.

The study concludes that tailored soil management strategies focusing on SOM enhancement and balanced macronutrient levels can optimize grape composition and quality. In this regard, limits for optimal berry traits were calculated—e.g., SOM between 2 and 3% and total nitrogen (TN) above 0.1%—thereby offering actionable guidelines to improve phenolic richness and acidity balance in wine grapes [Contribution 5].

In addition to the management of soil, soil water, and nutrients, canopy management plays a vital role in developing sustainable viticultural systems [Contribution 10]. Particularly, pruning stands out as one of the most influential management practices on grapevine health and performance by impacting disease susceptibility, vigor, yield, and grape quality [28,31].

In this issue, Sánchez et al. [Contribution 9] examined how pruning severity, i.e., the number of buds left per cane, at two levels, namely, short (spur) and mixed (a combination of spur and cane pruning), affects the duration of phenological stages, thermal requirements, and ripening progression in three grape juice cultivars ('Bordô', 'BRS Cora', and

'BRS Violeta'). This primary experimental, field-based study was conducted on irrigated grapevines over two growing seasons in a vineyard located in São Paulo, southeastern Brazil, under a humid subtropical climate. The results showed that there were no significant differences between short pruning (one to three buds left per cane) and mixed pruning (a combination of one to three and four to twelve buds left) with regard to the duration of phenological stages, thermal requirements, or the progression of ripening with the exception of the ripeness index, i.e., the ratio of TSSs to titratable acidity (TA), which was overall significantly lower for mixed pruning.

The diminishing effect of mixed pruning on the ripeness index warrants further investigation, particularly in wine growing, as it may help to counteract the climate change-driven trend of increasing the TSS-to-TA ratio in grapes, which runs counter to the preferences of wine consumers [32]. Moreover, severity is only one of several parameters that define the pruning of grapevines. Another key factor is, e.g., the pruning aggressiveness, which is determined by the amount of protective wood left between the pruning cut and the vine's permanent structures, as well as whether the vine's preferential sap flow pathways are preserved [31]. Therefore, future studies should explore both the individual and combined effects of the factors that define the pruning technique on vine yield and grape quality across diverse training systems, cultivars, trellis designs, and climates. The goal is to refine pruning practices to better align with specific growing conditions, grape yield and quality objectives, and vineyard health priorities.

2.3. *Post-Harvest Treatments for Grape Quality*

Complementing previous research on soil properties, water availability, nutrient dynamics, and canopy management, Kaya et al. [Contribution 4] investigated how post-harvest treatments, differing in environmental sustainability, impact grape quality. Specifically, they examined the effects of two alternative shade-drying methods on the phytochemical composition of grapes destined for raisin production. This primary experimental study, conducted in Hadim and Karaman (southern–central Turkey), focused on the phenolic profiles of 'Gök Üzüm' grapes subjected to hot dipping in either oak ash or potassium carbonate (K_2CO_3) aqueous solutions prior to shade drying.

Oak ash, a by-product of oak wood burning, represents an a priori more sustainable alternative to K_2CO_3, which is an industrially produced chemical. Using advanced analytical techniques (HPLC-DAD-MS and UHPLC-MS/MS), the authors quantified key grape constituents, including anthocyanins, flavonoids, phenolic acids, and phytoalexins.

The results revealed that grapes treated with oak ash exhibited significantly higher levels of most phenolic compounds. For instance, anthocyanidins reached 215 mg/kg under oak ash treatment, compared to 154 mg/kg with K_2CO_3. Similar enhancements were observed for flavonols (68 vs. 50 mg/kg), flavanones (15.1 vs. 11.0 mg/kg), catechins (61.6 vs. 39.8 mg/kg), and resveratrol (33.8 vs. 23.2 mg/kg). Delphinidin-3-O-glycoside, a key anthocyanin, was also significantly more abundant in grapes subjected to oak ash treatment. Principal Component Analysis further supported these compositional differences, showing clear clustering of samples based on the pre-treatment applied.

While both treatments altered the berry composition, oak ash consistently enhanced attributes linked to antioxidant capacity and color stability—critical factors for raisin quality and consumer appeal. In contrast, K_2CO_3 treatment was associated with elevated levels of certain hydroxybenzoic acids (gallic, vanillic, etc.), which feature antioxidant, anti-inflammatory, cardioprotective, neuroprotective, anticancer, and antimicrobial potential [33,34], although overall phenolic retention was lower. These findings highlight the sensitivity of grape composition to brief chemical pre-treatments and underscore the importance of aligning processing methods with desired nutritional and functional outcomes in grape products.

Future research might also explore the potential introduction of harmful substances into the food chain resulting from the use of wood ashes in post-harvest treatments. These ashes may be enriched with heavy metals and some xenobiotic organic compounds, depending on the wood's composition and its combustion conditions [35]. Furthermore, upcoming studies might incorporate comprehensive Life Cycle Assessments (LCAs) and Carbon Footprint analyses to ensure a thorough evaluation of environmental impacts. By addressing these critical aspects of grape, raisin, grape juice, and wine production, holistic "One Health" assessments can be developed, ultimately maximizing viticultural and societal benefits.

3. Conclusions

As demonstrated by the studies in this Special Issue, modern viticulture is inherently multifaceted. The effective management of soil, water, nutrients, pruning, and post-harvest practices is essential not only for optimizing grape yield and quality, but also for promoting environmental sustainability.

To frame the subject, Visconti et al. [Contribution 10] examined the pivotal role of soil health in sustaining vineyard productivity and terroir expression. Their work highlights the detrimental impacts of conventional viticultural practices on soil properties, the benefits of enhancing SOM and soil biological activity, and the pressing need for sustainable management strategies to ensure long-term viticultural and environmental viability.

In this context, SOM consistently emerges as a key determinant of vineyard health throughout the contributions to this Special Issue. As shown by Mpanga et al. [Contribution 2], SOM plays a crucial role in improving soil physical, chemical, and biological fertility. However, as Pérez-Rodríguez et al. [Contribution 7] illustrate, SOM may also contribute to undesirable outcomes, particularly in vineyards affected by heavy metal pollution (e.g., Zn contamination), emphasizing the need for site-specific management approaches.

Addressing soil biological fertility, Blanco et al. [Contribution 1] demonstrated the potential of MSW compost in mitigating soil temperature fluctuations, thereby enhancing soil microbial activity. These effects may bolster plant resilience and support environmental sustainability in grapevines, particularly when exposed to warm climates. In parallel, Vázquez-Blanco et al. [Contribution 8] explored the use of cover crops in vineyards with coarse-textured soils and acidic to mildly alkaline pH values, showing their potential to improve soil and environmental health, particularly where elevated soil Cu levels have become a concern.

Beyond soil management, several studies emphasized the critical role of soil water availability. Martínez-Vidaurre et al. [Contribution 3] demonstrated how soil water status influences vine health and berry composition, underscoring the importance of sustainably enhancing the soil water-holding capacity and precisely managing plant water stress during key phenological stages to balance yield and grape quality. Advancing the field, Andrade et al. [Contribution 6] showcased the application of machine learning algorithms to predict grape yield in support of early marketing decisions, illustrating the potential of data-driven approaches in viticultural management. Similarly, Li et al. [Contribution 5] examined the influence of soil characteristics on grape yield and quality, further emphasizing the importance of site-specific soil management.

In addition to soil, water, and nutrients, pruning practices also play a significant role in vineyard sustainability. Sánchez et al. [Contribution 9] investigated how different pruning methods affect berry chemistry, particularly in subtropical climates, providing insights into how canopy management can influence grape maturation and quality. Finally, Kaya et al.

[Contribution 4] explored the modulation of grape phytochemical composition through post-harvest treatments, particularly comparing two shade-drying methods of differing environmental sustainability for raisin production.

Collectively, these studies reflect a clear and ongoing shift towards sustainable viticultural practices, particularly regarding soil fertility, plant nutrition, and grape quality. We believe that this collection of articles provides both valuable insights and practical guidance for viticulturists, agronomists, and researchers striving to enhance vineyard performance while fostering resilient and eco-friendly viticultural systems.

Conflicts of Interest: The authors declare no conflicts of interest.

List of Contributions

1. Blanco, I.; Cardinale, M.; Domanda, C.; Pappaccogli, G.; Romano, P.; Zorzi, G.; Rustioni, L. Mulching with Municipal Solid Waste (MSW) Compost Has Beneficial Side Effects on Vineyard Soil Compared to Mulching with Synthetic Films. *Horticulturae* **2024**, *10*, 769. https://doi.org/10.3390/horticulturae10070769
2. Mpanga, I.K.; Ijato, T.; Dapaah, H.K.; Tronstad, R. Fishpond Water Potential on Vineyard Soil Health: An Exploratory Study of a Circular System. Horticulturae 2024, 10, 390. https://doi.org/10.3390/horticulturae10040390
3. Martínez-Vidaurre, J.M.; Pérez-Álvarez, E.P.; García-Escudero, E.; Ramos, M.C.; Peregrina, F. Differences in Soil Water Holding Capacity and Available Soil Water along Growing Cycle Can Explain Differences in Vigour, Yield, and Quality of Must and Wine in the DOCa Rioja. Horticulturae 2024, 10, 320. https://doi.org/10.3390/horticulturae10040320
4. Kaya, O.; Delavar, H.; Ates, F.; Yilmaz, T.; Sahin, M.; Keskin, N. Fine-Tuning Grape Phytochemistry: Examining the Distinct Influence of Oak Ash and Potassium Carbonate Pre-Treatments on Essential Components. Horticulturae 2024, 10, 95. https://doi.org/10.3390/horticulturae10010095
5. Li, Y.; Xiao, J.; Yan, Y.; Liu, W.; Cui, P.; Xu, C.; Nan, L.; Liu, X. Multivariate Analysis and Optimization of the Relationship between Soil Nutrients and Berry Quality of *Vitis vinifera* cv. Cabernet Franc Vineyards in the Eastern Foothills of the Helan Mountains, China. Horticulturae 2024, 10, 61. https://doi.org/10.3390/horticulturae10010061
6. Andrade, C.B.; Moura-Bueno, J.M.; Comin, J.J.; Brunetto, G. Grape Yield Prediction Models: Approaching Different Machine Learning Algorithms. Horticulturae 2023, 9, 1294. https://doi.org/10.3390/horticulturae9121294
7. Pérez-Rodríguez, P.; Nóvoa-Muñoz, J.C.; Arias-Estévez, M.; Fernández-Calviño, D. Soil Abandonment as a Trigger for Changes in Zn Fractionation in Afforested Former Vineyard Acidic Soils. Horticulturae 2023, 9, 1121. https://doi.org/10.3390/horticulturae9101121
8. Vázquez-Blanco, R.; Arias-Estévez, M.; Fernández-Calviño, D.; Arenas-Lago, D. Early Growth Assessment of *Lolium perenne* L. as a Cover Crop for Management of Copper Accumulation in Galician Vineyard Soils. Horticulturae 2023, 9, 1029. https://doi.org/10.3390/horticulturae9091029
9. Sánchez, C.A.P.C.; Callili, D.; Carneiro, D.C.d.S.; Silva, S.P.S.d.; Scudeletti, A.C.B.; Leonel, S.; Tecchio, M.A. Thermal Requirements, Phenology, and Maturation of Juice Grape Cultivars Subjected to Different Pruning Types. Horticulturae 2023, 9, 691. https://doi.org/10.3390/horticulturae9060691
10. Visconti, F.; López, R.; Olego, M.Á. The Health of Vineyard Soils: Towards a Sustainable Viticulture. Horticulturae 2024, 10, 154. https://doi.org/10.3390/horticulturae10020154

References

1. International Organisation of Vine and Wine (OIV). Resolution CST 1/2004: General Principles of Sustainable Vitiviniculture-Environmental-Social-Economic. 2004. Available online: https://www.oiv.int/public/medias/2074/cst-1-2004-en.pdf (accessed on 6 June 2025).
2. Robinson, J. (Ed.) *The Oxford Companion to Wine*, 4th ed.; Oxford University Press: Oxford, UK, 2015.
3. Çakıtlı, G. Sustainability in viticulture. *Acta Hortic.* **2024**, *1385*, 157–164. [CrossRef]

4. Calistri, P.; Iannetti, S.; Danzetta, M.L.; Narcisi, V.; Cito, F.; Di Sabatino, D.; Bruno, R.; Sauro, F.; Atzeni, M.; Carvelli, A.; et al. The Components of 'OneWorld-One Health' Approach. *Transbound. Emerg. Dis.* **2013**, *60*, 4–13. [CrossRef] [PubMed]
5. Corino, L.; Calò, A. Sustainable Viticulture: Current Practices and Future Developments. *Agric. Conspec. Sci.* **2001**, *66*, 3–11.
6. Litskas, V.; Mandoulaki, A.; Vogiatzakis, I.N.; Tzortzakis, N.; Stavrinides, M. Sustainable Viticulture: First Determination of the Environmental Footprint of Grapes. *Sustainability* **2020**, *12*, 8812. [CrossRef]
7. Ferreira, C.S.; Veiga, A.; Caetano, A.; Gonzalez-Pelayo, O.; Karine-Boulet, A.; Abrantes, N.; Keizer, J.; Ferreira, A.J. Assessment of the Impact of Distinct Vineyard Management Practices on Soil Physico-Chemical Properties. *Air Soil Water Res.* **2020**, *13*, 1178622120944847. [CrossRef]
8. García-Navarro, F.J.; Jiménez-Ballesta, R.; Perales, J.A.L.; Perez, C.; Amorós, J.A.; Bravo, S. Sustainable Viticulture in the Valdepeñas Protected Designation of Origin: From Soil Quality to Management in *Vitis vinifera*. *Sustainability* **2023**, *15*, 9339. [CrossRef]
9. Soil Survey Staff. *Keys to Soil Taxonomy*, 12th ed.; USDA-Natural Resources Conservation Service: Washington, DC, USA, 2014.
10. Komárek, M.; Čadková, E.; Chrastný, V.; Bordas, F.; Bollinger, J.-C. Contamination of vineyard soils with fungicides: A review of environmental and toxicological aspects. *Environ. Int.* **2010**, *36*, 138–151. [CrossRef]
11. Droz, B.; Payraudeau, S.; Rodríguez Martín, J.A.; Tóth, G.; Panagos, P.; Montanarella, L.; Borrelli, P.; Imfeld, G. Copper Content and Export in European Vineyard Soils Influenced by Climate and Soil Properties. *Environ. Sci. Technol.* **2021**, *55*, 7327–7334. [CrossRef]
12. Neaman, A.; Schoffer, J.-T.; Navarro-Villarroel, C.; Pelosi, C.; Peñaloza, P.; Dovletyarova, E.A.; Schneider, J. Copper contamination in agricultural soils: A review of the effects of climate, soil properties, and prolonged copper pesticide application in vineyards and orchards. *Plant Soil Environ.* **2024**, *70*, 407–417. [CrossRef]
13. Vukicevich, E.; Lowery, T.; Bowen, P.; Úrbez-Torres, J.R.; Hart, M. Cover crops to increase soil microbial diversity and mitigate decline in perennial agriculture. *A review. Agron. Sustain. Dev.* **2016**, *36*, 48. [CrossRef]
14. Blanco-Canqui, H.; Ruis, S.J. Cover crop impacts on soil physical properties: A review. *Soil Sci. Soc. Am. J.* **2020**, *84*, 1527–1576. [CrossRef]
15. Koudahe, K.; Allen, S.C.; Djaman, K. Critical review of the impact of cover crops on soil properties. *Int. Soil Water Conserv. Res.* **2022**, *10*, 343–354. [CrossRef]
16. Tiecher, T.L.; Tiecher, T.; Ceretta, C.A.; Ferreira, P.A.A.; Nicoloso, F.T.; Soriani, H.H.; De Conti, L.; Kulmann, M.S.S.; Schneider, R.O.; Brunetto, G. Tolerance and translocation of heavy metals in young grapevine (*Vitis vinifera*) grown in sandy acidic soil with interaction of high doses of copper and zinc. *Sci. Hortic.* **2017**, *222*, 203–212. [CrossRef]
17. Lo, K.S.L.; Yang, W.F.; Lin, Y.C. Effects of organic matter on the specific adsorption of heavy metals by soil. *Toxicol. Environ. Chem.* **1992**, *34*, 139–153. [CrossRef]
18. Salati, S.; Quadri, G.; Tambone, F.; Adani, F. Fresh organic matter of municipal solid waste enhances phytoextraction of heavy metals from contaminated soil. *Environ. Pollut.* **2010**, *158*, 1899–1906. [CrossRef]
19. Prosdocimi, M.; Tarolli, P.; Cerdà, A. Mulching practices for reducing soil water erosion: A review. *Earth-Sci. Rev.* **2016**, *161*, 191–203. [CrossRef]
20. Visconti, F.; Peiró, E.; Pesce, S.; Balugani, E.; Baixauli, C.; de Paz, J.M. Straw mulching increases soil health in the inter-row of citrus orchards from Mediterranean flat lands. *Eur. J. Agron.* **2024**, *155*, 127115. [CrossRef]
21. Qadir, M.; Wichelns, D.; Raschid-Sally, L.; McCornick, P.G.; Drechsel, P.; Bahri, A.; Minhas, P.S. The challenges of wastewater irrigation in developing countries. *Agric. Water Manag.* **2010**, *97*, 561–568. [CrossRef]
22. Rosenbaum, P.R. *Observational Studies*, 2nd ed.; Springer: New York, NY, USA, 2002; pp. 1–375.
23. Hernán, M.A.; Robins, J.M. *Causal Inference: What If*; Chapman & Hall/CRC: Boca Raton, FL, USA, 2020.
24. White, R.; Balachandra, L.; Edis, R.; Chen, D. The soil component of terroir. *J. Int. Sci. Vigne Vin* **2007**, *41*, 9–18. [CrossRef]
25. Van Leeuwen, C. Terroir: The effect of the physical environment on vine growth, grape ripening and wine sensory attributes. In *Managing Wine Quality: Viticulture and Wine Quality*; Reynolds, A.G., Ed.; Woodhead Publishing: Cambridge, UK, 2010; pp. 273–315. [CrossRef]
26. Morgani, M.B.; Fanzone, M.; Peña, J.E.P.; Sari, S.; Gallo, A.E.; Tournier, M.G.; Prieto, J.A. Late Pruning Modifies Leaf to Fruit Ratio and Shifts Maturity Period, Affecting Berry and Wine Composition in *Vitis vinifera* L. cv. 'Malbec' in Mendoza, Argentina. *Sci. Hortic.* **2023**, *313*, 111861. [CrossRef]
27. Cameron, W.; Petrie, P.R.; Bonada, M. Effects of Vineyard Management Practices on Winegrape Yield Components: A Review Using Meta-Analysis. *Am. J. Enol. Vitic.* **2024**, *75*, 0750007. [CrossRef]
28. Ghiglieno, I.; Facciano, L.; Valenti, L.; Amari, F.; Cola, G. Evaluation of the Impact of Vine Pruning Periods on Grape Production and Composition: An Integrated Approach Considering Different Years and Cultivars. *OENO One* **2025**, *59*, 8239. [CrossRef]
29. Uriarte, D.; Santesteban, L.G.; Mirás-Avalos, J.M.; Buesa, I.; Cancela, J.J.; Chacón, J.L.; Escalona, J.M.; Intrigliolo, D.S.; Lampreave, M.; Montoro, A.; et al. Quantifying the Effects of Water Status on Grapevine Vegetative Growth, Yield, and Grape Composition Through a Collaborative Analysis. *Aust. J. Grape Wine Res.* **2025**, *2025*, 1588228. [CrossRef]

30. Theocharis, S.; Nikolaou, N.; Zioziou, E.; Kyraleou, M.; Kallithraka, S.; Kotseridis, Y.; Koundouras, S. Effects of post-veraison irrigation on the phenolic composition of *Vitis vinifera* L. cv. 'Xinomavro' grapes. *OENO One* **2021**, *55*, 173–189. [CrossRef]
31. Galar-Martínez, M.; Torres, N.; Sebastián, B.; Palacios, J.; Arzoz, I.; Juanena, N.; Villa-Llop, A.; Loidi, M.; Dewasme, C.; Roby, J.P.; et al. Respectful Pruning Improves Grapevine Development: A Case Study in Young Vineyards. *Aust. J. Grape Wine Res.* **2024**, *2024*, 8448405. [CrossRef]
32. Olego, M.Á.; Quiroga, M.J.; Cuesta-Lasso, M.D.; Visconti Reluy, F.; Garzón-Jimeno, E. Auxins seem promising as a tuning method for balancing sugars with acidity in grape musts from cv. Tempranillo, but not defoliation or application of magnesium to leaves. *OENO One* **2023**, *57*, 70–83. [CrossRef]
33. Hadidi, M.; Liñán-Atero, R.; Tarahi, M.; Christodoulou, M.C.; Aghababaei, F. The Potential Health Benefits of Gallic Acid: Therapeutic and Food Applications. *Antioxidants* **2024**, *13*, 1001. [CrossRef]
34. Kaur, J.; Gulati, M.; Singh, S.K.; Kuppusamy, G.; Kapoor, B.; Mishra, V.; Gupta, S.; Arshad, M.F.; Porwal, O.; Jha, N.K.; et al. Discovering Multifaceted Role of Vanillic Acid beyond Flavours: Nutraceutical and Therapeutic Potential. *Trends Food Sci. Technol.* **2022**, *122*, 187–200. [CrossRef]
35. Smołka-Danielowska, D.; Jabłońska, M. Chemical and Mineral Composition of Ashes from Wood Biomass Combustion in Domestic Wood-Fired Furnaces. *Int. J. Environ. Sci. Technol.* **2022**, *19*, 5359–5372. [CrossRef]

horticulturae

MDPI

Article

Mulching with Municipal Solid Waste (MSW) Compost Has Beneficial Side Effects on Vineyard Soil Compared to Mulching with Synthetic Films

Ileana Blanco [1], Massimiliano Cardinale [1,*], Corrado Domanda [1,2], Gianluca Pappaccogli [1], Piergiorgio Romano [1], Gianni Zorzi [3] and Laura Rustioni [1]

1 Department of Biological and Environmental Sciences and Technologies (DiSTeBA), University of Salento, SP6 Lecce-Monteroni, 73100 Lecce, Italy; ileana.blanco@unisalento.it (I.B.); c.domanda@phd.uniss.it (C.D.); gianluca.pappaccogli@unisalento.it (G.P.); piergiorgio.romano@unisalento.it (P.R.); laura.rustioni@unisalento.it (L.R.)

2 Department of Agriculture Sciences, University of Sassari, Viale Italia 39, 07100 Sassari, Italy

3 Dipartimento Ambiente, Centro Ricerca Fondazione Edmund Mach, 38098 San Michele all'Adige (TN), Italy; gianni.zorzi@yahoo.it

* Correspondence: massimiliano.cardinale@unisalento.it

Abstract: Municipal solid waste (MSW) compost represents a sustainable alternative to plastic film for mulching in viticulture. This study investigated the effects of MSW compost on vineyard soil properties, specifically focusing on side effects such as soil temperature and microbial decomposition activity, independently from its role in weed control. The experiment was conducted in a vineyard located in the Mediterranean region (Southern Italy), with six different mulching treatments: black polyethylene (PE) film, black and white biodegradable film, three different amounts of MSW compost (8, 15, and 22 kg plant^{-1}), and a control without mulching. Weed growth was monitored to determine the optimal compost application amount. The 15 kg plant^{-1} treatment was selected for further analyses, as it did not significantly impact weed growth compared to the control. Results indicated that MSW compost mulching maintained lower soil temperatures compared to other treatments (up to 5 °C in the warmest hours) and reduced the amplitude of the thermal wave up to 50% compared to the non-mulched soil and even more compared to black film mulched soil, particularly during the warmest periods. This suggests that MSW compost can mitigate heat stress on plant roots, potentially enhancing plant resilience and preserving crop production also in stressful growing conditions. Microbial decomposition activity, assessed using the tea bag index, was higher in the MSW compost treatment during spring compared to the control, indicating temperature as a key driver for organic matter decomposition, but this effect disappeared during summer. These findings highlight the potential of MSW compost to support sustainable viticulture by reducing reliance on synthetic mulching materials and promoting environmental sustainability through the recycling of organic municipal waste.

Keywords: mediterranean region; organic fraction of municipal solid waste; agricultural plastic films; microbial decomposition activity; tea bag index (TBI); soil temperature; weed control

1. Introduction

Mulching is a key soil management practice based on the application of a layer of material on top of the soil surface [1]. It effectively controls weeds, enhances soil moisture retention [2], and regulates soil temperature [3]. Additionally, it prevents soil erosion caused by abiotic factors [4] and reduces soil compaction from heavy agricultural machinery [5].

Several studies have highlighted the role of mulching in viticulture. Steinmaus et al. [6] and Mairata et al. [7] reported mulching as an alternative method to conventional weed control methods that include mechanical cultivation, herbicide application, and manual control [8]. Mechanical cultivation involves using machinery to till the soil and remove weeds, which can be labor-intensive and potentially harmful to soil structure. Herbicide application involves the use of chemical substances to kill or inhibit weed growth, which can raise environmental and health concerns. Manual control, while effective, is time consuming and not always feasible for large-scale operations [9]. Various authors discussed the contribution of mulching to improving the water use efficiency of grapevine [10] in semi-arid conditions [11–13]. Prosdocimi et al. [14] focused on the role of mulching in reducing soil erodibility and surface runoff. Ferrara et al. [15] observed positive impacts on grapevine physiology and soil fertility using exhausted olive pomace as an organic mulch. Alternative organic mulches, such as compost, not only improve grapevine health and soil texture but also contribute to soil fertility, and are particularly beneficial in the context of circular economy and sustainable practices [16]. Their relevance is even more pronounced in arid climates, where challenges include both soil salinization and the limited availability of organic matter [17].

Various materials can serve as mulch, including vegetative residues, plastic films, biological geotextiles, gravel, and crushed stones [18]. Therefore, mulches are generally categorized into inorganic, organic, and special-type materials. Inorganic mulches include plastic film, with or without holes—widely used in agriculture—landscape fabric—used alone or below other mulches—and biodegradable or photodegradable plastic film [19]. Polyethylene (PE) mulch films, produced using non-renewable petroleum-based polymers, can have negative effects on the environment if not properly recovered and disposed of after use [20]. Thus, considerable research efforts have been expended in developing and investigating biodegradable mulch alternatives to PE mulching film in the last decades [21] and in investigating the use of organic mulch like MSW compost as a sustainable alternative [18]. Biodegradable and photodegradable plastic films were developed to address the environmental issues caused by low-density PE accumulation and plastic waste in soil and further problems resulting from improper disposal [22]. Biodegradable mulches, now widely available, consist of various polymers or additives like starch, cellulose, polyhydroxyalkanoates (PHA), and polylactic acid (PLA) [4]. The functionality of bio-based films employed in agriculture is still a matter of major concern [23]. Organic mulches, which should be weed-free, easy to apply, and readily available, often consist of on-site or off-site plant residues left on the soil surface after cropping [19]. Further organic materials like shredded grass, litter, compost, and small branches can also be utilized. In particular, MSW compost enhances soil carbon and nitrogen content and enzyme activities, providing a cost-effective alternative to expensive chemical amendments [24,25]. However, ensuring the safe use of MSW compost in agriculture requires careful monitoring for potential contaminants and heavy metals [26]. Special mulches include living mulch, reflective mulch, or biodegradable mulch [18,27]. Each type of mulch meets different needs and suits different environments. Organic mulches improve soil health and fertility over time as they get decomposed, but this decomposition requires a regular replenishment. In contrast, inorganic and specialized mulches are more durable and require less maintenance; however, they are typically more costly and often impractical for large-scale agricultural use due to both their higher initial investment and potential environmental impact [28]. The choice of mulch depends on factors such as crop specific needs, local climate conditions, the availability of materials, and the farmers' goals [29].

In agriculture, a variety of application techniques are utilized for mulching, each with distinct benefits. For example, traditional flat mulching involves covering with organic, inorganic, or a combination of materials to maintain a specific thickness. Plastic mulching with holes, a variation of flat mulching, partially covers the soil to improve rainfall infiltration and aeration [4,18]. Another technique, ridge-shaped mulching, involves covering ridges with plastic to direct rainwater into furrows, thus reducing surface runoff and improving water use efficiency. This method is especially effective in rainwater harvesting and reducing evaporation. Additionally, the ridge-furrow system, utilizing both plastic film and organic mulches, optimizes water availability for crops and increases soil productivity, demonstrating significant environmental benefits in semi-arid regions [30].

To assess the effectiveness of mulching practices, weed suppression is generally assessed by analyzing the percentage of weed coverage in a selected area, for instance, using smartphone applications [31]. Furthermore, several other parameters could be monitored in order to evaluate the additional beneficial effects of mulching. For example, soil moisture monitoring is useful to assess water conservation [18,32]; soil and atmospheric temperature allow evaluating the insulative effects of mulching against temperature fluctuations [18]. A further critical parameter is the soil decomposition activity, which provides insights into the contribution of mulching to soil fertility and ecosystem health by influencing the microbial degradative functions [33]. Decomposition of organic matter (OM) is one of the most important soil functions in agroecosystems; therefore, it has been studied extensively, and one of the most useful techniques is the litterbag method [34]. Among these methods, the "tea bag index" uses readily available green tea and red rooibos [33] to estimate OM weight loss due to microbial decomposition.

The aim of the present study was to evaluate the effects of MSW compost mulching on vineyard soil properties, specifically focusing on soil temperature and microbial decomposition activity. Additionally, the research attempted to identify the optimal compost application amount for effective weed control. By using MSW compost as a mulch material, the study aims to support sustainable agricultural practices by reducing reliance on synthetic mulching materials, aligning with circular economy principles, and promoting environmental sustainability through the recycling of organic municipal waste.

2. Materials and Methods

2.1. Vineyard Localization and Climatic Conditions

The 5-year-old experimental vineyard (*Vitis vinifera* cv. 'Primitivo', rootstock M4) is located in Salento (Apulia region, Southern Italy), 40°27′50.6″ N–17°58′18.7″ E, and belongs to the company "Cantina Due Palme" (Cellino San Marco-BR-, Italy) (Figure S1A,B).

Soil physico-chemical analyses were performed by the winery "Cantina Due Palme" in 2020. The soil had about 55 g kg^{-1} of skeleton, 61% of sand, 26% of silt and 13% of clay. The pH was 7.6, and the soil conductibility was 0.194 dS m^{-1}. Ca, Mg, K, and Na concentrations were 2367 g kg^{-1}, 158 g kg^{-1}, 404 g kg^{-1}, and 20 g kg^{-1}, respectively. Total limestone was 41 g kg^{-1}. The soil was poor in organic matter (1%), as well as in nitrogen (0.5 g kg^{-1}).

The site is on flat land, 50–60 m a.s.l, with a typically Mediterranean climate, hot and dry in summer and mild and humid in winter, belonging to the class Csa in the Köppen-Geiger classification [35]. During summertime, intense heat waves are prevalent over the Salento peninsula due to the dominance of a subtropical high-pressure system for several days. In contrast, mild and wet winters exhibit moderate and variable temperatures [36]. In the Salento peninsula, the average annual precipitation amounts are approximately 650 mm [37], primarily occurring from January to March and throughout the autumn season [38]. Summers are typically characterized by dry conditions, where short and intense convective rainstorms generate mostly low rainfall. The average annual temperature is around 16–17 °C [39].

Table 1 shows an overview of the main weather parameters observed throughout the year 2023 in the territory of the experimental study. Data was obtained from a weather station (iMetos 3.3, Pessl Instruments GmbH, Weiz, Austria) positioned at ~4 km from the experimental site. The climatic parameters were measured with a frequency of 5 min, averaged, and recorded by the data logger every 60 min. Air relative humidity was measured by a Hygromer® IN-1 sensor (Rotronic, Zurich, Switzerland). Air temperature was measured by a PT1000 sensor and precipitation by a double-tipping bucket rain gauge (Pessl Instruments GmbH, Weiz, Austria). The maximum and minimum temperature values (T_{max} and T_{min}, respectively) were calculated on a daily basis, and averaged over the month (T_{avg}). Precipitation is represented as a monthly cumulative value. During the winter season, T_{avg} ranged from 8.1 °C to 11.5 °C, with the lowest daily minimum temperature recorded in February (7.6 °C). Relative humidity consistently remained high during this period, fluctuating between 78.9% and 81.9%, indicative of humid conditions. By May, T_{avg} rose to 18.4 °C, while relative humidity remained relatively stable (83.7%). During the summer season, T_{avg} peaked at 27.5 °C in July and 26.3 °C in August, coinciding with a noticeable drop in relative humidity, which fell to 63.8%. The precipitation pattern remained relatively consistent until May, experiencing a significant spike in April with 155.6 mm of rainfall. A transition to a drier period occurred from June onwards, resulting in no precipitation in August, before returning to higher values for the remaining months.

Table 1. Overview of the main weather parameters during the test year 2023.

Month	Air Temperature (°C)			Air Relative Humidity (%)	Precipitation (mm)
	T_{avg}	T_{max}	T_{min}		
Jan-23	9.6	10.2	9.1	90.4	77.4
Feb-23	8.1	8.7	7.6	78.9	13.4
Mar-23	11.5	12.1	10.9	81.9	50.2
Apr-23	12.9	13.5	12.3	83.4	155.6
May-23	18.4	18.9	18.0	83.7	60.6
Jun-23	22.8	23.4	22.1	77.9	29.0
Jul-23	27.5	28.1	26.9	63.8	0.0
Aug-23	26.3	26.8	25.8	67.7	1.2
Sep-23	24.0	24.5	23.4	73.8	7.4
Oct-23	19.9	20.5	19.3	83.2	49.4
Nov-23	14.5	15.1	14.2	84.5	154.2
Dec-23	10.2	10.8	9.4	89.5	113.8

2.2. Experimental Plan

The vineyard implantation consists of rows, with a spacing of 1 m intra-row and 2.2 m inter-row between the plants, a density of around 4500 plants per hectare, and a vertical shoot positioning with Guyot pruning. Five rows were included in the experiment (Figure 1): the first and last row (rows 1 and 5, respectively) were considered borders and were neither treated nor monitored. In the three middle rows (rows 2, 3 and 4), grapevine plants were divided into eight blocks of six plants each; the first and last blocks (blocks A and H) were also considered borders and were neither treated nor monitored. Each of the six internal blocks (blocks B to G) of rows 2, 3, and 4 was then subjected to one of the following treatments:

1. Film-black: mulching with a traditional black PE film;
2. Film-b/w: mulching with a black and white biodegradable film;
3. Compost (8): mulching with municipal solid waste compost (MSW compost)—8 kg plant^{-1}, 4 cm thick;
4. Compost (15): mulching with MSW compost—15 kg plant^{-1}, 6.5 cm thick;
5. Compost (22): mulching with MSW compost—22 kg plant^{-1}, 10 cm thick;
6. Control (no mulching).

Compost quantities were selected based on the previous experience of the authors. The mulch films, black plastic film (Film-black) and biodegradable and bi-color film (Film-b/w), were chosen as conventional mulching techniques, for comparison to the MSW compost treatments. In particular, the innovative Film-b/w was chosen with the purpose of assessing the feasibility of its application in vineyards.

Therefore, each treated row contained one block per treatment, and the treatments were applied according to a completely randomized design to finally obtain three replicates per treatment randomly positioned within each row (Figures 1 and S1C). During the experiment, inter-row spaces were left untreated and were neither tilled nor unweeded.

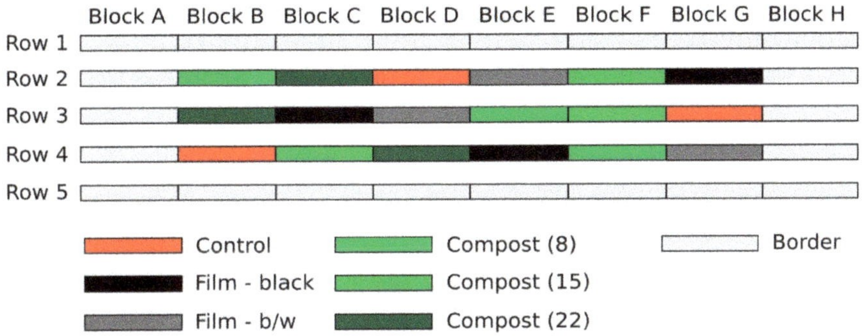

Figure 1. Experimental design. Three randomized replicates per each treatment were created, one in each treated row (rows 2, 3, and 4). Each block contained six *Vitis vinifera* cv. "Primitivo" plants. Rows 1 and 5, and blocks A and H were considered as borders.

Immediately after an initial tillage, common to all treatments, mulching was applied on 24 March 2023 and the trial was monitored until September 2023, to include the complete grapevine vegetative season. The compost used in this work as mulching was produced and provided by the company Heracle S.r.l. (Erchie-BR, Italy); it consists of a composted mixture of green residues and the organic fraction of municipal solid waste (MSW) (Table S1), collected through separate waste collection at source and subjected to strictly controlled biological processes. The compost was weighed in a bucket, and the same amount (8, 15, or 22 kg, according to the treatment) was distributed around each grapevine plant by keeping a width of ~50 cm (~25 cm to both sides of the vine line). Thus, keeping the treatment width constant, the thickness of the compost layer changed accordingly to the amount of compost used.

As a comparison with MSW compost mulching, soil was covered with black mulching film in UV-stabilized PE, hereafter "Film-black" (Tenax Black Cover, Tenax SpA, Lecco, Italy; technical data from manufacturer: thickness 60 micron, weight 37 g m^{-2}) or with black/white biodegradable mulching film made with the bioplastic ecovio® (BASF), hereafter "Film-b/w" (Ecotelo, Filnova S.r.l., Milano, Italy; technical data from manufacturer: thickness 18 micron, weight 24.84 g m^{-2}). The same wideness of the coverage made with MSW compost mulching (~50 cm) was maintained.

2.3. Canopy Cover/Weed Control

To identify the mulching strategy most suitable for the experimental purpose, weed growth was measured during springtime. In 18 April, 29 May and 23 June 2023, the percentage of grass cover was measured using the smartphone application "Canopy Cover Free" [40]. In detail, the application calculates the percentage of green area in a photo of the soil (as shown in Figure S2). In each block, 3 photos were collected (between the 2nd and 3rd plants; between the 3rd and 4th plants; and between the 4th and the 5th plants). Thus, on each date, 9 measurements were collected for each treatment. As explained in the next paragraphs, the treatment "Compost (15)" (mulching with MSW compost—15 kg $plant^{-1}$, 6.5 cm thick) was selected for further analyses (soil temperature and decomposition of organic matter) due to the similarity to the control treatment in the weed growth dynamics.

2.4. Soil Temperature

The soil and the atmospheric temperature were logged continuously in the period July–September 2023 by using portable WatchDog 1000 Series Micro Stations (Spectrum Technologies Instruments, Aurora, IL, USA) placed at 170 cm height inside the vineyard at the top of the support poles (Figure S3). The atmospheric temperature measurement was carried out using the internal sensor of one of the micro-stations. The station was adequately shielded from solar radiation (Figure S3). The soil temperature was detected by the SMEC 300 sensor (Spectrum Technologies Instruments, Aurora, IL, USA) placed at 20 cm soil depth under the treatments Film-black, compost (15), and control. One soil temperature sensor was used for each treatment. The stations stored the values with a 10 min measurement interval.

The thermal performance of the different treatments was assessed by the temperature mean values of the soil under the black film (T_{Fb}), under the compost layer (T_{C15}), and of the untreated soil used as a control (T_{Ctrl}). Air temperature (T_{air}) was also recorded.

2.5. Decomposition of Organic Matter (Tea Bag Index)

The tea bag index (TBI) method is a special litterbag method that was used to assess the soil microbial activity through the estimation of the organic matter decomposition rate [33]. The green tea used was the Lipton green tea (EAN: 8711327515765); the red tea used was the Lipton rooibos tea (EAN: 8711327514348). Both of them were purchased at Dutchsupermarket (Gelderland, NL; article codes: THEE049 and THEE056, respectively). Despite the changes in tea bag net manufacturers that occurred in the last ten years, the method was proven to still be valid [41]. The content of green tea comprises a higher fraction of easily degradable organic compounds, while rooibos tea contains a higher fraction of more recalcitrant organic compounds. Tea bags were weighed and then buried at a depth of ~12 cm, one green and one red tea bag per block of the treatments Film-black, compost (15), and control, halfway between two vines. According to the original protocol [33], after a burial time of 3 months, tea bags were recovered, dried at 65 °C for 48 h, and then re-weighted to calculate the mass loss (excluding the weight of the bag, the cord and the label). The measure was repeated twice, in the periods April–June (spring) and July–September (summer). This method was already applied in vineyards [42–44]. Green tea and red tea mass loss were used as a reliable and intuitive proxy to estimate the degradation of labile organic matter fraction and recalcitrant organic matter fraction, respectively [45–47].

2.6. Statistical Analyses

Data were analyzed using SPSS v.20 software (IBM Corporation, Armonk, NY, USA) or MATLAB v.2023b (The MathWorks Inc., Natick, USA). For weed control and tea bag analysis (Sections 2.3 and 2.5, respectively), one-way ANOVA was used to compare means between treatments for each measurement date. For soil temperatures (Section 2.4), one-way analysis of variance (ANOVA) was carried out to identify significant differences between T_{Fb}, $T_{C1}5$, T_{Ctrl}, and T_{air}. One-way ANOVA analyses were also performed for comparing the mean temperature difference between the soil (under the black film, the compost layer, and the control soil) and the air (ΔT_{Fb}, ΔT_{C15}, ΔT_{Ctrl}, respectively); the analyses were carried out for every value of the air temperature (1 °C intervals) in the studied summer period. Additionally, to evaluate the effects of treatments across the whole monitored period, a repeated-measures ANOVA was used. All ANOVAs were followed by a post hoc test (Tukey HSD or Duncan's) at $p \leq 0.05$.

3. Results

3.1. Weed Control and Treatment Selection

The trend of weed growth during the first 3 months of the experiment is reported in Figure 2. To inhibit weed growth, with an effect similar to the one obtained using mulching films, a high quantity of compost was necessary (22 kg plant^{-1}). However, this effect lasted less than that obtained with the mulching films. In fact, the compost mulching (22 kg plant^{-1}) favored weed growth later on. On the contrary, a thin compost layer (8 kg plant^{-1}) stimulated the weed growth earlier, inducing a significant increase in the covered surface already 2 months after the application. It is worth noting that, three months after the application, any of the compost mulching produced significant differences from the non-mulched control.

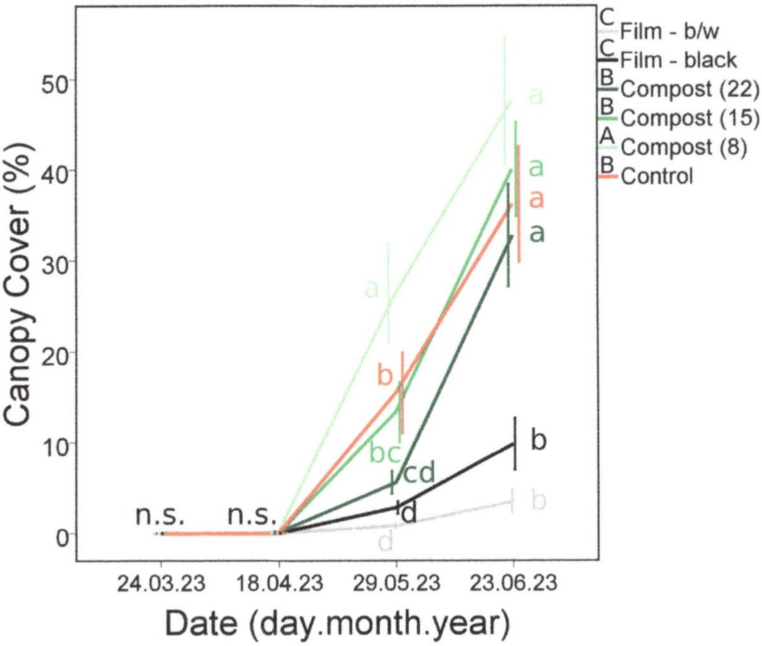

Figure 2. Trend in weed growth, expressed as percentage of the soil covered by green canopies. The whisker indicates the standard error. Different lowercase letters indicate significantly different means between the treatments for each date (Duncan post hoc test, $p < 0.05$). Different capital letters indicate significantly different means between the treatments with repeated measures for the entire duration of the surveys (Duncan post hoc test, $p < 0.05$). n.s.= not significant.

The intermediate quantity of compost (15 kg plant^{-1}) did not produce any significant effect on the percentage of surface covered by weeds with respect to the non-mulched control during the springtime. Thus, this treatment was selected to highlight the effects of compost mulching on the soil properties, excluding weed interference.

Both mulching films were able to significantly reduce weed growth, but differences among them were not significant. Thus, the most commonly used PE black film was selected as a conventional reference mulching film product.

3.2. Soil Temperature

Statistically significant differences were always found when comparing the values of T_{Fb}, T_{C15}, T_{Ctrl}, and T_{air} in each month and over the entire period of July–September 2023 (Figure 3). The average soil temperature followed the trend of the air temperature related to the advancement of the months. Interestingly, the lowest average soil temperature was recorded in each month as well as in the whole period in the compost treatment, followed by the control, and then the Film-black. Regardless of the different weather conditions, the adoption of the compost layer allowed for maintaining the average soil temperature at values lower than T_{Ctrl}, demonstrating the cooling effect of the MSW compost during the analyzed summer period. The lowest mean soil temperature (27.5 °C) was recorded in the compost treatment in September, while the black film kept the soil temperature over 29.1 °C. During July, the month with the highest mean air temperature (28.6 °C), the compost layer allowed for maintaining the mean T_{C15} under 31.1 °C, significantly lower than the mean T_{Ctrl} (31.7 °C) and T_{Bf} (32.4 °C).

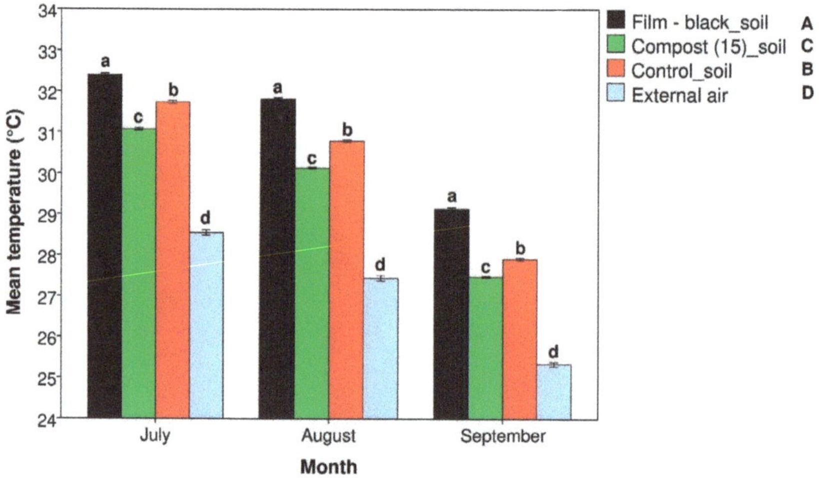

Figure 3. Mean values ± standard errors (whiskers) of the temperature of the soil treated with the black film (T_{Fb}) and the compost layer (T_{C15}), of the control soil (T_{Ctrl}) and of the air (T_{air}), during July–September 2023. Different lowercase letters indicate significant differences between the means for each month (Duncan post hoc test, $p < 0.05$). Different capital letters indicate significant differences between the means for the entire duration of the surveys (Duncan post hoc test, $p < 0.05$).

Figure 4 shows the mean daily patterns of air and soil temperature from July to September 2023. As expected, minimum temperatures were typically observed for both air and ground temperatures during the early morning hours. The air temperature exceeded all ground temperatures from 09:00 until 14:00; the pattern reversed from 18:00 to 09:00. In comparison to the temperature of the control soil, the Film-black showed significantly higher values at all times. This difference notably increased between 11:00 and 17:00. In contrast, the soil temperature under the compost was higher than the control between 05:00 and 11:00 and was lower between 13:00 and 01:00; it was not different from control between 02:00 and 04:00. Note that the flexure between 13:00 and 14:00 is due to the shading of the vineyard.

Significant differences were shown for the three treatments over the entire monitored period; in particular, the Film-black showed the highest values, the control intermediate values, and the compost the lowest values (Figure 4). Significant differences between the treatments at almost all times, except between Film-black and compost from 05:00 to 09:00 and between compost and control from 02:00 to 04:00, were also revealed.

The damping of thermal waves at 20 cm was on average higher in compost mulched soil with a mean amplitude of 3.1 °C than in the control and black film mulched soil with mean amplitude values of 6.0 °C and 7.4 °C, respectively, during the whole summer period (Figure 4). Monthly mean amplitudes in the soil covered with the compost were lower than about 47–49% of those calculated for the control soil, while increments of 9–35% were associated with the black film treatment. The compost allowed for a reduction in the daily fluctuation of soil temperature, due to the relevant lowering of the daily maximum temperature value. It is worth considering that temperature peaks are not detected through the analysis of average data.

A more in-depth analysis of the mean temperature difference between the soil under the different treatments and the air (ΔT_{Fb}, ΔT_{C15}, ΔT_{Ctrl}) was carried out (Figure 5). The results showed that, for air temperature values above 23.0 °C, the compost on average allowed for maintaining the soil temperature slightly cooler than the control and fairly cooler than the black film, with a more pronounced effect for air temperature values above 28.1 °C.

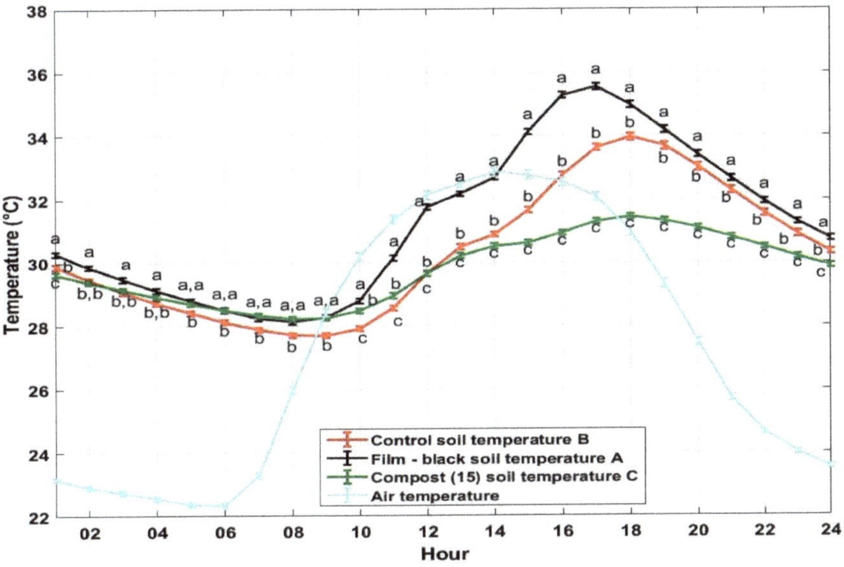

Figure 4. Diurnal pattern of the soil and air temperature. The whisker indicates the standard error. Different lowercase letters indicate significant differences between the three selected treatments for each hour (hsd post hoc test, $p < 0.05$). Different capital letters indicate significant differences between treatments with repeated measures for the entire monitored period (hsd post hoc test, $p < 0.05$). The air temperature pattern is shown for reference only.

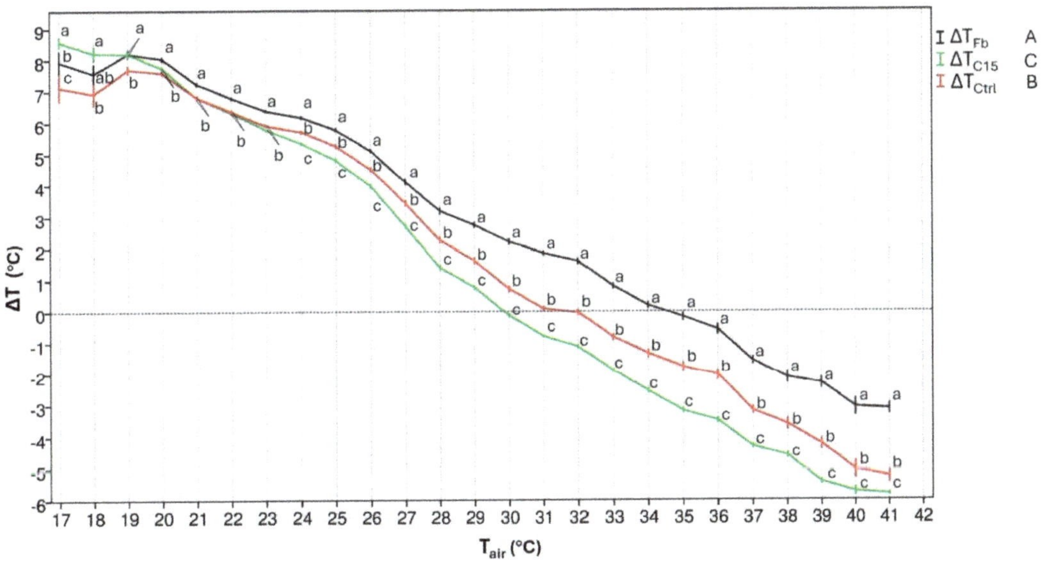

Figure 5. Mean values ± standard errors (whiskers) of the temperature difference between the soil under the black film, the compost layer and the control treatments and the air temperature, in relation to T_{air} (1 °C intervals). Different lowercase letters indicate significant differences between ΔT_{Fb}, ΔT_{C15} and ΔT_{Ctrl} means for each T_{air} value (Duncan post hoc test, $p < 0.05$). Different capital letters indicate significant differences between ΔT_{Fb}, ΔT_{C15} and ΔT_{Ctrl} means for the entire range of T_{air} values (Duncan post hoc test, $p < 0.05$).

3.3. Decomposition of Organic Matter (Tea Bag Index)

Tea bag index, a method used as a proxy for microbial decomposition activity in soil, was measured for the treatments "Compost (15)", "Film-black" and "Control" in two consecutive 90-day periods (spring and summer); it showed that percentages of mass losses due to microbial decomposition ranged between 40% and 60% for the labile organic fraction (green tea) and between 18% and 40% for the recalcitrant organic fraction (rooibos red tea) (Figure 6A,B). The decomposition of recalcitrant organic fraction (red tea) was higher in spring (April–June) than in summer (July–September) (*t*-test, $p = 0.007$), while labile organic fraction (green tea) did not show significant differences between the two periods. MSW compost acted similarly to the Film-black in spring, stimulating the microbial degradation activity of the labile organic fraction (green tea) of the soil (Figure 6A), while in summer this effect disappeared and the MSW compost values were similar to the untreated control (Figure 6B). The same trend was observed for the recalcitrant organic fraction (rooibos red tea), although no statistical significance was observed in this case due to a higher variability between the three replicates (Figure 6A,B). Over the two periods (ANOVA repeated measures), the mulching treatment significantly affected the weight loss of green tea ($p = 0.016$), while no significance was found for red tea ($p > 0.05$) (Figure 6).

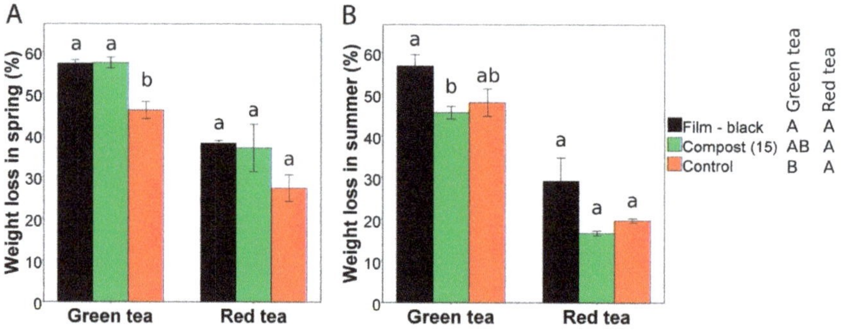

Figure 6. Means ± standard errors (whiskers) of the tea bag index, measured in two periods: April–June (panel (**A**)) and July–September (panel (**B**)), for the treatments "Film-black", "Compost" (15) and "Control". Different lowercase letters indicate significantly different means between treatments for each type of tea (Duncan post hoc test, $p < 0.05$). Different capital letters indicate significantly different means between treatments with repeated measures over the two periods for each type of tea separately (Duncan post hoc test, $p < 0.05$).

4. Discussion

It is well known that MSW compost could be used in viticulture as a fertilizer and that its application as organic mulch on the vine row could inhibit weed growth, modulating the competition among plants. However, other "side-effects" could be exploited: application of MSW compost can maintain a lower soil temperature than the soil under the black PE mulch film and also be compared to uncovered soil, which is beneficial for the plant root system. Furthermore, besides chemical nutrition, soil biological fertility could play an important role in plant resilience. In this work, we explored these side effects of compost mulching using MSW-derived compost.

The most widely used mulch in commercial agricultural systems is PE plastic film [48], because of its relatively modest cost, availability, and physical and mechanical properties that enable achieving several benefits [49]. MSW compost, during the summer, kept on average lower soil temperature values than in the control and Film-black treatments, particularly with lower peak values experienced as a consequence of high levels of solar radiation intensity. An analogous effect on soil temperature was observed by Chan et al. [50] regarding the use of composted mulch in vineyards in Australia; the application of composted mulch significantly changed the soil temperature at 10 cm depth, leading to a shorter daily temperature range, lowering daily maximum temperatures, and rising daily minimums. Similar results on soil temperature were evaluated in vineyards with organic mulch under Mediterranean conditions during the summer months, where the monthly maximum soil temperature, at a depth of 7 cm, was found to be lower in mulched soil [51]. By maintaining lower soil temperatures during summertime, compost can mitigate heat stress on plant roots, enhancing their health and overall function. This is crucial, as extreme temperatures can reduce root activity, hinder water and nutrient uptake, and ultimately affect plant growth and resilience. Maintaining root zone temperatures below 30 °C is recommended to ensure normal growth and functionality. A study carried out in China on grapevines showed that temperatures above 30 °C can reduce root and shoot growth, decrease chlorophyll content in leaves, and increase lignin in roots and shoots; additionally, high root zone temperatures can shift root respiration to cyanide-resistant respiration, releasing more heat to alleviate stress damage [52]. Thus, using compost to stabilize soil temperature can be very useful for maintaining healthy root systems during the hottest part of the day. In fact, as shown in Figure 5, the cooling effect of the compost mulching has a higher magnitude at higher, thus more stressful, temperatures (>28 °C).

The values of weight loss for both green and red tea observed in our work are coherent with those reported previously in vineyards [45–47]. The effect of MSW compost on microbial decomposition activity was different in the two different periods investigated (spring and summer): in spring, the compost enhanced the soil decomposition activity with respect to the untreated control, showing values similar to those of the treatment with Film-black, while in summer this effect disappeared. Soil microbial decomposition activity is mainly dependent on humidity and temperature [53]; thus, our results can be explained by observing the temperature trend, which was significantly affected by the mulching treatments in our experiment. In fact, soil temperature, measured in summer, showed exactly the same trend as soil decomposition activity (compare Figure 3 with Figure 6B). Therefore, temperature appears to be the main driver for the decomposition of organic matter, both labile and recalcitrant, in summer. However, other factors may play a role in this seasonal variation of the decomposition activity. For example, the moisture content obviously reduces in summer, and this negatively influences both the growth rate and the metabolism of microbe cells.

In spring, the compost stimulated soil decomposition activity. Organic matter decomposition is a task performed by soil microbes that depends not only on microbial abundance but also on microbial diversity [54,55]; keeping this in mind, besides soil temperature, a reason for the increase in decomposition activity in spring (Figure 6A) could be a direct "microbial inoculation effect" by the MSW compost applied. In the following period, the microbial community likely undergoes physiological changes, and the initial inoculation effect is expected to reduce. In general, active soil microbes decompose organic matter, releasing essential nutrients that promote robust plant growth. This activity also enhances soil aeration and water retention, supporting better root development and resilience to environmental stresses like salinity, drought, and nutrient deficiency. Additionally, beneficial microbes can increase nutrient mobility and availability in the soil and can also outcompete harmful pathogens, thus reducing disease incidence and improving general plant fitness [56,57]. However, confirming this would require thorough microbiological analyses of both soil and MSW compost used [58], to assess the increment in soil microbial load as well as a shift of the microbiota towards an enrichment of heterotrophic species particularly efficient in decomposition activity.

On the contrary, in summertime, the organic matter decomposition was kept lower with respect to the one observed with plastic films. Soil degradation, resulting in low organic matter contents, is a key problem in hot environments (such as the Mediterranean regions), and it could be worsened by climate change [59,60]. Thus, the soil temperature decreases, and the decomposition activity consequently decelerates. This could be an important advantage in the choice of MSW compost as mulching material, as it determines the conservation of organic matter in the soil in hot climates.

Although in this work we did not collect data about the effects the treatments had on vine nutrition and grape quality, the results obtained (especially with tea bags) indicate that MSW compost increases the turnover of organic matter into the soil. This could benefit the plants in terms of available nutrients, which could in turn impact the quality of the grapes. However, this requires further dedicated research.

Further studies could implement the knowledge, including applications in different environmental conditions (e.g., different sites and different climates). Our analyses did not highlight any negative effects of the MSW compost; however, we cannot exclude that further investigations could find different results.

5. Conclusions

The application of MSW compost maintained lower soil temperature values than the soil under the black PE mulch film and was also compared to the control soil. Thus, using MSW compost to stabilize soil temperature is helpful for maintaining healthy root systems, particularly during the hottest hours of the day. We also found that MSW compost stimulates the soil biological activity in the first period after addition, but it reduces the organic matter lost during the hottest period with respect to the plastic film. However, more studies are needed to disentangle the effects of compost-derived microbes from those of environmental factors.

The application of MSW compost as a mulch material supports sustainable practices in viticulture and, in general, in tree crop management, by recycling municipal waste and reducing the use of synthetic materials like PE films. This aligns with the principles of the circular economy and promotes environmental sustainability in agriculture.

Supplementary Materials: The following supporting information can be downloaded at: https://www.mdpi.com/article/10.3390/horticulturae10070769/s1, Table S1: Physico-chemical parameters of MSW compost. Values derived from five subsamples of the municipal solid waste (MSW) compost used in this work, as provided by the producing company; Figure S1: Map and photo of the experimental site. (A) Location of the Apulia region in south-east Italy; (B) zoomed-in area (background map from Google Earth) with the red rectangle indicating the location of the experimental vineyard; (C) detail of the experimental vineyard with the different mulching treatments; Figure S2: Example of the Canopy Cover Free smartphone application. (A) Original photo of weeds. (B) Identification and measurement of the green area (leaves); Figure S3: Micro station localized on the vineyard row connected to sensors for recording both soil and air temperature.

Author Contributions: Conceptualization, I.B., M.C., G.Z. and L.R.; methodology, I.B., M.C. and L.R.; software, I.B., M.C., P.R., G.P. and L.R.; validation, I.B., M.C., G.Z. and L.R.; formal analysis, I.B., M.C., P.R., G.P. and L.R.; investigation, I.B., M.C., C.D., P.R., G.P. and L.R.; resources, M.C., I.B. and L.R.; writing—original draft preparation, I.B., M.C., P.R., G.P. and L.R.; writing—review and editing, I.B. and M.C.; supervision, I.B., M.C. and L.R.; project administration, M.C. and L.R.; funding acquisition, M.C. and L.R. All authors have read and agreed to the published version of the manuscript.

Funding: PR received a PhD grant in "Biological and environmental sciences and technologies" (co-funded by PNRR-DM_117/2023; CUP number: F83C23001130009). LR and MC received financial support from the company "Heracle S.r.l." in the form of generic contribution to the research topic concerning the use of compost in viticulture.

Data Availability Statement: The original contributions presented in the study are included in the article/supplementary material, further inquiries can be directed to the corresponding author/s.

Acknowledgments: The authors are grateful to the company "Cantina Due Palme" (Cellino San Marco-BR-, Italy) for the access to the vineyard and for the technical support.

Conflicts of Interest: The authors declare no conflicts of interest. The Company "Heracle S.r.l.", which provided the compost and gave financial support, was not involved in any way in the study conceiving, experimental design, data collection and analysis, and results interpretation.

References

1. Prosdocimi, M.; Tarolli, P.; Cerdà, A. Mulching practices for reducing soil water erosion: A review. *Earth-Sci. Rev.* **2016**, *161*, 191–203. [CrossRef]
2. Bodner, G.; Nakhforoosh, A.; Kaul, H.-P. Management of crop water under drought: A review. *Agron. Sustain. Dev.* **2015**, *35*, 401–442. [CrossRef]
3. Jiang, L.; Dami, I.; Doohan, D. Effects of Mulching on Soil Temperature, Scion Rooting, and Soil Moisture of Mounded Grapevines. *Int. J. Fruit Sci.* **2015**, *16*, 182–190. [CrossRef]
4. El-Beltagi, H.S.; Basit, A.; Mohamed, H.I.; Ali, I.; Ullah, S.; Kamel, E.A.R.; Shalaby, T.A.; Ramadan, K.M.A.; Alkhateeb, A.A.; Ghazzawy, H.S. Mulching as a Sustainable Water and Soil Saving Practice in Agriculture: A Review. *Agronomy* **2022**, *12*, 1881. [CrossRef]
5. Gaiotti, F.; Marcuzzo, P.; Belfiore, N.; Lovat, L.; Fornasier, F.; Tomasi, D. Influence of compost addition on soil properties, root growth and vine performances of Vitis vinifera cv Cabernet sauvignon. *Sci. Hortic.* **2017**, *225*, 88–95. [CrossRef]
6. Steinmaus, S.; Elmore, C.L.; Smith, R.J.; Donaldson, D.; Weber, E.A.; Roncoroni, J.A.; Miller, P.R.M. Mulched cover crops as an alternative to conventional weed management systems in vineyards. *Weed Res.* **2008**, *48*, 273–281. [CrossRef]
7. Mairata, A.; Labarga, D.; Puelles, M.; Huete, J.; Portu, J.; Rivacoba, L.; Pou, A. The organic mulches in vineyards exerted an influence on spontaneous weed cover and plant biodiversity. *Eur. J. Agron.* **2023**, *151*, 126997. [CrossRef]
8. Bajwa, A.A.; Mahajan, G.; Chauhan, B.S. Nonconventional Weed Management Strategies for Modern Agriculture. *Weed Sci.* **2017**, *63*, 723–747. [CrossRef]
9. Abbas, T.; Zahir, Z.A.; Naveed, M.; Kremer, R.J. Limitations of Existing Weed Control Practices Necessitate Development of Alternative Techniques Based on Biological Approaches. In *Advances in Agronomy*; Sparks, D.L., Ed.; Elsevier: Amsterdam, The Netherlands, 2018; Volume 147, pp. 239–280.
10. Zhang, Q.; Wang, S.; Li, L.; Inoue, M.; Xiang, J.; Qiu, G.; Jin, W. Effects of mulching and sub-surface irrigation on vine growth, berry sugar content and water use of grapevines. *Agric. Water Manag.* **2014**, *143*, 1–8. [CrossRef]
11. Gil, P.M.; Lobos, P.; Durán, K.; Olguín, J.; Cea, D.; Schaffer, B. Partial root-zone drying irrigation, shading, or mulching effects on water savings, productivity and quality of 'Syrah' grapevines. *Sci. Hortic.* **2018**, *240*, 478–483. [CrossRef]
12. Phogat, V.; Pitt, T.; Stevens, R.M.; Cox, J.W.; Šimůnek, J.; Petrie, P.R. Assessing the role of rainfall redirection techniques for arresting the land degradation under drip irrigated grapevines. *J. Hydrol.* **2020**, *587*, 125000. [CrossRef]
13. Romero, P.; Navarro, J.M.; Ordaz, P.B. Towards a sustainable viticulture: The combination of deficit irrigation strategies and agroecological practices in Mediterranean vineyards. A review and update. *Agric. Water Manag.* **2022**, *259*, 107216. [CrossRef]
14. Prosdocimi, M.; Jordan, A.; Tarolli, P.; Keesstra, S.; Novara, A.; Cerda, A. The immediate effectiveness of barley straw mulch in reducing soil erodibility and surface runoff generation in Mediterranean vineyards. *Sci. Total Environ.* **2016**, *547*, 323–330. [CrossRef]
15. Ferrara, G.; Fracchiolla, M.; Chami, Z.A.; Camposeo, S.; Lasorella, C.; Pacifico, A.; Aly, A.; Montemurro, P. Effects of Mulching Materials on Soil and Performance of cv. Nero di Troia Grapevines in the Puglia Region, Southeastern Italy. *Am. J. Enol. Vitic.* **2012**, *63*, 269–276. [CrossRef]
16. Fang, X.; Gao, B.; Zhong, D.; Wang, L.; Borrion, A.; Huang, W.; Xu, S.; Cui, S. Closing the food waste loop: Analysis of the agronomic performance and potential of food waste disposal products. *J. Clean. Prod.* **2023**, *382*, 135174. [CrossRef]
17. Chakraborty, D.; Nagarajan, S.; Aggarwal, P.; Gupta, V.K.; Tomar, R.K.; Garg, R.N.; Sahoo, R.N.; Sarkar, A.; Chopra, U.K.; Sarma, K.S.S.; et al. Effect of mulching on soil and plant water status, and the growth and yield of wheat (*Triticum aestivum* L.) in a semi-arid environment. *Agric. Water Manag.* **2008**, *95*, 1323–1334. [CrossRef]
18. Kader, M.A.; Senge, M.; Mojid, M.A.; Ito, K. Recent advances in mulching materials and methods for modifying soil environment. *Soil Tillage Res.* **2017**, *168*, 155–166. [CrossRef]
19. Jordán, A.; Zavala, L.M.; Muñoz-Rojas, M. Mulching, Effects on Soil Physical Properties. In *Encyclopedia of Agrophysics*; Springer: Berlin/Heidelberg, Germany, 2011; pp. 492–496.
20. Blanco, I.; Loisi, R.V.; Sica, C.; Schettini, E.; Vox, G. Agricultural plastic waste mapping using GIS. A case study in Italy. *Resour. Conserv. Recycl.* **2018**, *137*, 229–242. [CrossRef]
21. Mansoor, Z.; Tchuenbou-Magaia, F.; Kowalczuk, M.; Adamus, G.; Manning, G.; Parati, M.; Radecka, I.; Khan, H. Polymers Use as Mulch Films in Agriculture-A Review of History, Problems and Current Trends. *Polymers* **2022**, *14*, 5062. [CrossRef]
22. Serrano-Ruiz, H.; Martin-Closas, L.; Pelacho, A.M. Biodegradable plastic mulches: Impact on the agricultural biotic environment. *Sci. Total Environ.* **2021**, *750*, 141228. [CrossRef]
23. Briassoulis, D.; Giannoulis, A. Evaluation of the functionality of bio-based plastic mulching films. *Polym. Test.* **2018**, *67*, 99–109. [CrossRef]
24. Meena, M.D.; Yadav, R.K.; Narjary, B.; Yadav, G.; Jat, H.S.; Sheoran, P.; Meena, M.K.; Antil, R.S.; Meena, B.L.; Singh, H.V.; et al. Municipal solid waste (MSW): Strategies to improve salt affected soil sustainability: A review. *Waste Manag.* **2019**, *84*, 38–53. [CrossRef] [PubMed]
25. Crecchio, C.; Curci, M.; Mininni, R.; Ricciuti, P.; Ruggiero, P. Short-term effects of municipal solid waste compost amendments on soil carbon and nitrogen content, some enzyme activities and genetic diversity. *Biol. Fertil. Soils* **2001**, *34*, 311–318. [CrossRef]
26. Hargreaves, J.; Adl, M.; Warman, P. A review of the use of composted municipal solid waste in agriculture. *Agric. Ecosyst. Environ.* **2008**, *123*, 1–14. [CrossRef]

27. Touchaleaume, F.; Martin-Closas, L.; Angellier-Coussy, H.; Chevillard, A.; Cesar, G.; Gontard, N.; Gastaldi, E. Performance and environmental impact of biodegradable polymers as agricultural mulching films. *Chemosphere* **2016**, *144*, 433–439. [CrossRef]
28. Guerra, B.; Steenwerth, K. Influence of Floor Management Technique on Grapevine Growth, Disease Pressure, and Juice and Wine Composition: A Review. *Am. J. Enol. Vitic.* **2012**, *63*, 149–164. [CrossRef]
29. Winter, S.; Bauer, T.; Strauss, P.; Kratschmer, S.; Paredes, D.; Popescu, D.; Landa, B.; Guzman, G.; Gomez, J.A.; Guernion, M.; et al. Effects of vegetation management intensity on biodiversity and ecosystem services in vineyards: A meta-analysis. *J. Appl. Ecol.* **2018**, *55*, 2484–2495. [CrossRef] [PubMed]
30. Gan, Y.; Siddique, K.H.M.; Turner, N.C.; Li, X.-G.; Niu, J.-Y.; Yang, C.; Liu, L.; Chai, Q. Ridge-Furrow Mulching Systems—An Innovative Technique for Boosting Crop Productivity in Semiarid Rain-Fed Environments. *Adv. Agron.* **2013**, *118*, 429–476.
31. Jàuregui, J.M.; Delbino, F.G.; Brance Bonvini, M.A.I.; Berhongaray, G. Determining yield of forage crops using the Canopeo mobile phone app. *J. N. Z. Grassl.* **2019**, *81*, 41–46. [CrossRef]
32. Mulumba, L.N.; Lal, R. Mulching effects on selected soil physical properties. *Soil Tillage Res.* **2008**, *98*, 106–111. [CrossRef]
33. Keuskamp, J.A.; Dingemans, B.J.J.; Lehtinen, T.; Sarneel, J.M.; Hefting, M.M.; Muller-Landau, H. Tea Bag Index: A novel approach to collect uniform decomposition data across ecosystems. *Methods Ecol. Evol.* **2013**, *4*, 1070–1075. [CrossRef]
34. Chassain, J.; Vieublé Gonod, L.; Chenu, C.; Joimel, S. Role of different size classes of organisms in cropped soils: What do litterbag experiments tell us? A meta-analysis. *Soil Biol. Biochem.* **2021**, *162*, 108394. [CrossRef]
35. Rubel, F.; Kottek, M. World Map of the Köppen-Geiger Climate Classification. Available online: https://koeppen-geiger.vu-wien.ac.at/ (accessed on 24 April 2024).
36. Giorgi, F.; Lionello, P. Climate change projections for the Mediterranean region. *Glob. Planet. Change* **2008**, *63*, 90–104. [CrossRef]
37. Martano, P.; Elefante, C.; Grasso, F. Ten years water and energy surface balance from the CNR-ISAC micrometeorological station in Salento peninsula (southern Italy). *Adv. Sci. Res.* **2015**, *12*, 121–125. [CrossRef]
38. Donateo, A.; Conte, M.; Grasso, F.M.; Contini, D. Seasonal and diurnal behaviour of size segregated particles fluxes in a suburban area. *Atmos. Environ.* **2019**, *219*, 117052. [CrossRef]
39. Donateo, A.; Palusci, O.; Pappaccogli, G.; Esposito, A.; Martilli, A.; Santiago, J.L.; Buccolieri, R. Analysis of urban heat island and human thermal comfort in a Mediterranean city: A case study of Lecce (Italy). *Sustain. Cities Soc.* **2023**, *98*, 104849. [CrossRef]
40. Mendes, J.; Pinho, T.M.; Neves dos Santos, F.; Sousa, J.J.; Peres, E.; Boaventura-Cunha, J.; Cunha, M.; Morais, R. Smartphone Applications Targeting Precision Agriculture Practices—A Systematic Review. *Agronomy* **2020**, *10*, 855. [CrossRef]
41. Middelanis, T.; Pohl, C.M.; Looschelders, D.; Hamer, U. New directions for the Tea Bag Index: Alternative teabags and concepts can advance citizen science. *Ecol. Res.* **2023**, *38*, 690–699. [CrossRef]
42. Costantini, E.A.C.; Castaldini, M.; Diago, M.P.; Giffard, B.; Lagomarsino, A.; Schroers, H.J.; Priori, S.; Valboa, G.; Agnelli, A.E.; Akca, E.; et al. Effects of soil erosion on agro-ecosystem services and soil functions: A multidisciplinary study in nineteen organically farmed European and Turkish vineyards. *J. Environ. Manag.* **2018**, *223*, 614–624. [CrossRef]
43. Guzmán, G.; Cabezas, J.M.; Sánchez-Cuesta, R.; Lora, Á.; Bauer, T.; Strauss, P.; Winter, S.; Zaller, J.G.; Gómez, J.A. A field evaluation of the impact of temporary cover crops on soil properties and vegetation communities in southern Spain vineyards. *Agric. Ecosyst. Environ.* **2019**, *272*, 135–145. [CrossRef]
44. Andrés, P.; Doblas-Miranda, E.; Silva-Sánchez, A.; Mattana, S.; Font, F. Physical, Chemical, and Biological Indicators of Soil Quality in Mediterranean Vineyards under Contrasting Farming Schemes. *Agronomy* **2022**, *12*, 2643. [CrossRef]
45. Ferrante, M.; Lamelas-López, L.; Nunes, R.; Monjardino, P.; Lopes, D.J.H.; Soares, A.O.; Lövei, G.L.; Borges, P.A.V. A simultaneous assessment of multiple ecosystem services and disservices in vineyards and orchards on Terceira Island, Azores. *Agric. Ecosyst. Environ.* **2022**, *330*, 107909. [CrossRef]
46. Pingel, M.; Reineke, A.; Leyer, I. Disentangling the mixed effects of soil management on microbial diversity and soil functions: A case study in vineyards. *Sci. Rep.* **2023**, *13*, 3568. [CrossRef] [PubMed]
47. Steiner, M.; Pingel, M.; Falquet, L.; Giffard, B.; Griesser, M.; Leyer, I.; Preda, C.; Uzman, D.; Bacher, S.; Reineke, A. Local conditions matter: Minimal and variable effects of soil disturbance on microbial communities and functions in European vineyards. *PLoS ONE* **2023**, *18*, e0280516. [CrossRef] [PubMed]
48. Ilyas, M.A.; Rehman, F.U.; Ilyas, I.; Kalsoom, M.; Bilal, M.T.; Gull, M.; Ilyas, I.; Iqbal, R.U.; Ilahi, H.; Shakeel, M.A.; et al. A detailed review of mulching: An important technique in agricultural crop production. *Asian J. Adv. Res.* **2021**, *4*, 492–499.
49. Vox, G.; Loisi, R.V.; Blanco, I.; Mugnozza, G.S.; Schettini, E. Mapping of Agriculture Plastic Waste. *Agric. Agric. Sci. Procedia* **2016**, *8*, 583–591. [CrossRef]
50. Chan, K.Y.; Fahey, D.J.; Newell, M.; Barchia, I. Using Composted Mulch in Vineyards—Effects on Grape Yield and Quality. *Int. J. Fruit Sci.* **2010**, *10*, 441–453. [CrossRef]
51. Cabrera-Pérez, C.; Llorens, J.; Escolà, A.; Royo-Esnal, A.; Recasens, J. Organic mulches as an alternative for under-vine weed management in Mediterranean irrigated vineyards: Impact on agronomic performance. *Eur. J. Agron.* **2023**, *145*, 126798. [CrossRef]
52. Su, Y.; Li, X.; Cao, Z.; Gao, Z.; Du, Y. Effects of Long-Term High Temperatures in the Root Zone on the Physiological Characteristics of Grapevine Leaves and Roots: Implications for Viticulture Practices. *Horticulturae* **2024**, *10*, 245. [CrossRef]
53. Prescott, C.E. Litter decomposition: What controls it and how can we alter it to sequester more carbon in forest soils? *Biogeochemistry* **2010**, *101*, 133–149. [CrossRef]

54. Wardle, D.; Yeates, G.; Barker, G.; Bonner, K. The influence of plant litter diversity on decomposer abundance and diversity. *Soil Biol. Biochem.* **2006**, *38*, 1052–1062. [CrossRef]
55. Eilers, K.G.; Lauber, C.L.; Knight, R.; Fierer, N. Shifts in bacterial community structure associated with inputs of low molecular weight carbon compounds to soil. *Soil Biol. Biochem.* **2010**, *42*, 896–903. [CrossRef]
56. Darriaut, R.; Lailheugue, V.; Masneuf-Pomarede, I.; Marguerit, E.; Martins, G.; Compant, S.; Ballestra, P.; Upton, S.; Ollat, N.; Lauvergeat, V. Grapevine rootstock and soil microbiome interactions: Keys for a resilient viticulture. *Hortic. Res.* **2022**, *9*, uhac019. [CrossRef] [PubMed]
57. Vimal, S.R.; Singh, J.S.; Arora, N.K.; Singh, S. Soil-Plant-Microbe Interactions in Stressed Agriculture Management: A Review. *Pedosphere* **2017**, *27*, 177–192. [CrossRef]
58. Daebeler, A.; Petrová, E.; Kinz, E.; Grausenburger, S.; Berthold, H.; Sandén, T.; Angel, R. Pairing litter decomposition with microbial community structures using the Tea Bag Index (TBI). *Soil* **2022**, *8*, 163–176. [CrossRef]
59. Hamidov, A.; Helming, K.; Bellocchi, G.; Bojar, W.; Dalgaard, T.; Ghaley, B.B.; Hoffmann, C.; Holman, I.; Holzkamper, A.; Krzeminska, D.; et al. Impacts of climate change adaptation options on soil functions: A review of European case-studies. *Land Degrad. Dev.* **2018**, *29*, 2378–2389. [CrossRef]
60. Sahoo, G.; Wani, A.M. Effect of Climate Change on Land Degradation. *Int. J. Innov. Eng. Manag. Res.* **2020**, *9*, 483–494. [CrossRef]

horticulturae

MDPI

Article

Fishpond Water Potential on Vineyard Soil Health: An Exploratory Study of a Circular System

Isaac Kwadwo Mpanga [1,2,*], Toyosi Ijato [3], Harrison Kwame Dapaah [4] and Russell Tronstad [5,*]

1. Circular Planet Institute LLC, 5900 Bacons Drive, Suite 100, Austin, TX 78731, USA
2. College of Agriculture and Life Sciences, Cooperative Extension, The University of Arizona, 2830 N Commonwealth Dr, #103, Camp Verde, AZ 86322, USA
3. The United States Department of Agriculture, 1201 W. Gregory Drive, 197 ERML, Urbana, IL 61801, USA; babtoy1@gmail.com
4. Faculty of Agricultural Sciences, Akenten Appiah-Menka University of Skills Training and Entrepreneurial Development, Mampong-Ashanti P.O. Box 40, Ghana; hkdapaah@yahoo.com
5. Department of Agricultural and Resource Economics & Cooperative Extension, College of Agriculture and Life Sciences, The University of Arizona, McClelland Park 308, 650 N. Park Ave, Tucson, AZ 85721, USA

* Correspondence: mpangaisaac75@gmail.com (I.K.M.); tronstad@arizona.edu (R.T.)

Abstract: Climate variabilities continue to hinder sustainable food production with an increasing need to use resources such as water and soil efficiently. The quest for these efficiencies in agriculture systems drives innovations among farmers. However, limited data on farm practices, such as irrigating with fishpond water and their effects on soil health, hinder their adoption as climate-smart innovations. In a nearly twenty-year-old vineyard with two distinct irrigation practices (irrigation with recycled fishpond water and irrigation with ditch water), this study was carried out as an exploratory study to investigate the influence of recycling fishpond water on soil health parameters and yield. Soil samples were taken from two different irrigation fields in summer and winter for lab analysis on soil health parameters (organic matter and carbon, nitrogen, phosphorus, microbial biomass, and microbial respiration). Averages over the two seasons of field measurements indicate that long-term irrigation using recycling fishpond water increased the measured soil health parameters (organic matter (13%), organic carbon (30%), nitrogen (17%), phosphorus (46%), microbial biomass (18%), and microbial respiration (56%)) in both summer and winter months when compared to fields receiving just ditchwater irrigation. Using water in a way that can improve soil health increases biodiversity and improves the efficiency of our limited water resources in semi-arid agricultural lands, and this strategy is a climate-smart tool that can help reduce water risks in dry agricultural regions such as Arizona.

Keywords: circular agriculture; fishpond water; irrigation; soil health; regenerative agriculture

1. Introduction

Water, soil health, and biodiversity are major issues in semi-arid agriculture systems, especially as climate variabilities continue to increase and hinder sustainable food production. Nitrogen and phosphorus play key roles in plant development, and they have been harnessed in agricultural systems to improve crop yield and profitability [1,2] through synthetic and/or organic fertilizer source applications. In most agricultural systems, synthetic fertilizers are used to meet crop nutrient requirements. The efficient use of both organic and synthetic fertilizer sources is important to avoid excessive use in agriculture [3], environmental risks such as eutrophication, the leaching of nutrients into our aquifers, and more [4–8]. The need for these efficiencies in agriculture is shifting many practitioners' approaches toward circular agriculture systems.

Circular agriculture is a re-emerging ancient practice that was prominent among preindustrial societies but was pushed aside by modern large-scale farming based on monoculture [9] and highly intensive practices that prioritized farm-level profits over environmental concerns. The focus of circular agriculture is on the minimal use of external inputs, closing nutrient loops, and regenerating soils with practices that have lesser negative impacts on human health and the environment. Circular agriculture emphasizes promoting farming practices such as mixed cropping, crop rotations, waste reuse, and agroforestry practices [9] that are regenerative in nature for both plant and soil health improvement. In practice, a wider-scale acceptance and use of these regenerative and circular agricultural practices is common among smallholder farms, and circular practices can use resources more efficiently and reduce our agricultural ecological footprint [10], especially in resource-constraint areas such as arid and semi-arid regions.

In semi-arid regions, irrigation water, soil organic matter, and carbon are key drivers for improving soil health and providing sustainable crop production regimes. Soil organic matter and carbon are linear to soil nitrogen and phosphorus available to plants [11] and are good soil health indicators. Soil health is a concept that refers to the functionality of soil and its ability to maintain a balanced ecosystem with high biodiversity and productivity that is responsive to management practices [12,13]. Also, soil health includes the role of soil on water quality, climate change, and human health [14]. Techniques such as intercropping, cover crops, reduced tillage, no-till, nutrient management, and manure applications are high-ranking sustainable approaches that have been recommended for soil health improvement in the context of circular agriculture [15,16].

Animal waste products can also be incorporated into crop production systems as farm inputs, providing a nitrogen source with additional benefits such as improved soil structure, water holding capacity, and soil carbon, as well as the activity of beneficial microorganisms in the soil [17]. The proper management of all nitrogen sources is key to minimizing leaching into drinking water aquifers, whereas organic sources of fertilizers such as manure, plant waste material, and compost are challenging to manage well due to their slow and unpredicted release [18,19]. Also, the processing, handling, and transportation of manures has been a major limitation, but the concept of the "manure shed" (which has been identified as a major solution to the logistic problem of transporting manure) can be advantageous if maximized [20].

Exploring the interplays between an agroecosystem is also crucial, as nutrient pools that have been considered waste in some agricultural systems can be easily reutilized in other systems, thereby improving N use efficiencies within and between agroecosystems. Elsewhere, the irrigation of crop plants with water from fishponds has been demonstrated to improve crop yields [21,22]. They can also be used directly in hydroponic and/or aeroponic farming systems as nutritional sources for plants. Alternatively, fishpond effluent could be a good irrigation water source for pasture and crop fields as water scarcity continues to increase around the globe due to warmer temperatures, growing populations, and climate variability. For instance, in the United States and across the globe, the use of irrigation water has increased significantly in recent decades due to multiple extended drought conditions with a need for the sustainable use of irrigation water in agriculture. Urgency persists in semi-arid regions that have record low water levels such as the Southwest United States. The use of integrated approaches that combine technology with best management practices is needed to develop sustainable approaches to tackle our water crises [23,24] with additional benefits such as soil health. To contribute to the concept of sustainable water use in a circular agroecosystem that improves soil health and biodiversity for a resilient food system in Northern Arizona, this study was conceived. Here, we investigate the effect of irrigating with fishpond vs. ditchwater (control) on soil (organic matter, carbon, nitrogen, and phosphorus), microbial activities (respiration and biomass), and grape crop yields after two decades.

This exploratory study aims to understand how long-term furrow irrigation with recycling fishpond water may influence soil health (organic matter and carbon, nitrogen, phosphorus, and soil microbial biomass and respiration) in a semi-arid vineyard soil compared to furrow irrigation with just ditchwater from a mountain-fed creek.

2. Materials and Methods

2.1. Study Location, Field Layout, and Soil Sampling

Location: This exploratory study was carried out at Clear Creek Vineyard, Camp Verde, AZ, USA, where the irrigation system currently practiced was installed over two decades ago (Figure 1).

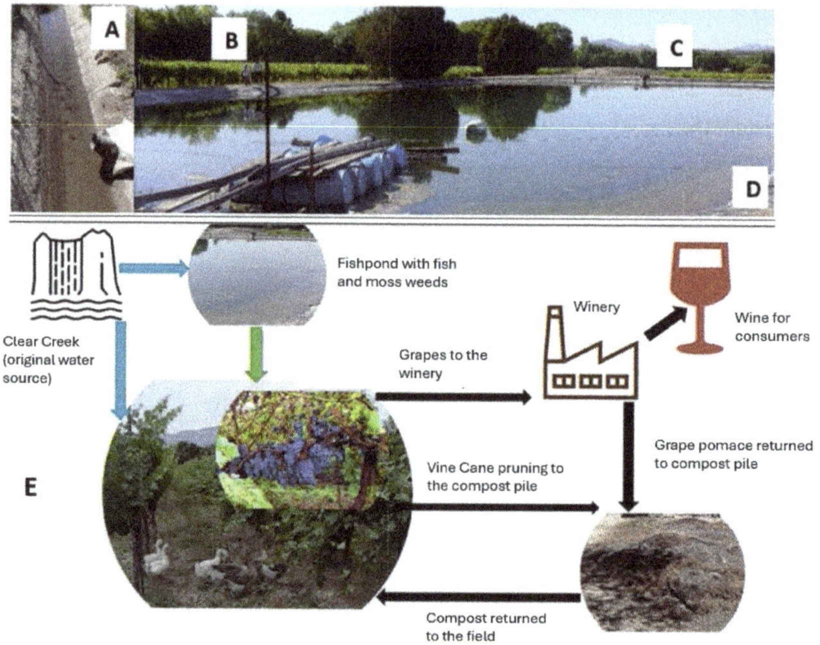

Figure 1. Clear Creek Vineyard, Camp Verde, Arizona, United States. The vineyard (**B**) is irrigated with either direct ditchwater (**A**) from Clear Creek near Camp Verde or stored in a fishpond (**D**) for 3–4 weeks before using it for the vineyard (**C**) in a circular agricultural setup (**E**).

History of vineyard: Clear Creek Vineyard and Winery was originally a horse and cattle pasture dating back to 1875, which changed after the current owner bought the property in 1999 and started a vineyard and winery. The vineyard is set up as a circular farm as all waste is composted and reused on the farm and only the wine processed from the grapes leaves the farm (Figure 1E).

The fishpond and creek: The fishpond (Figure 1D) was constructed and stocked with fish in 2005 (Figure 1A). The pond has fish species such as smallmouth bass, largemouth bass, bluegill, catfish, and trout in the winter. The pond also has moss weed, which serves as feedstock for the fish. The moss weed population is controlled every two years by manually pulling some out into the field. The chemical properties of both creek and pond water are presented in Table 1.

Planting distance of vines: The planting distance of the vines is 2 m between vines in the row and 3 m between the vine rows with a metal trellis system for training the vines.

Irrigation: The farm applies furrow irrigation from either the ditch or fishpond water at 100 cubic meters per hectare every three weeks depending on the weather.

Soil amendment and weed control: The compost made on the farm with waste from the winery and poultry manure sourced from Hickman Farm (Phoenix, AZ, USA) is applied to both fields at the same rate of two tons per hectare in spring. Weed Wacker, hand pulling, and geese are used alternatively as needed to control weeds and pests in the vineyard on both sides of the fields.

Pruning: Vine canes are pruned from February to March before bud break, and the management of excessive growth and fruit set is carried out in June after fruits are set. The June pruning is performed to have adequate airflow in the vineyard to prevent fungal infections and have quality fruit clusters with high sugar content.

Table 1. Chemical properties of the ditch/creek and pond water at Clear Creek Vineyard, Arizona.

	Ditchwater	Fishpond Water
pH	8.5	8.1
Sodium absorption ratio	0.2	0.3
Electrical conductivity (mmho/cm)	0.4	0.3
Cations (me/L)	4.0	4.1
Anions (me/L)	4.1	3.4
Sodium (mg/L)	7.0	8.0
Calcium (mg/L)	36.1	30.8
Magnesium (mg/L)	22.0	26.0
Potassium (mg/L)	3.0	4.0
Total hardness (mg/L)	182.0	186.0
Nitrate (mg/L)	0.1	0.1
Sulfur (mg/L)	2.0	1.0
CO_3 (mg/L)	3.7	<1.0
HCO_3 (mg/L)	227.0	185.0
Chloride (mg/L)	4.0	10.0
Total alkalinity (mg/L)	192.0	154.0

Field layout and Experimental design: The vineyard has historically been irrigated in two fields (side-by-side along a slight slope gradient) based on the type of water (ditch or fishpond) used for irrigation. The field irrigated with ditchwater has received water only from the ditch since 1999, while the field irrigated from the fishpond has received irrigation water only from the fishpond since 2005. To replicate our collection of samples, each field was subdivided into three sections along the slope gradient.

Soil sampling: Soil samples were collected one foot from vine trunks at 0–15 cm depth in the summer (June) and winter (December) of 2021. On each replicate, ten composite samples were collected within the grape plants using a soil probe. The soil samples were collected on ice in a chest and frozen immediately after sampling at −20 °C for one week before shipping with ice packs to the laboratory (Ward Laboratories, Inc., Kearney, NE, USA) for soil health indicator measurements.

2.2. Soil Analysis

The Haney test was used to analyze N, P, organic carbon, organic matter, and soil respiration at Ward Laboratories (website at www.wardlab.com), as described below.

Soil preparation: Each soil sample received at Ward Laboratories was dried at 50 °C, ground to pass a 2 mm sieve, and weighed into two 50 mL Erlenmeyer flasks (4 g each) and one 50 mL plastic beaker (40 g) that is perforated to allow water infiltration.

Soil NH_4-N and PO_4-P: Two 4 g samples were extracted with 40 mL of DI water and 40 mL of N-(3-chlorophenyl)-2-(3-methoxyphenyl) acetamide (H3A) (Sigma-Aldrich, Inc., St. Louis, MO, USA), respectively. The soil extractant mixtures were shaken for 10 min, centrifuged for 5 min, and filtered through Whatman 2V filter paper. The water and H3A extracts were analyzed on a Lachat 8000 flow injection analyzer (Hach Company, Loveland, CO, USA) for NO_3-N, NH_4-N, and PO_4-P [25].

Organic carbon and matter: The water extract from the 4 g soil samples was analyzed with Teledyne Tekmar Torch C: N analyzer (LabX Media Group, Midland, ON, Canada) for water-extractable organic C and N, which represent water-extractable organic C (WEOC) or the amount of organic C in the soil, which is roughly 80 times smaller than the total soil organic C pool (% organic matter) and reflects the energy source feeding soil microbes [25].

Total soil microbial biomass: The phospholipid fatty acid (PLFA) analytical method was used to determine soil microbial biomass. Frozen soil samples were shipped on dry ice to Ward Laboratories (WARD Laboratories, Inc., Kearney, NE, USA) for processing and analysis. The PLFA analyses in Ward Laboratories were based on the method described in [26], where total soil lipid fractions were extracted by shaking a mixture of 2 g frozen soil in 9.5 mL of a liquid buffer of dichloromethane, methanol, and citrate buffer (1:2:0.8 v/v) consistently for 1 h at 240 rpm. After lipid-class separation and creating fatty acid methyl esters, samples were then analyzed using an Agilent 7890A GC (OGMI, Ramsey, NJ, USA) equipped with a 7693 autosampler instrument and a flame ionization detector, which quantified the total microbial biomass.

Soil microbial respiration: The 40 g soil sample was analyzed in a 24 h incubation test at 24 °C. The soil was wetted through capillary action with 20 mL of deionized (DI) water in a 236.6 mL glass jar, capped, and incubated for 24 h. After 24 h, the gas inside the jar was analyzed using an infrared gas analyzer (IRGA) Li-Cor 840A (LI-COR Biosciences, Lincoln, NE, USA) for CO^2-C [25].

Statistical analysis: Data were first checked for normality using Proc GLM with plot = diagnostics followed by a t-test at p equal or less than 0.05 using two-tailed distributions and homogeneity of variances (homoscedasticity).

3. Exploratory Results

3.1. Grape Yields

Irrigation with fishpond water increased *Cabernet sauvignon* grape fruit cluster numbers, length (Figure 2a,b), and weight per plant and hectare when compared to the control treatment of just ditchwater (Figure 2c,d).

Figure 2. Average grape (Cabernet sauvignon variety) yields based on the two fields (ditchwater and fishpond irrigation water) (* = different at $p = 0.05$).

3.2. Soil pH, Organic Matter, and Carbon

Interestingly, ditchwater stored in the fishpond with aquatic life before using it to irrigate the vineyards seems to have increased organic matter by 17% in the summer and 8% in the winter soil samples. Also, in fishpond irrigation, the total organic carbon was higher in both summer and winter soil samples at 40% and 18%, respectively, when compared to irrigating the ditchwater coming directly from Clear Creek (Table 2). Generally, summer month soil samples had higher organic matter and carbon, and rather minimal differences were observed in organic matter and carbon in the winter soil samples (Table 2). Also, no differences were observed in the soil pH between sampling seasons (Table 2).

3.3. Soil Nitrate and Ammonium

The use of fishpond water increased organic total N in both summer and winter soil samples by 36% and 10%, respectively, when compared to ditchwater irrigation. Inorganic N was also slightly higher in the fishpond-irrigated section with a 13% increase in both seasons. Nitrate analysis showed that fishpond irrigation had a minimal advantage over ditchwater irrigation in both summer and winter at 9%, with lower values in the winter samples. Ammonium forms between the treatments in the two sampling periods showed relatively high values in the fishpond-irrigated field for the summer (Table 3).

Table 2. Effects of fishpond irrigation water on semi-arid vineyard soil pH and organic matter.

	Soil pH	Organic Matter (%)	Total Organic C (mg C/kg Soil)
		Summer soil samples	
Ditchwater-irrigated field	8.1 [a]	3.9 [b]	138.7 [b]
Fishpond-water-irrigated field	8.1 [a]	4.6 [a]	193.7 [a]
t-test ($p = 0.05$)	na	0.0	0.0
		Winter soil samples	
Ditchwater-irrigated field	8.1 [a]	3.7 [a]	123.3 [a]
Fishpond-water-irrigated field	8.0 [a]	4.0 [a]	145.7 [a]
t-test ($p = 0.05$)	na	0.6	0.8

Same letter = no difference at $p = 0.05$. na = not applicable.

Table 3. Effects of recycling fishpond and direct ditch irrigation water on semi-arid vineyard soil nitrogen.

	Organic Nitrogen (mg N/k Soil)	Inorganic Nitrogen (mg N/kg Soil)	Nitrate (mg N/kg Soil)	Ammonium (mg N/kg Soil)
		Summer soil samples		
Ditchwater-irrigated field	24.0 [a]	47.6 [a]	46.5 [a]	1.1 [a]
Fishpond-water-irrigated field	32.7 [a]	53.7 [a]	50.7 [a]	3.0 [a]
t-test ($p = 0.05$)	0.1	0.1	0.1	0.1
		Winter soil samples		
Ditchwater-irrigated field	19.2 [a]	30.9 [a]	30.2 [a]	1.3 [a]
Fishpond-water-irrigated field	21.2 [a]	34.9 [a]	33.0 [a]	1.9 [a]
t-test ($p = 0.05$)	0.3	0.1	0.1	0.4

Same letter = no difference at $p = 0.05$.

3.4. Soil Phosphorus

The fishpond irrigation water had higher effects on inorganic phosphorus at 47%; however, a 4% increase for organic phosphorus was observed in favor of irrigation with just ditchwater in the summer soil samples. Considering the winter soil samples, when compared to irrigating with only ditchwater, fishpond irrigation seems to have the advantage of increasing both organic P (17%) and inorganic P (19%) (Table 4).

Table 4. Effects of recycling fishpond and direct ditch irrigation water on semi-arid vineyard soil phosphorus.

	Inorganic Phosphorus (mg P/kg Soil)	Organic Phosphorus (mg P/kg Soil)
	Summer soil samples	
Ditchwater-irrigated field	23.3 [b]	9.0 [a]
Fishpond-water-irrigated field	34.3 [a]	8.6 [a]
t-test ($p = 0.05$)	0.0	0.8
	Winter soil samples	
Ditchwater-irrigated field	14.4 [b]	7.4 [b]
Fishpond-water-irrigated field	17.1 [a]	8.7 [a]
t-test ($p = 0.05$)	0.00	0.0

Same letter = no difference at $p = 0.05$.

3.5. Soil Microbial Respiration and Total Microbial Biomass

After almost two decades of applying irrigation water from the fishpond, soil microbial respiration and biomass have increased when compared to direct irrigation with only ditchwater (Table 5). In the summer months soil samples, the total microbial biomass, total bacterial biomass, total fungal biomass, and soil respiration increased by 13%, 27%, 29%, and 59%, respectively, under fishpond irrigation when compared to ditchwater irrigation. In the winter soil samples, the total microbial biomass, total bacterial biomass, total fungal biomass, and soil respiration increased at 12%, 11%, 30%, and 47%, respectively, under fishpond irrigation when compared to ditchwater irrigation. Comparatively, summer samples had higher microbial activities than the winter samples (Table 5 and [27]).

Table 5. Effects of ditch and fishpond irrigation water on semi-arid vineyard soil's total soil microbial biomass, bacterial biomass, fungal biomass, and microbial respiration.

	Total Microbial Biomass (PLFA ng/g)	Total Bacterial Biomass (PLFA ng/g)	Total Fungal Biomass (PLFA ng/g)	Soil Respiration (mg CO_2-C /kg Soil)
		Summer soil samples		
Ditchwater-irrigated field	1807.0 ^{a}	652.7 ^{a}	101.5 ^{a}	88.7 ^{b}
Fishpond-water-irrigated field	2041.2 ^{a}	826.1 ^{a}	131.0 ^{a}	141.0 ^{a}
t-test ($p = 0.05$)	0.6	0.1	0.1	0.0
		Winter soil samples		
Ditchwater-irrigated field	1064.1 ^{a}	340.1 ^{a}	5.9 ^{a}	38.2 ^{b}
Fishpond-water-irrigated field	1189.9 ^{a}	378.4 ^{a}	7.7 ^{a}	56.2 ^{a}
t-test ($p = 0.05$)	0.5	0.5	0.8	0.0

Same letter = no difference at $p = 0.05$.

4. Discussion

Irrigation water and soil health are major challenges for sustainable and profitable crop production in the Southwest United States and semi-arid regions in general, especially among small-to-medium-sized farms due to increasing temperatures, drought, and reduced aquifer and reservoir levels. The Southwest United States can be more resilient in food production using circular agricultural systems by using waste streams more efficiently, closing nutrient loops, regenerating soils, and minimizing environmental impacts [9]. This study highlights the effects of recycling fishpond irrigation water as a catalyst for improving soil health and biodiversity in a circular agricultural model vineyard in Northern Arizona.

The study results reveal that watering with fishpond water for almost two decades has positively affected grape yields compared to irrigating with just ditchwater (Figure 2), which is consistent with similar studies from other locations and systems. In different studies, fishpond water increased paddy rice [21,22,28] and wheat [29] yields. The positive yield effects from fishponds relate to soil fertility and health improvements, as observed by others [28,29], which is also consistent with the findings of this study. Irrigating with fishpond water improved soil health compared to irrigating with just ditchwater (Tables 1–4).

Yuan et al. reported higher paddy rice yields [28], while Abdelraouf and Ragab reported a yield increase in wheat [29] from using fishpond water compared to freshwater, which was associated with increases in nitrogen and phosphorus and was also confirmed by this study (Tables 3 and 4). Recall that both ditchwater- and fishpond-water-irrigated fields receive compost and manure from the geese on the farm. Thus, the additional nitrogen and phosphorus from fishpond-water-irrigated fields are mainly related to droppings and organic matter from fish and other living organisms such as moss, as well as the migratory birds visiting the pond during the winter. Therefore, these systems represent a classical integration of fish farming into semi-arid agriculture, where nutrients are sustainably recycled from biodiversity and integrated into agroecosystems with the extra benefits of harvesting fish from the pond for food or household income.

Similarly, the highest amount of organic and inorganic phosphorous recorded was in the soil irrigated with fishpond water (Table 4). Phosphorus in the soil system is mainly available as non-labile P, labile P, and solution P [30]. However, the highest amount of phosphorous was in the summer for all treatments, which is related to organic matter and carbon that is mostly in the mineralized form during the summer (Table 4). Not only did this integration benefit major nutrient cycles from the increased biodiversity, but the organic matter content of the soil also increased in the fishpond-irrigated fields (Table 2). Organic matter and carbon have a linear correlation with nitrogen and phosphorus availability [11], which could be why more nitrogen and phosphorus were observed in the field irrigated with fishpond water. The higher organic matter and carbon in the soils of fishpond-irrigated fields could be attributed to the fish and bird feces and moss plants in the pond, which supports soil health for more biomass in those areas. Though no differences were observed in the soil pH, the slightly lower pH of the pond water (Table 1) could support the higher P availability observed in Table 4.

The activity of soil microorganisms is of great importance to soil health [31]. Therefore, irrigation from the pond water not only added substrate in the form of organic matter but also induced respiration and microbial activities such as microbial, fungal, and bacterial biomass. The increase in microbial activities in the fishpond treatment could be related to lower EC values for the fishpond water when compared to the ditchwater's slightly higher values (Table 1). According to Rietz and Haynes, irrigation water with higher EC and salt reduces microbial activities [32], which is similar to the findings of this study, except that the fishpond water in this study seems to have fairly high salt values, which makes it difficult to relate the positive soil health activities to low EC or salt values. Also, Chen et al. reported a higher diversity in the bacterial community they studied in aquaculture-wastewater-irrigated soils compared to freshwater-irrigated soils [33].

Irrigation water sources from fishponds could be a driving tool for healthy semi-arid soils [34] with added benefits such as increased productivity, long-term profitability, and environmental safety. The use of fishpond water for irrigation comes with additional benefits to the soil, plants, farmer's harvest and income, and support to biodiversity and the environment (Figure 3).

Figure 3. Pros and cons of using ditchwater versus fishpond water for crop irrigation.

Fishponds provide greater habitat for birds, especially during the winter, and aquatic life compared to irrigation ditches that are always flowing fast and may not have water during winter months. The location under study is in a somewhat tempered climate compared to the low desert or more northern climates where water temperatures can be too hot or cold for good year-round fish growth and production. Nonetheless, in areas where geographic and climate conditions are conducive to constructing a fishpond that can water relatively modest-sized fields, a fishpond should be considered for enhanced water, soil, and environmental benefits, even though the initial pond construction costs will be high (Figure 3). The Clear Creek Vineyard owner loves the fishpond not just for the enhanced water, soil, and environmental amenities it provides but also because it makes their property more attractive to tourists, wine tasters, and future potential bidders.

5. Conclusions

In semi-arid regions, integrating circular agricultural practices such as fishponds into cropping systems can enhance the resource use efficiency of water and improve environmental indicators (soil health and biodiversity). Since these ancient practices are regaining attention within agricultural production systems as climate-smart solutions, more research and extension program collaborations should be developed to make these systems more sustainable and resilient for safe food production. Our results suggest that more formal research is warranted on this subject for longer time periods with randomization of more trials to further prove the practice of using fishponds. The limitations associated with the use of fishponds are initial costs and concerns about pond surface evaporations, which we have not addressed. Resource conservation policies need to consider these costs along with the expected benefits before implementation.

Author Contributions: I.K.M., conceptualization, design, data collection, analysis, manuscript drafting, and final editing; T.I., data analysis, first draft writing, and final editing; R.T., critical review, the editing of the first draft, and final draft completion; H.K.D., critical review and the editing and review of the final draft. All authors have read and agreed to the published version of the manuscript.

Funding: This research received no external funding.

Data Availability Statement: The data presented in this study are available upon request from the corresponding author (accurately indicate status).

Acknowledgments: The authors thank Ignacio Mesa who made his farm available for this research.

Conflicts of Interest: Isaac Kwadwo Mpanga was employed by the Circular Planet Institute LLC. The remaining authors declare that the research was conducted in the absence of any commercial or financial relationships that could be construed as a potential conflict of interest.

References

1. Albornoz, F. Crop responses to nitrogen overfertilization: A review. *Sci. Hortic.* **2016**, *205*, 79–83. [CrossRef]
2. Hendricks, G.; Shukla, S.; Roka, F.; Sishodia, R.; Obreza, T.; Hochmuth, G.; Colee, J. Economic and environmental consequences of overfertilization under extreme weather conditions. *J. Soil Water Conserv.* **2019**, *74*, 160–171. [CrossRef]
3. Innes, R. Economics of Agricultural Residuals and Overfertilization: Chemical Fertilizer Use, Livestock Waste, Manure Management, and Environmental Impacts. In *Encyclopedia of Energy, Natural Resource, and Environmental Economics*; Elsevier Science: Waltham, MA, USA, 2013; pp. 50–57. [CrossRef]
4. Andersen, J.H.; Conley, D.J.; Hedal, S. Palaeoecology, reference conditions and classification of ecological status: The EU Water Framework Directive in practice. *Mar. Pollut. Bull.* **2004**, *49*, 283–290. [CrossRef]
5. Bouwman, L.; Goldewijk, K.K.; Van Der Hoek, K.W.; Beusen, A.H.W.; van Vuuren, D.P.; Willems, J.; Rufino, M.C.; Stehfest, E. Exploring global changes in nitrogen and phosphorus cycles in agriculture induced by livestock production over the 1900–2050 period. *Proc. Natl. Acad. Sci. USA* **2013**, *110*, 20882–20887. [CrossRef]
6. Fowler, D.; Coyle, M.; Skiba, U.; Sutton, M.A.; Cape, J.N.; Reis, S.; Sheppard, L.J.; Jenkins, A.; Grizzetti, B.; Galloway, J.N.; et al. The global nitrogen cycle in the twenty-first century. *Philos. Trans. R. Soc. B Biol. Sci.* **2013**, *368*, 20130164. [CrossRef]
7. Yang, X.E.; Wu, X.; Hao, H.L.; He, Z.L. Mechanisms and assessment of water eutrophication. *Zhejiang Da Xue Xue Bao Li Xue Ban* **2008**, *9*, 197–209. [CrossRef]
8. Shah, T.; Lateef, S.; Noor, M.A. Carbon and Nitrogen Cycling in Agro-ecosystems: An Overview. In *Carbon and Nitrogen Cycling in Soil*; Datta, R., Meena, R.S., Pathan, S.I., Ceccherini, M.T., Eds.; Summerer Singapore: Singapore, 2020; Volume 348, pp. 1–15.

9. Helgason, K.S.; Iversen, K.; Julca, A. Circular Agriculture for Sustainable Rural Development. UN/DESA Policy, 2021, Brief #105. Available online: https://www.un.org/development/desa/dpad/publication/un-desa-policy-brief-105-circular-agriculture-for-sustainable-rural-development/ (accessed on 10 December 2021).
10. Mpanga, I.K.; Schuch, U.K.; Schalau, J. Adaptation of resilient regenerative agricultural practices by small-scale growers towards sustainable food production in north-central Arizona. *Curr. Res. Environ. Sustain.* **2021**, *3*, 100067. [CrossRef]
11. Yang, X.; Chen, X.; Yang, X. Effect of organic matter on phosphorus adsorption and desorption in a black soil from Northeast China. *Soil Tillage Res.* **2019**, *187*, 85–91. [CrossRef]
12. Cardoso, E.J.B.N.; Vasconcellos, R.L.F.; Bini, D.; Miyauchi, M.Y.H.; dos Santos, C.A.; Alves, P.R.L.; de Paula, A.M.; Nakatani, A.S.; Pereira, J.d.M.; Nogueira, M.A. Soil health: Looking for suitable indicators. What should be considered to assess the effects of use and management on soil health? *Sci. Agricola* **2013**, *70*, 274–289. [CrossRef]
13. Lal, R. Soil health and carbon management. *Food Energy Secur.* **2016**, *5*, 212–222. [CrossRef]
14. Lehmann, J.; Bossio, D.A.; Kögel-Knabner, I.; Rillig, M.C. The concept and future prospects of soil health. *Nat. Rev. Earth Environ.* **2020**, *1*, 544–553. [CrossRef] [PubMed]
15. Mpanga, I.K.; Tronstad, R.; Guo, J.; LeBauer, D.S.; Idowu, O.J. On-farm land management strategies and production challenges in United States organic agricultural systems. *Curr. Res. Environ. Sustain.* **2021**, *3*, 100097. [CrossRef]
16. Wang, M.; Pendall, E.; Fang, C.; Li, B.; Nie, M. A global perspective on agroecosystem nitrogen cycles after returning crop residue. *Agric. Ecosyst. Environ.* **2018**, *266*, 49–54. [CrossRef]
17. Zhang, Q.; Zhou, W.; Liang, G.; Wang, X.; Sun, J.; He, P.; Li, L. Effects of Different Organic Manures on the Biochemical and Microbial Characteristics of Albic Paddy Soil in a Short-Term Experiment. *PLoS ONE* **2015**, *10*, e0124096. [CrossRef] [PubMed]
18. Rashid, M.T.; Voroney, R.P. Recycling Soil Nitrate Nitrogen by Amending Agricultural Lands with Oily Food Waste. *J. Environ. Qual.* **2003**, *32*, 1881–1886. [CrossRef]
19. Adjei, E.; Santo, K.G.; Poku, I.A.; Alexander, D.-A.; Antwi, A.O. Garden Egg (*Solanum melongena* L.) Performance under Different Sources of Animal Manure as a Sustainable Alternative Fertilizer for Farmers. *Agric. Sci.* **2023**, *14*, 1053–1067. [CrossRef]
20. Spiegal, S.; Kleinman, P.J.; Endale, D.M.; Bryant, R.B.; Dell, C.; Goslee, S.; Meinen, R.J.; Flynn, K.C.; Baker, J.M.; Browning, D.M.; et al. Manuresheds: Advancing nutrient recycling in US agriculture. *Agric. Syst.* **2020**, *182*, 102813. [CrossRef]
21. Tsuruta, T.; Yamaguchi, M.; Abe, S.-I.; Iguchi, K. Effect of fish in rice-fish culture on the rice yield. *Fish. Sci.* **2011**, *77*, 95–106. [CrossRef]
22. Oehme, M.; Frei, M.; Razzak, M.A.; Dewan, S.; Becker, K. Studies on nitrogen cycling under different nitrogen inputs in integrated rice-fish culture in Bangladesh. *Nutr. Cycl. Agroecosystems* **2007**, *79*, 181–191. [CrossRef]
23. Mpanga, I.K.; Idowu, O.J. A Decade of Irrigation Water use trends in Southwestern USA: The Role of Irrigation Technology, Best Management Practices, and Outreach Education Programs. *Agric. Water Manag.* **2021**, *243*, 106438. [CrossRef]
24. Mpanga, I.K.; Gaikpa, D.S.; Koomson, E.; Dapaah, H.K. Innovations in Water Management: Agriculture. In *The Palgrave Handbook of Global Sustainability*; Palgrave Macmillan: Cham, Switzerland, 2022. [CrossRef]
25. Haney, R.L.; Haney, E.B.; Smith, D.R.; Harmel, R.D.; White, M.J. The soil health tool—Theory and initial broad-scale appli-cation. *Appl. Soil Ecol.* **2018**, *125*, 162–168. [CrossRef]
26. Hamel, C.; Hanson, K.; Selles, F.; Cruz, A.F.; Lemke, R.; McConkey, B.; Zentner, R. Seasonal and long-term resource-related variations in soil microbial communities in wheat-based rotations of the Canadian prairie. *Soil Biol. Biochem.* **2006**, *38*, 2104–2116. [CrossRef]
27. Luo, S.-P.; He, B.-H.; Zeng, Q.-P.; Li, N.-J.; Yang, L. Effects of seasonal variation on soil microbial community structure and enzyme activity in a Masson pine forest in Southwest China. *J. Mt. Sci.* **2020**, *17*, 1398–1409. [CrossRef]
28. Zhou, Y.; Zhu, J.Q.; Li, G. The nutrients in the fertile water from fish pond assimilated and utilized by paddy field. In Proceedings of the 2011 International Symposium on Water Resource and Environmental Protection (ISWREP), Xi'an, China, 20–22 May 2011; pp. 2695–2698.
29. Abdelraouf, R.; Ragab, R. The Benefit of Using Drainage Water of Fish Farms for Irrigation: Field and Modelling Study Using the SALTMED Model. *Irrig. Drain.* **2017**, *66*, 758–772. [CrossRef]
30. Penn, C.J.; Camberato, J.J. A Critical Review on Soil Chemical Processes that Control How Soil pH Affects Phosphorus Availability to Plants. *Agriculture* **2017**, *9*, 120. [CrossRef]
31. Tahat, M.M.; Alananbeh, K.M.; Othman, Y.A.; Leskovar, D.I. Soil Health and Sustainable Agriculture. *Sustainability* **2020**, *12*, 4859. [CrossRef]
32. Rietz, D.; Haynes, R. Effects of irrigation-induced salinity and sodicity on soil microbial activity. *Soil Biol. Biochem.* **2003**, *35*, 845–854. [CrossRef]
33. Chen, L.; Feng, Q.; Li, C.; Wei, Y.; Zhao, Y.; Feng, Y.; Zheng, H.; Li, F.; Li, H. Impacts of aquaculture wastewater irrigation on soil microbial functional diversity and community structure in arid regions. *Sci. Rep.* **2017**, *7*, 11193. [CrossRef] [PubMed]
34. Mitchell, J.P.; Ferris, H.; Shrestha, A.; Larney, F.; Sposito, G. Managing irrigation for soil health in arid and semi-arid regions. In *Managing Soil Health for Sustainable Agriculture*; Reicosky, D., Ed.; Burleigh Dodds Series in Agricultural Science; Burleigh Dodds Science Publishing: Cambridge, UK, 2018; Volume 2, pp. 149–162.

horticulturae

MDPI

Article

Differences in Soil Water Holding Capacity and Available Soil Water along Growing Cycle Can Explain Differences in Vigour, Yield, and Quality of Must and Wine in the DOCa Rioja

José María Martínez-Vidaurre [1], Eva Pilar Pérez-Álvarez [1], Enrique García-Escudero [1], María Concepción Ramos [2] and Fernando Peregrina [3,*]

[1] Instituto de Ciencias de la Vid y del Vino-ICVV, Universidad de La Rioja, Gobierno de La Rioja, CSIC, 26007 Logroño, Spain; jmvidaurre@larioja.org (J.M.M.-V.); evapilar.perez@icvv.es (E.P.P.-Á.); egescudero@larioja.org (E.G.-E.)

[2] Departamento de Química, Física y Ciencias Ambientales y del Suelo, Universitat de Lleida-Agrotecnio Center CERCA, 25198 Lleida, Spain; mariaconcepcion.ramos@udl.cat

[3] Área de Edafología, Departamento Producción Agraria, E. T. S. I. Agronómica, Alimentaria y de Biosistemas, Universidad Politécnica de Madrid, 28040 Madrid, Spain

* Correspondence: fernando.peregrina@upm.es; Tel.: +34-910671046

Abstract: Soil water availability during the vine growth cycle can affect yield and grape quality. The objective was to evaluate the effect of soil water holding capacity (AWC) and available soil water (ASW) throughout the growing cycle on the nutritional status, vigor, production, and composition of grapes and wine. The study was conducted in the municipality of Uruñuela in the DOCa Rioja (Spain). The soils of four rainfed vineyard plots were characterized to determine AWC and its impacts on vine, grape, and wine composition. The N, P, and K foliar content, vigor, grape yield, berry weight, and composition of must and wine were analyzed in those vineyard plots during the period 2010–2014. The ASW was simulated in each plot and each year analyzed, considering the soil properties and the weather conditions, after model calibration in one plot in which soil water content was registered. The results showed that AWC influenced ASW along the growing cycle, so vines suffered from water stress in some periods of the vegetative cycle. Plots with higher AWC had higher ASW from fruit set to ripening and lower water stress during this period, which explains the higher N, P, and K foliar content, vigor and grape yield, and lower polyphenol and anthocyanin content in grapes and wines. The period where water availability had the most influence on the quality of the grapes was from veraison to ripening, during which ASW increased berry weight and acidity and decreased anthocyanins and polyphenolic compounds.

Keywords: soil rooting depth; polyphenols; anthocyanins; hydric stress

1. Introduction

Within the same climatic zone, the soil characteristics determine vine development and grape composition [1,2]. The depth of the soil, its texture, and porosity regulate the amount of soil accessible to roots, affecting the movement of water and nutrients, as well as soil water storage [3–5]. Furthermore, they impact water availability, subsequently influencing the water status of the vine [6,7]. Therefore, the water availability in the vineyard soil could be used to estimate the water stress on grapevines.

Water availability can affect the uptake by the vine of principal nutrients such as N, P, and K from the soil. Because the ionic forms NO_3^-, $H_2PO_4^-$, and K^+ are actively absorbed and require an energy input in the form of ATP, under hydric stress, the vine reduces its photosynthetic activity as a result of the stomata closure [8]. The reduction in the assimilation of these nutrients has a great impact on the vineyard, especially N, due to its great influence on vine vigor [9] and on the polyphenol and anthocyanin contents of the grape [10]. While K affects the flow of water in the plant and the acidity conditions of the grape, which are crucial for obtaining a quality wine [11], water stress can affect grapevines at different times during the growth cycle. Additionally, shoot growth is reduced under any level of water stress [8]. Furthermore, deficits in water availability, occurring during both the early and late stages of the growing season, lead to a reduction in yields [12].

Berry composition is also affected by soil properties, vine management, cultivar, and climate [13]. Thus, water deficits of up to 50% of crop evapotranspiration (ETc) have a minimal effect on yield but yield decreases if this threshold is exceeded [14]. So, Ramos and Martínez-Casasnovas [15] found that this effect on yield occurred in years when the water deficit was higher than 50% ETc during the growing season.

Berry growth is affected by water deficit, which also has an impact on grape quality [16–20]. This impact depends on the intensity of water stress and also on when this stress appears [21–23]. Sugar accumulation in berries is affected by a late water deficit, as this water stress reduces leaf photosynthesis [24]. The increase in irrigation rates provokes an increase in total acidity, while the malic acid concentration decreases under both regulated irrigation and prolonged deficit irrigation, resulting in an increase in the tartaric/malic acid ratio [25]. In addition, water stress before veraison negatively affects titratable acidity, malic acid, and aromatic quality [26]. But this hydric stress can have a positive effect by increasing the concentration of anthocyanin and polyphenols [27].

Thus, in irrigation treatments, the total polyphenol content decreased, with the lowest anthocyanin and phenol concentrations under the irrigation rate corresponding to 100% of crop evapotranspiration (ETc) [28].

According to these results, Cooley et al. [25] and Lizama et al. [29] observed that wine color density, anthocyanin, and polyphenolic levels decrease with increasing irrigation rates [30]. Bucchetti et al. [30] reported that water deficits increase the anthocyanin content per berry, reduce fruit growth, and consistently increase anthocyanin concentration.

In areas where vines are usually cultivated under rainfed conditions because of water scarcity, the vine response mainly depends on soil water content, which will be driven by the rainfall amount, its distribution along the cycle, and the soil properties. Thus, high variability can be recorded from year to year.

This research aims to analyze the variability in the response of grapevines cultivated in rainfed conditions to various water stress scenarios and their impact on plant development. The research was conducted in Rioja DOCa on the Tempranillo (*Vitis vinifera* L.) cultivar, which, due to this appellation, is the largest producer of that variety worldwide. The aim was to analyze the variability in the response of grapevine to water stress under rain-fed conditions over five years (2010–2014). Focused on: (i) nutritional status, vigor, and yield; (ii) grape and wine composition; and (iii) relationships between soil water content and vine, grape, and wine parameters throughout the grapevine growth cycle.

2. Materials and Methods

2.1. Area of Study

The plots used in the study are in the municipality of Uruñuela, belonging to the DOCa Rioja, and more specifically to the Rioja Alta subzone. The description of the geomorphology and soils of this area is given in [31].

The study area has a Mediterranean climate, according to the UNESCO aridity index, with a slight oceanic influence. Weather conditions throughout the study period were recorded on a daily scale using an agro-climatic station operated by the Government of La Rioja (www.larioja.org/siar, accessed 20 February 2024), situated within the same mesoclimate area (latitude: 42°27′43″ N; longitude: 2°42′46″ W, altitude: 465 m a.s.l.). Figure 1 presents data on annual average temperature, annual average precipitation, annual solar radiation (MJ m^{-2}), and annual potential evapotranspiration, as well as precipitation recorded at intervals throughout the vine cycle in the studied years.

Figure 1. Climatic data of years 2010 to 2014 and the historical series 2004 to 2018 (**a**) precipitation and evapotranspiration, and (**b**) annual solar radiation and annual average temperature.

2.2. Experimental Design and Vineyard Plot Description

Four commercial vineyards of the Tempranillo variety (*Vitis vinifera* L.) grafted on Richter-110 rootstock were selected for this study. The vineyards were located on platforms with slopes less than 2% (less than one km apart) and were selected for this study. In consequence, climatic conditions can be considered homogenous. The study was conducted over a five-year period (from 2010 to 2014). Within each vineyard, three adjacent rows of 50 vines (plots) were selected. Therefore, each row serves as a repetition for sampling and determinations on vines and berries.

The grapevines were between 20 and 35 years old, with a planting density of 3086 to 3137 grapevines per hectare (Table 1), with vines and row spacing of 1.20 × 2.70 m (vine and row spacing). Double cordon and gobelet were the training systems. Chemical weed control was performed in the row under the vines. In addition, in the interrow, the soil was tilled with a cultivator to eliminate weeds every four to six weeks from February to August.

Table 1. Soil characteristics and plant density for each vineyard plot.

Plot	Plant Density ha^{-1}	Soil Classification (USDA, 2006)	pH (H_2O)	E.C. [1] dS/m	O.M. %	Clay %	Silt %	Sand %	$CaCO_3$ %	Available Water Capacity mm
			Ap Horizon							Section Control
P21	3086	Fluventic Haploxerept	8.15	0.15	1.00	20.6	38.8	40.6	0.5	128.5
P22	3086	Fluventic Haploxerept	8.20	0.13	1.05	24.1	28.7	47.2	1.3	146.5
P53	3137	Typic Calcixerept	8.35	0.14	0.97	18.5	52.5	29.0	3.5	56.3
P63	3086	Petrocalcic Palexerolls	8.40	0.15	1.87	22.9	40.0	37.1	14.7	59.9

[1] E.C.: electric conductivity; O.M.: organic matter.

2.3. Soil Description and Soil Analysis

On 24 May 2010, two soil pits were excavated in each plot for the purpose of describing and sampling the soil horizons. A composite sample of 3 sub-samples (approximately 2 kg) was sampled from each horizon. The soil samples were subsequently air-dried, sieved at 2 mm, and subjected to various analyses. The analyses included pH and electrical conductivity in water (at a ratio of 1:5 soil/solution), organic matter determined by dichromate oxidation [32], soil texture assessed by laser diffraction particle size analyzer (Diffractometer LS^{TM} 13 320, Beckman Coulter, Brea, CA, USA), total carbonate measured by infrared (EQUILAB CO-202, Equilab, Jakarta, Indonesia), active lime content determined through volumetric analysis (Bernard Calcimeter), cation exchange capacity (CEC) using the cobaltihexamine method [33], and extractable potassium analyzed via the Mehlich 3 method [34].

The equations formulated by Saxton and Rawls [35] were used to calculate the soil water holding capacity on each horizon. This process involved incorporating parameters such as electrical conductivity, organic matter content, particle size distribution, and the volume percentage of coarse elements into the calculations. The collective soil water holding capacity for each soil was determined by aggregating the water holding capacities of individual horizons.

Table 1 presents the soil classification and main soil physicochemical characteristics of each plot.

2.4. Calculation of Available Soil Water

Soil water was simulated for each plot and year, considering the respective soil properties and weather conditions recorded in each year. The AWC was simulated for each plot based on the climatic conditions in each study year. The simulation was conducted using the Vineyard-Soil Irrigation Model (VSIM—https://www.pvts.net/pdfs/VSIM_Models1/VSIM_5_03.PDF, accessed 20 February 2024). The model performs a daily water balance to determine the water content of the entire soil profile from water inputs, crop evapotranspiration, and drainage. The model also considers the spacing between vines and rows and the cover crop. In a vineyard in the same area, soil moisture data were determined using TDT (Time Domain Transmissometry) GroGraph Moisture Solution probes (ESI Environmental Sensors Inc., Sidney, BC, Canada) at depths of 30, 60, and 100 cm. With this data, the model was calibrated and validated as described in [36].

Following the validation and calibration of the model, the available soil water (ASW) difference between field capacity water and water content at the wilting point was assessed throughout the entire growth cycle for each plot and each year.

According to the Baggiolini scale [37], phenological stages were determined through observations in the experimental plots.

Water stress conditions in grapevines were analyzed by considering the percentage of ASW relative to AWC. According to Pellegrino et al. [38] and van Leeuwen et al. [20], water deficit levels (weak, moderate, and severe) were established with thresholds in terms of the percentage of the AWC as follows: weak water deficit: 32–20% of AWC; weak to moderate water deficit: 20–8% of AWC; moderate to severe: 8–2% of AWC; and severe water deficit for values <2% of AWC.

2.5. Grapevine Nutritional Status

In each row of each plot, 60 leaves were sampled from leaves opposite the second bunch at the veraison stage [39]. Leaf blades and petioles were separated from each leaf, washed with tap water, and rinsed with distilled water. The plant materials were dried at 60 °C in a forced-air oven for 72 h. Then, they were ground through a 0.5 mm sieve using an ultracentrifugal mill (ZM1 Retsch, Haan, Germany).

Nitrogen concentration in petioles (% N) was determined with a CNS elemental analyzer (TruSpec CN, LECO, St. Joseph, MI, USA). Phosphorus (% P) and K (% K) concentrations in petioles were determined by microwave hydrogen peroxide digestion and ICP–optical emission spectroscopy (ICP-3300 DV, PerkinElmer, Shelton, CT, USA). Concentrations were expressed on a dry weight basis as g 100 g^{-1}.

2.6. Grapevine Agronomic Performance

All plots were harvested when the grapes had reached 13% v/v probable alcoholic degree, which is the technical maturity for cv. Tempranillo grapes in the DOCa Rioja.

Harvest dates had an interval of 7 to 10 days. At harvest, in each plot, the number of bunches per vine and yield were recorded to calculate grape yield as kg grape $vine^{-1}$ (GrY) and the average bunch weight as kg (BuW).

At postharvest (late November or early December), 20 vines were randomly selected in each row of each plot, and weight of the pruning wood and the number of shoots were determined to calculate the average shoot weight as g (ShW) and the average pruning wood per vine as kg (PrW).

2.7. Grape Sampling and Analytical Parameters of Must

One day before harvest, random samples consisting of 400 berries were collected from six clusters of 20 grapevines randomly dispersed in each row of each experimental plot. Five berries were selected from each cluster, including two from opposite sides at the top of the cluster, two from opposite sides in the middle, and one from the tip of the cluster. In the laboratory, 200 berries were meticulously separated, counted, and weighed to calculate the average weight of each individual berry. Subsequently, a masticator (IUL Instruments GmbH, Königswinter, Germany) was used to crush the berries and extract the must. Probable volumetric alcoholic degree (PVAD), pH (pHM), total acidity (AcTM), malic acid (AcM), and must potassium (KM) were determined using methods outlined by the International Organization of Vine and Wine (OIV) [40]. The total polyphenol index (TPIM) was determined through spectroscopy, measuring the absorbance at 280 nm with a UV–visible spectrophotometer PU8720 (Philips, Eindhoven, The Netherlands). The sample was previously diluted with distilled water to obtain an absorbance within the range. The sample absorbance multiplied by the dilution corresponds to TPIM [41].

An additional set of 200 berries was treated with 1% HCl by heating at 40 °C for 30 min. From this extract, the anthocyanin content in the skin of the berries (AntM) was determined following the method proposed by Ribéreau-Gayon and Stonestreet [42]. Finally, color intensity (CIM) was determined in accordance with the European Community Official Methods. For this purpose, the absorbances at 410 nm, 520 nm, and 620 nm were measured using a spectrophotometer (PU8720, Philips, Eindhoven, The Netherlands). The color intensity corresponds to the sum of these absorbances multiplied by ten [42].

2.8. Vinification

Small-scale winemaking was carried out identically for all plots. At harvest, 30 to 45 kg of fruit were taken from each row of each plot. First, the harvested grapes per plot were weighed, crushed, and the stems removed. Subsequently, 3.5 kg of crushed material from each plot were weighed and placed in a glass container adapted for vinification.

Alcoholic fermentation was carried out using a selected yeast strain of *Saccharomyces cerevisiae* (VRB® Uvaferm SIV-Lallemand, Blagnac, France). The fermentation proceeded in a temperature-controlled chamber at 26 °C for 15 days. The final alcoholic grade was approximately 13%. Winemaking was carried out following the protocol described by Sampaio et al. [43].

2.9. Wine Analysis

Oenological parameters such as alcohol grade (AlG), pH (pHW), total acidity (AcTW), K (KW), tonality (TnW), and color intensity (CIW) were determined according to OIV methods [44]. Anthocyanin content (AntW) was determined by the Ribéreau-Gayon and Stonestreet method [41], and the total polyphenol index (TPIW) was determined by measuring the absorbance at 280 nm.

2.10. Statistical Analysis

The plot factor was selected as the fixed factor and the year factor as the random factor. An ANOVA analysis was conducted with these two factors. When one factor is fixed and the other is random, this type is called a mixed-effects ANOVA. In addition, to determine pair-wise differences by post hoc tests (least significant differences), a one-way ANOVA was also performed.

Pearson correlations were calculated between grapevine, must, and wine parameters with mean ASW in the budbreak–bloom, bloom–fruit set, fruit set–veraison, and veraison–maturity periods. Partial Least Squares regression (PLS regression) analysis between the grape parameters and ASW from one week before stage I2 (week 1) to one week before stage N (week 15) was performed, considering all plots and years.

All statistical procedures were conducted using the STATGPHICS Centurion XV Version 16.1.03 software (Statgraphics Technologies, Inc., The Plains, VA, USA).

3. Results

3.1. Weather Conditions Recorded during the Period of the Study

The maximum, minimum, mean temperature (TmaxGS, TminGS, and TmGS) and precipitation (PGS) recorded during the growing season (April–October) at the weather station for years from 2010 to 2014 are presented in Table 1. In 2013, precipitation levels were above the average recorded between 2004 and 2018. Although total precipitation in 2012 was below average, autumn precipitation levels exceeded the typical monthly averages. However, the total precipitation for 2012 did not surpass the average of the period considered (2004–2018).

3.2. Simulated Soil Water Contents and Available Soil Water for the Selected Plots and Years

Due to the differences in soil characteristics, the maximum AWC that each soil can retain in the profile ranged from 146.5 mm in P22 to 56.3 mm in P63. In the remaining plots, the values were 128.5 and 60.2 mm, respectively, for P53 and P21 (Table 1).

Figure 2 shows the ASW throughout the crop cycle for each plot in the analyzed years, and Figure 3 shows the % of AWC throughout the crop cycle for each plot.

It can be seen that, at the beginning of the crop cycle (in phase D, week ending 25 April), the ASW reached its maximum capacity in almost all plots and years. However, two weeks before the I2 stage (bloom), the ASW decreased rapidly, reaching 50% of the maximum holding capacity within a few weeks, although this varied among plots and years (Figures 2 and 3).

In 2010, plots P21 and P22 reached 50% of their maximum AWC six weeks after bloom, while P53 or P63 reached this level three weeks after bloom. The minimum percentage of the AWC recorded in the vegetative cycle for plots P21 and P22 was 29% and 31%, respectively, reached three weeks after the M1 stage (veraison). For plots P53 and P63, ASW decreased below 20% of AWC (considered as the threshold to define a moderate to weak water deficit) four weeks after the J stage (fruit set). Additionally, in both cases, lower percentages of the AWC (a minimum value of 11 and 14%, respectively) were reached one week before veraison. Finally, ASW for plots P53 and P63 increased, reaching about 20% of AWC four weeks after veraison (Figures 2 and 3).

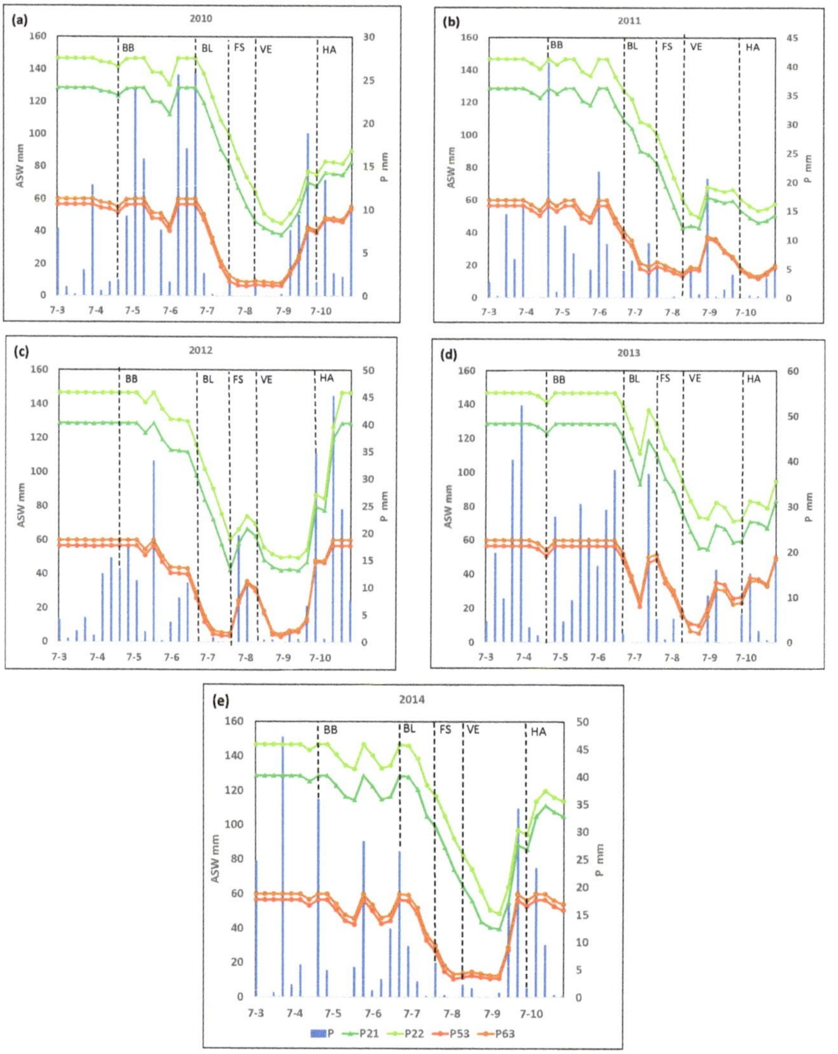

Figure 2. Available soil water (ASW, mm) and rainfall (P, mm) along the growing cycle in the selected plots P2, P22, P53, and P63 in years (**a**) 2010, (**b**) 2011, (**c**) 2012, (**d**) 2013, and (**e**) 2014. BB, budbreak; BL, bloom; FS, fruit set; VE, veraison; HA, harvest.

In 2011, P21 and P22 reached 50% of the AWC six and seven weeks, respectively, after the bloom, while plots P53 and P63 reached this threshold two weeks later. In plot P21, the minimum percentage of the AWC about 33%) was reached in veraison, and in plot P22, this minimum percentage of AWC (about 34%) was reached two weeks after veraison, while the minimum value reached in the other two plots (P53 and P63) was about 24 and 25% of the AWC, respectively, and was reached in veraison (Figures 2 and 3).

In 2012, 50% of the AWC was reached three weeks after bloom in P21 and P22, while it occurred at bloom in P53 and P63. The minimum percentages of the AWC in P21 and P22 were 33% and 34%, respectively, and were reached three weeks after veraison. In P53 and P63, 20% of the AWC was reached two weeks after bloom, and the minimum (6 and 9%, respectively) was reached in fruit set. Plots P53 and P63 reached a level above 20% AWC again for four weeks after fruit set. After that, they decreased again to below 20% of AWC until the week of harvest (Figures 2 and 3).

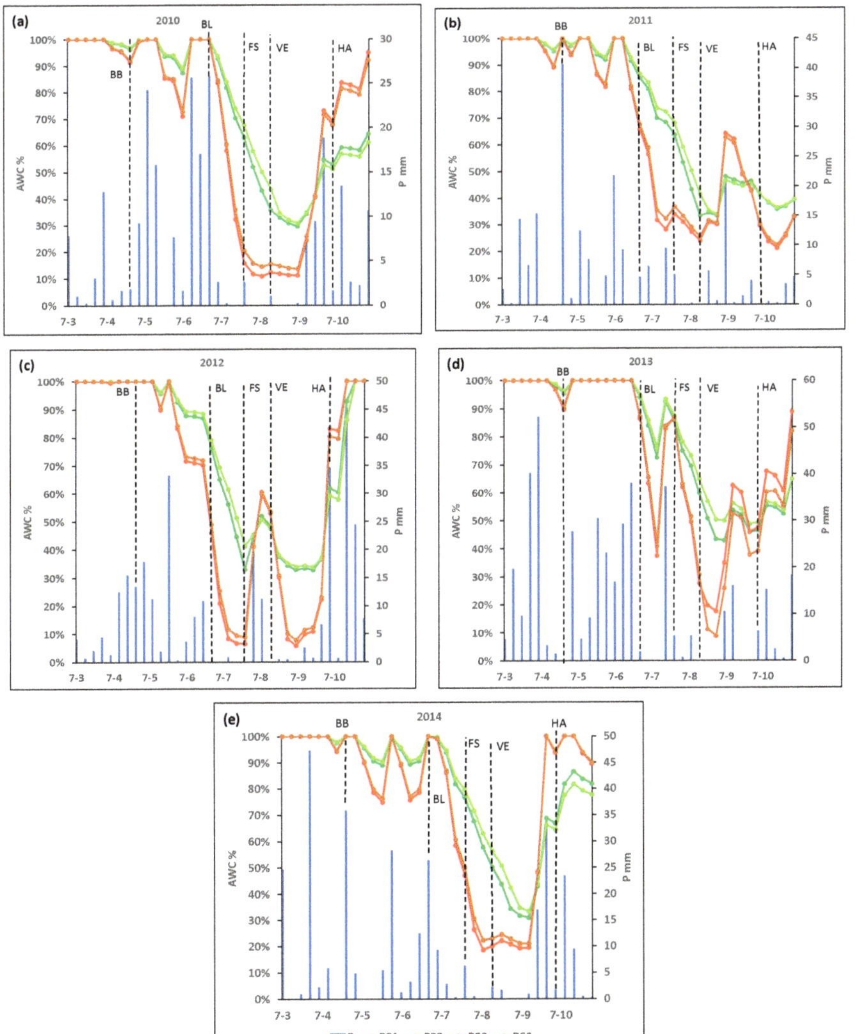

Figure 3. Percentage of available water holding capacity (AWC, mm) and rainfall (P, mm) along the growing cycle in the selected plots P21, P22, P53, and P63, in years (**a**) 2010, (**b**) 2011, (**c**) 2012, (**d**) 2013 and (**e**) 2014. BB, budbreak; BL, bloom; FS, fruit set; VE, veraison; HA, harvest.

In 2013, 50% of the AWC was reached two weeks after veraison in P21 and P22, while in P53 and P63, that threshold was reached one week before the veraison. The lowest percentage of the AWC in plots P53 and P63 (18% and 9% of the AWC, respectively) were reached two weeks after veraison. These plots, P53 and P63, recovered values above 20% of the AWC three weeks after veraison (Figures 2 and 3).

Finally, in 2014, both P21 and P22 reached 50% of the AWC one week after veraison, while for P53 and P63, this threshold was reached two and three weeks after bloom, respectively, and one week before fruit set. The minimum percentage of the AWC in P21 and P22 (31% and 33%, respectively) was reached four weeks after veraison, while in plots P53 and P63, the minimum values were 21% and 22% of the AWC, respectively, reached one week before veraison (Figures 2 and 3).

3.3. Nutritional Status, Vigor, and Yield

The average values for each plot and year analyzed and the ANOVA pertaining to the nutritional status and vine response are presented in Table 2.

Table 2. Comparison of means for foliar contents and vegetative grapevine parameters in each plot (P21, P22, P53, and P63) and year (2010–2014). Results of a mixed-effects ANOVA (plot as a fixed factor and year as a random factor).

	Petiole % N	Petiole % P	Petiole % K	Shoot Weight g	100 Berries Weight g	Grape Weight per Vine kg vine^{-1}	Pruning Wood per Vine kg vine^{-1}
				2010			
P21	0.59b [1]	0.29c	2.30c	161.77b	244b	5.54c	1.55b
P22	0.56b	0.28c	1.23b	133.89b	222ab	4.30ab	1.32b
P53	0.45a	0.18b	0.78a	71.17a	213a	4.69bc	0.71a
P63	0.45a	0.05a	0.81a	75.85a	208a	3.55a	0.71a
				2011			
P21	0.63c	0.24c	2.50c	183.41c	312b	6.10c	1.75c
P22	0.56b	0.28c	1.18b	121.83b	266a	4.97b	1.22b
P53	0.47a	0.17b	0.81a	76.55a	277ab	3.42a	0.75a
P63	0.45a	0.05a	0.78a	70.09a	245a	2.89a	0.67a
				2012			
P21	0.55b	0.33c	2.64c	158.59c	314d	7.81c	1.47c
P22	0.48a	0.31bc	1.66b	129.37b	264c	5.69b	1.17b
P53	0.45a	0.22b	1.11a	59.50a	183b	2.84a	0.59a
P63	0.49a	0.09a	1.11a	48.91a	147a	2.02a	0.46a
				2013			
P21	0.67b	0.50c	2.72c	186.24b	270a	3.31a	1.70c
P22	0.61b	0.45c	1.04b	157.20b	258a	3.87a	1.34b
P53	0.45a	0.19b	0.71ab	78.46a	269a	4.45a	0.80a
P63	0.41a	0.05a	0.47a	74.84a	245a	3.29a	0.69a
				2014			
P21	0.65c	0.19b	2.28b	142.60c	339c	7.04b	1.42c
P22	0.48b	0.28c	1.09a	83.42b	303b	7.09b	0.88b
P53	0.42ab	0.25bc	0.87a	60.34a	301b	3.64a	0.68a
P63	0.38a	0.06a	0.58a	57.92a	281a	3.81a	0.67a
				Average value per year			
2010	0.51	0.20	1.28a	110.70bc	221.58a	4.52	1.08b
2011	0.52	0.18	1.32a	112.97bc	275.14b	4.34	1.10b
2012	0.49	0.24	1.63b	99.09ab	226.98a	4.59	0.92a
2013	0.53	0.30	1.24a	124.18c	260.58b	3.73	1.13b
2014	0.48	0.19	1.20a	86.07a	305.81c	5.39	0.91a
				Average value per plot			
P21	0.62c	0.31c	2.49c	166.52c	295.83b	5.96c	1.58c
P22	0.54b	0.32c	1.24b	125.14b	262.67ab	5.18bc	1.19b
P53	0.45a	0.20b	0.86a	69.21a	248.45a	3.81ab	0.71a
P63	0.44a	0.06a	0.75a	65.52a	225.11a	3.11a	0.64a
				p values			
Year	0.4088	0.1436	0.0099	0.0069	0.0063	0.4377	0.0221
Soil	0.0001	0.0001	0.0000	0.0000	0.0129	0.0100	0.0000
Soil × Year	0.0007	0.0000	0.3596	0.1850	0.000	0.0000	0.2815

[1] Different letters indicate significant differences within soil plots (L.S.D., >95%).

Of all the vine parameters studied, only % K, ShW, BW, and PrW were affected by the year factor. Berry weight was lower in 2010 and 2012 compared to the other years. Pruning weight was also lower in 2012 and 2014 compared to other years. Finally, the % K in the petiole was higher in 2012 than in the other years.

As for the plot factor, this affected all the vine parameters analyzed. So, plots P21 and P22 had higher N, P, and K content, ShW, and PrW than plots P53 and P63, while plot P21 had higher BW, BuW, and GrY than plots P53 and P63.

The interaction between the year and plot was significant for all vine parameters except K content, ShW, and PrW. The significant differences between plots for each year are shown in Table 3 and are not described in detail to avoid excessive length of the text.

Table 3. Must analysis for each plot (P21, P22, P53, and P63) and year (2010–2014). Results of a mixed-effects ANOVA (plot as a fixed factor and year as a random factor).

	Probable Volumetric Alcoholic Degree	pH	Total Acidity g L^{-1}	Malic Acid mg L^{-1}	K mg L^{-1}	IPT	Color Intensity	Anthocyanins mg g^{-1}
2010								
P21	13.3a [1]	3.44ab	6.52c	3.80d	1794b	13.23a	3.70a	1.36a
P22	13.2a	3.47b	5.99bc	3.09c	1670b	15.17b	4.27ab	1.62ab
P53	13.3a	3.36a	5.87ab	2.56b	1376a	19.01c	4.86b	1.88b
P63	13.2a	3.38a	5.36a	1.99a	1296a	18.79c	4.72b	1.83b
2011								
P21	12.8ab	3.47a	5.85c	2.74c	2005b	13.07a	3.32a	1.19a
P22	12.6a	3.49a	4.93b	2.32bc	1793a	12.38a	2.78a	1.11a
P53	13.5b	3.44a	5.11b	2.05b	1862ab	18.90b	4.44b	1.80b
P63	13.2ab	3.58b	4.24a	1.53a	1724a	19.22b	4.88b	1.64b
2012								
P21	12.8a	3.60a	5.47c	2.81c	2050b	14.83b	3.42a	0.99a
P22	12.5a	3.60a	4.72b	2.12b	1837a	12.82a	3.30a	1.17a
P53	13.6b	3.53a	4.78b	1.71a	1944ab	26.36d	7.08b	2.08b
P63	13.6b	3.75b	4.04a	1.54a	1946ab	22.66c	6.27b	2.47c
2013								
P21	12.0a	3.32a	9.04b	5.01b	2038c	17.32a	5.70a	1.28a
P22	11.8a	3.28a	8.48b	4.49b	1824b	16.15a	6.33a	1.61b
P53	13.1b	3.21a	7.11a	3.25a	1474a	23.45b	10.32b	2.08c
P63	13.2b	3.26a	6.50a	2.65a	1373a	23.76b	9.93b	1.64b
2014								
P21	12.2b	3.41b	5.90b	2.90c	1858c	12.67a	3.74a	1.23a
P22	11.4a	3.28a	5.69b	2.39bc	1565b	12.33a	3.68a	1.32a
P53	14.3c	3.41b	5.36ab	2.22b	1663b	18.33b	5.92b	1.98b
P63	13.8c	3.38b	4.85a	1.63a	1400a	17.67b	5.56b	2.05b
Mean value per year								
2010	13.23	3.41bc	5.94c	2.86b	1533.75a	16.55ab	4.39a	1.67
2011	13.02	3.49c	5.03ab	2.16a	1845.92b	15.89a	3.85a	1.43
2012	13.13	3.62d	4.75a	2.04a	1944.42b	19.16bc	5.01a	1.68
2013	12.52	3.27a	7.78d	3.85c	1677.25a	20.17c	8.07b	1.65
2014	12.92	3.37b	5.45bc	2.28a	1621.58a	15.25a	4.72a	1.65
Mean value per plot								
P21	12.60a	3.45	6.56c	3.45d	1949.07c	14.22a	3.98a	1.21a
P22	12.29a	3.42	5.96b	2.88c	1737.80b	13.77a	4.07a	1.37a
P53	13.57b	3.39	5.65b	2.36b	1663.67ab	20.42b	6.52b	1.97b
P63	13.40b	3.47	5.00a	1.87a	1547.80a	21.21b	6.27b	1.93b
p values								
Year	0.3708	0.0000	0.0000	0.0000	0.0046	0.0037	0.0001	0.5773
Soil	0.0046	0.2574	0.0001	0.0000	0.0027	0.0000	0.0003	0.0004
Soil × Year	0.0003	0.0011	0.0074	0.0050	0.0000	0.0000	0.0000	0.0000

[1] Different letters indicate significant differences within soil plots (L.S.D., >95%).

3.4. Must Composition

The mean values of the variables analyzed in the must as well as the ANOVA results are shown in Table 3.

All the must parameters analyzed were affected by the year factor except for the probable volumetric alcoholic degree (PVAD) and AntM. The year 2013 was the year with the lowest pHM, the highest AcTM, and AcM acid.

Regarding the plot factor, it affected all parameters except pH. Plots P22 and P23 exhibited lower PVAD, TPIM, CIM, and AntM and higher AcM compared to plots P53 and P63. Additionally, plot P21 had higher KM and AcTM than plots P53 and P63.

The interaction among plot and year factors also affected all must variables, and the differences between plots for each year are shown in Table 4.

Table 4. Wine analysis for each plot (P21, P22, P53, and P63) and year (2010–2014). Results of a mixed-effects ANOVA (plot as a fixed factor and year as a random factor).

	Probable Alcoholic Degree	pH	K mg L^{-1}	Total Acidity g L^{-1}	Color Intensity	IPT 186	Anthocyanins mg L^{-1}
				2010			
P21	13.8a [1]	3.79a	1687b	6.63c	9.10a	47.10a	567a
P22	13.5a	3.80a	1540b	6.23b	10.44a	49.93a	614a
P53	13.3a	3.70a	1357a	6.3bc	15.69b	65.83b	836b
P63	13.1a	3.73a	1317a	5.63a	16.92b	68.87b	796b
				2011			
P21	13.5a	3.87bc	1713c	6.13c	7.89a	51.13a	596a
P22	13.5a	3.88c	1537b	5.43a	7.74a	50.13a	587a
P53	13.4a	3.74a	1280a	5.83b	13.87b	68.47b	845b
P63	13.5a	3.79ab	1377a	5.2a	14.63b	68.77b	779b
				2012			
P21	12.7ab	3.81a	1773c	5.97c	7.88a	49.27a	465a
P22	12.1a	3.75a	1540ab	5.67b	6.96a	46.77a	486a
P53	13.0b	3.83a	1423a	5.5b	13.45b	80.27b	996b
P63	13.3b	4.02b	1650bc	4.37a	15.25b	78.57b	1069b
				2013			
P21	14.0a	3.68c	1923c	8.83c	6.49a	63.43b	447a
P22	13.7a	3.60bc	1603bc	7.9b	7.05a	52.97a	508a
P53	13.5a	3.47ab	1263b	7.43b	11.82b	60.40b	699b
P63	13.8a	3.43a	882a	6.77a	12.80b	60.07b	630b
				2014			
P21	12.7b	3.80b	1522b	5.32a	4.35a	31.92b	542b
P22	11.6a	3.68ab	1201a	5.54a	3.50a	27.19a	439a
P53	14.9d	3.78ab	1437b	5.33a	8.81b	50.68d	778c
P63	14.3c	3.61a	1129a	5.58a	7.99b	44.82c	614b
			Mean values per year				
2010	13.43	3.75bc	1475.0	6.20b	13.04c	57.93b	703.26ab
2011	13.49	3.82cd	1476.7	5.65a	11.03bc	59.62b	701.79ab
2012	12.79	3.85d	1596.7	5.37a	10.88bc	63.72b	754.08b
2013	13.75	3.54a	1418.0	7.73c	9.54b	59.21b	571.08a
2014	13.39	3.72b	1322.3	5.42a	6.16a	38.67a	593.17a
			Mean values per plot				
P21	13.32	3.79	1723.7b	6.58b	7.14a	48.57a	523.47ᵃ
P22	12.90	3.74	1484.2ab	6.15b	7.14a	45.40a	526.65a
P53	13.63	3.70	1371.4a	6.08b	12.73b	64.22b	830.93b
P63	13.62	3.72	1251.6a	5.51a	13.52b	65.13b	777.63b
			p values				
Year	0.4655	0.0037	0.3180	0.0000	0.0000	0.0012	0.1012
Soil	0.3728	0.4786	0.0066	0.0095	0.0000	0.0005	0.0004
Soil × Year	0.0000	0.0001	0.0000	0.0000	0.2652	0.0000	0.0000

[1] Different letters indicate significant differences within soil plots (L.S.D., >95%).

3.5. Wine Composition

The mean values of the variables analyzed in the wine as well as the ANOVA results are shown in Table 4.

The year factor significantly affected the following wine parameters: pH, AcTW, ICW, TnW, and IPTW. The year 2013 had the lowest pH and the highest total acidity compared to the other years. Additionally, in 2014, the lowest CIW and IPTW were observed, while the highest TnW was recorded compared to the other years.

The soil factor also had a significant effect on KW, AcTW, CIW, TnW, IPTW, and anthocyanin content. Plots P21 and P22 had lower CIW, IPT, and AntW than plots P53 and P63, while plot P21 had higher CIW and KW than plots P53 and P63.

All wine parameters analyzed except CIW had interactions among year and plot factors, the differences between plots for each year are shown in Table 4.

3.6. Relationship between Mean Available Soil Water and Grapevine, Must, and Wine Parameters

Regarding the parameters analyzed in grapevine petiole, N, P, and K contents were significantly correlated with the mean ASW in every period studied (Table 5). The ShW and the PrW were correlated with the average ASW in the four periods studied. On the other hand, BW was correlated with mean ASW for periods bloom–fruit set and veraison–maturity, and BuW was correlated with mean ASW for periods budbreak–bloom, bloom–fruit set, and veraison–maturity. However, GrW was correlated with ASW for all four periods studied along the growth cycle.

Table 5. Pearson correlation coefficients of foliar contents and grapevine parameters with available soil water (ASW) in different periods of the growth cycle.

	Mean ASW mm in Period			
	Budbreak–Bloom	**Bloom–Fruit Set**	**Fruit Set–Veraison**	**Veraison–Maturity**
Petiole N %	0.7695	0.7282	0.7605	0.7508
p value	0.0001	0.0003	0.0001	0.0001
Petiole P %	0.7401	0.6896	0.7603	0.773
p value	0.0002	0.0008	0.0001	0.0001
Petiole K %	0.6191	0.5174	0.5261	0.5652
p value	0.0036	0.0195	0.0172	0.0094
Shoot weight	0.8242	0.7586	0.7618	0.7851
p value	0	0.0001	0.0001	0
Berry weight	0.4139	0.499	0.4214	0.4944
p value	0.0696	0.0251	0.0643	0.0267
Bunch weight	0.5753	0.5515	0.4427	0.5301
p value	0.008	0.0117	0.0506	0.0162
Grape yield	0.6273	0.6001	0.5044	0.5893
p value	0.0031	0.0052	0.0233	0.0063
Pruning weight	0.8306	0.7834	0.76	0.7872
p value	0	0	0.0001	0

With respect to the must parameters, PVAD was negatively correlated with the mean ASW for all four periods studied (Table 6). Must pH was not correlated with the mean ASW in any of the periods studied. Total acidity in the must was correlated with ASW in the budbreak–bloom, bloom–fruit set, and fruit set–veraison periods, with a higher r value for the bloom–fruit set, fruit set–veraison and veraison-maturity periods. Similarly, AcM was correlated with mean ASW for all four periods studied with the highest r value in the fruit set–veraison period, while KM was not correlated with mean ASW in any of the periods studied. With respect to TPIM and AntM, there was a negative correlation with the mean ASW for each period studied, while CI was negatively correlated with the mean ASW in the budbreak–bloom, bloom–fruit set, and veraison–maturity periods.

Table 6. Pearson correlation coefficients of must properties with available soil water (ASW) in different periods of the growth cycle.

	Mean ASW mm in Period			
	Budbreak–Bloom	**Bloom–Fruit Set**	**Fruit Set–Veraison**	**Veraison-Maturity**
Probable volumetric alcoholic grade	−0.7646	−0.7687	−0.844	−0.8254
p value	0.0001	0.0001	0	0
pH	−0.0338	−0.2645	−0.2301	−0.1316
p value	0.8874	0.2597	0.3291	0.5801
Acidity Total	0.3903	0.498	0.5642	0.4602
p value	0.0889	0.0255	0.0096	0.0412

Table 6. *Cont.*

	Mean ASW mm in Period			
	Budbreak–Bloom	Bloom–Fruit Set	Fruit Set–Veraison	Veraison-Maturity
Malic acid	0.5693	0.6288	0.6801	0.6037
p value	0.0088	0.003	0.001	0.0048
K	0.4214	0.2451	0.3798	0.4138
p value	0.0643	0.2976	0.0986	0.0697
TPI	−0.8148	−0.8111	−0.6739	−0.801
p value	0	0	0.0011	0
Color intensity	−0.5543	−0.4638	−0.3294	−0.5098
p value	0.0112	0.0394	0.1561	0.0217
Anthocyanins	−0.805	−0.7325	−0.6883	−0.7576
p value	0	0.0002	0.0008	0.0001

In the wine, AlG, pHW, and AcTW were not correlated with the mean ASW in any period studied (Table 7). Potassium in wine was correlated with mean ASW in the budbreak–bloom and veraison–maturity periods. The TnW was not correlated with the mean ASW for any period studied, while TPIW, AntW, and CIW were negatively correlated with the mean ASW in each period studied.

Table 7. Correlation coefficients of wine properties with available soil water (ASW) in different periods of the growth cycle.

	Mean ASW mm in Period			
	Budbreak–Bloom	Bloom–Fruit Set	Fruit Set–Veraison	Veraison–Maturity
Alcoholic degree	−0.365	−0.2455	−0.2869	−0.3281
p value	0.1136	0.2968	0.22	0.1578
pH	0.1435	−0.0094	−0.0282	0.0674
p value	0.5462	0.9687	0.9062	0.7776
Acidity total	0.4061	0.4311	0.2882	0.3287
p value	0.0757	0.0577	0.2178	0.1571
K	0.5328	0.3928	0.4397	0.4977
p value	0.0156	0.0867	0.0524	0.0256
Color intensity	−0.7649	−0.8197	−0.8071	−0.8302
p value	0.0001	0	0	0
Tonality	0.3125	0.3651	0.3606	0.4117
p value	0.1797	0.1135	0.1183	0.0713
IPT	−0.6578	−0.7649	−0.6381	−0.689
p value	0.0016	0.0001	0.0025	0.0008
Antochyanins	−0.8025	−0.8387	−0.7962	−0.8244
p value	0	0	0	0

3.7. Relationship between Available Soil Water and Grape Composition

Figure 4 shows the results of the PLS regression analysis, in which the coefficients for each week start one week before the I2 stage and end one week before maturity is reached. The graph represents the number of components that gave the best fit and lower error, with all variables significantly adjusted with a significance level of 95%. For the PVAD, two components were retained, and the percentage of the variance explained was 75.0%.

For the parameters AcTM, AcM, and pHM, three components were retained, and the percentage of variance explained was 67.6%, 65.4%, and 58.7%, respectively. In the variables associated with phenolic compounds, three components were retained, which explained 92.1%, 67.7%, and 86.8% of the variance for TPIM, AntM, and CIM, respectively. Finally, BW was adjusted by retaining three components, which explained 54.9% of the variance.

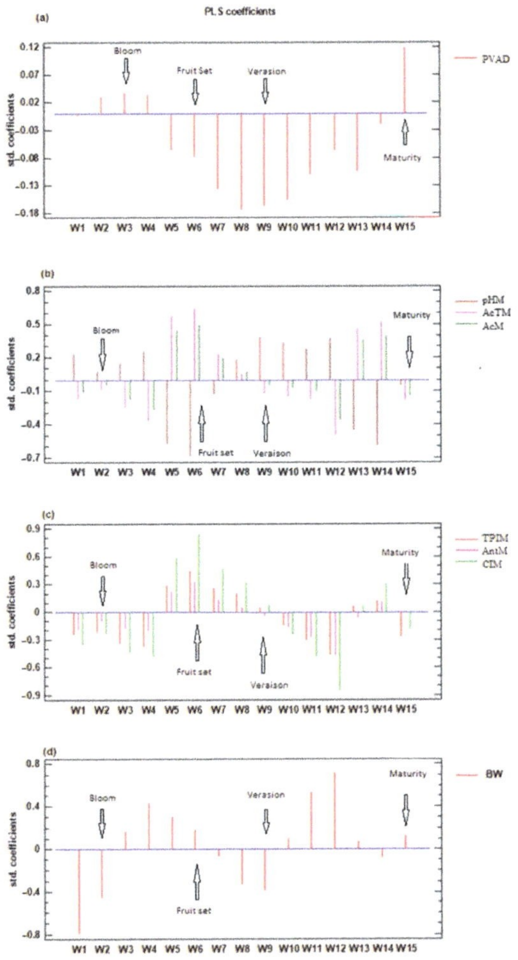

Figure 4. Coefficients of the PLS regression analysis were obtained for all analyzed plots and years between the grape parameters and ASW from stage H until reaching ripening (13°) and analyzed weekly (weeks 1–15; W1–W15) (**a**) probable volumetric alcoholic degree (PVAD) (**b**) must pH (pHM), total acidity (AcTM), malic acid (AcM) (**c**) total polyphenol index (TPIM), anthocyanins (AntM), and color index (CIM) (**d**) berry weight (BW).

4. Discussion

The properties of the soils in plots P21, P22, P53, and P63 closely resemble those of the vineyard soils in DOCa Rioja, as characterized by Peregrina et al. [45] The AWC values for these plots fall within the range (30–200 mm) for vine growing soils, as defined by van Leeuwen et al. [46].

Throughout the analyzed period (2010–2014), varying amounts and distributions of rainfall, along with temperature discrepancies, were recorded. These variations provided information on ASW under diverse weather conditions.

It was observed that the soils of P21, P22, P53, and P63 usually maintained the maximum ASW up to bloom (week of 27/6), and from that time, the ASW decreased rapidly. Available soil water levels below 50% of the AWC were reached in all plots during the growing season, usually between bloom and veraison. In a wet year, like 2013, the soil with higher AWC (P21 and P22) maintained soil water for a longer time. Thus, ASW below 50% of AWC was reached at the beginning of August, five weeks after bloom.

The minimum ASW values were usually reached at veraison, except in very wet years (e.g., 2013). Consequently, in P21 and P22, the minimum ASW values were above 20% of the AWC, while in P53 and P63, the ASW values were below or close to 20% of the AWC between 2 and 6 weeks after bloom. In 2013, this threshold was reached 1 week after veraison. Additionally, minimum ASW values were reached earlier in plots P53 and P63 than in plots P21 and P22, even in the wettest years, such as 2013. Plots P53 and P63 reached values below 10% of the AWC in 2012.

Other authors [36,37,47], reported the evolution of ASW and the minimum values similar to those found in our conditions. These authors expressed ASW in relation to AWC and found values below 0.1 by the end of the growing cycle in some of the years and locations analyzed. The vines experienced weak to moderate water stress at veraison and throughout most of the ripening period, as indicated by the low ASW values. This confirms that the diminished ASW towards the end of the growing cycle not only affected berry weight but also had an impact on berry composition.

Under our conditions, no water stress was observed until the bloom stage. However, moderate water stress (ASW below 20% of AWC) was recorded 2–6 weeks after the I2 stage in plots P53 and P63, with the highest water stress observed at veraison.

The N, P, and K contents in petioles were influenced by the plot factor, with higher contents in plots P21 and P22 than in plots P53 and P63. Additionally, N, P, and K contents were correlated with ASW in all the studied periods. These results could be due to a reduction in nutrient assimilation by mass flow when soil moisture decreases [48], making nutrient absorption by the vine more difficult [9,49]. Furthermore, these three nutrients (as NO_3^-, HPO_4^-, and K^{+}) are actively taken up by transport proteins against the electrochemical potential with ATP energy input [50]. Thus, a water deficit that promotes stomatal closure and reduced photosynthesis [8], which can affect ATP synthesis, can reduce the uptake of these nutrients. In the case of N, lower ASW would also contribute to a reduction in NO_3^- assimilation, as nitrate reductase activity decreases rapidly and reversibly at low leaf Ψ [51].

The increased N content in plots with higher AWC may contribute to lower concentrations of polyphenols and anthocyanins in musts and wines. This can be attributed to the heightened availability of N, which fosters greater vegetative development in the vines. This vegetative growth competes with the accumulation of sugars and pigments in grapes [52]. Similar findings were reported in other studies under low vine nutritional N status, resulting in elevated anthocyanin and polyphenol content in Cabernet Sauvignon [53] and Merlot [54] berries in the Bordeaux region (France), as well as in Tempranillo grapes in DOCa Rioja [10,55]. The response of the vine to reduced N uptake and lower water availability appears to be additive rather than interactive, as suggested by Keller [8]. In essence, the increased N uptake contributes to a reduced concentration of polyphenols and anthocyanins, aligning with the impact of water stress, as discussed later.

The positive correlations identified between ASW and foliar K content may be attributed to the fundamental role of K in vine water regulation, as noted by Keller [8], given that its assimilation is constrained under limited soil moisture conditions [16,56]. Similar findings were reported by King et al. [57], who observed that vines in areas with higher soil moisture levels exhibited elevated foliar K content. Furthermore, studies conducted in Tempranillo vineyards in La Rioja (Spain) by Zaballa and García-Escudero [58], as well as in Cabernet Sauvignon and Merlot vineyards in Israel by Klein et al. [59], documented a positive relationship between foliar K and increased moisture availability through irrigation.

The plot factor influences ShW and PrW, resulting in lower values observed in plots P53 and P63 compared to plots P21 and P22. This discrepancy may be attributed to the lower ASW after the bloom stage and the moderate water stress experienced in plots P53 and P63. Additionally, the year factor affected ShW and PrW, as evidenced by the highest shoot weight observed in 2013, which experienced the highest rainfall. Finally, ShW and PrW were found to be correlated with ASW across all periods studied; budbreak–bloom, bloom–fruit set, fruit set–veraison, and veraison–maturity. Our results indicate that shoot growth is highly sensitive to hydric stress, as it primarily involves cell expansion [60,61]. A reduction in shoot and leaf growth is typically the first visible sign of a water deficit in the vineyard [62,63]. Similar results, showing increased vine vigor in soils with higher AWC, have been reported by Tramontini et al. [7] in Bordeaux and by Tomasi et al. [64] in Italy. The weight of the berries exhibited a strong correlation with ASW, indicating higher values in wetter years compared to drier ones [65–68]. Coinciding with those results, in our study, ASW had a significant positive correlation with berry weight in the veraison–maturity period. Additionally, the PLS coefficients were high and positive during the weeks after veraison (W10 to W13). This effect after veraison is consistent with van Leeuwen et al. [20], who found smaller berry sizes under water limitations at ripening. In our conditions, BW showed high and negative PLS coefficients with ASW around bloom (W1 to W2) and veraison (W7 to W9). These results are according to those reported by Ramos et al. [36]. The explanation may be connected to the observation that, during these two periods, the balance between vegetative and reproductive growth becomes more crucial than in other phases. The increased ASW tends to promote vegetative growth over berry development in these specific periods.

Grape yield showed slight differences with BW, as it was affected by the plot factor but not by the year factor. Moreover, production correlated with ASW during the periods from budbreak to flowering and from veraison to ripening. A similar result with yield affected by water available during the budbreak–bloom period was reported by Ramos and Martínez-Casasnovas [15]. The relationship during the veraison to ripening period may be because hydric stress can affect photosynthesis, and lower yields may occur because not all berries reach full maturity [69].

In addition to berry development, a water deficit can also affect grape quality. Thus, PVAD was affected by the plot factor, with higher values in plots P53 and P63 compared to plots P21 and P22. These results would be due to the higher water stress in plots P53 and P63, as confirmed by the correlation between PVAD and ASW in all the periods studied. Moreover, PLS coefficients were negative during most of the period studied but were higher around veraison (W8–W10), showing that this period would be the most influential on the final PVAD. The positive PLS coefficients in the weeks prior to harvest agree with the results reported by Lasko and Pool [70], where water stress in the following weeks to veraison was not as critical as at the end of the cycle, when it can produce both a decrease in berry size and sugar accumulation. Thus, Bucelli et al. [71] and Ubalde et al. [72] reported similar results, indicating higher sugar content in grapes from vineyards with lower AWC soils and shallower soil rooting depth [71].

Finally, the alcoholic content of the wine (AlcG) was not affected by either the plot or the year factors. This would indicate that changes in the PVAD were not significant enough to impact the AlcG of the wines.

Must acidity is another parameter that can be directly affected by soil water availability. In our results, pH, total acidity, and malic acid were influenced by the year factor but not by the plot factor. These results confirm that climate has a greater influence on the acidity components than soil, as observed by van Leeuwen et al. [13].

The year 2013, the wettest of the analyzed years, had a lower pH and higher total acidity and malic acid levels than other years. Additionally, AcTM correlated with ASW during all periods studied except budbreak–bloom, while AcM showed a positive correlation in all the periods studied. Consistent with our results, higher acidity has been reported in wetter soils [1], in conditions of excessive soil moisture [73], and with well-watered grapes [74,75]. Additionally, Lopes et al. [76] reported a significant reduction in titratable acidity of the must when soil water content was reduced during spring. Likewise, Tomasi et al. [71] found higher acidity in soils with greater rooting depth. Furthermore, studies by Peyrot des Gachons et al. [77], de Souza et al. [78], and dos Santos et al. [79] have documented reduced total acidity attributed to water deficit and/or alterations in temperature in sun-exposed berries in rainfed vines.

As for the time of the cycle in which some effects on both the total acidity and malic acid in the must, the reasons may vary.

Enhanced water availability in the early phases of fruit growth (W5 to W8), especially when the berry is small and acid synthesis begins, could have a favorable impact on final acidity levels. This correlation corresponds to the positive PLS coefficients observed during this timeframe. Conversely, the negative PLS coefficients in the following period (W9 to W11) may be attributed to a dilution effect as the berry undergoes rapid expansion. Moreover, it was noted that elevated ASW during ripening contributes to heightened acidity levels.

Our result is according to Ramos and Martínez de Toda [67], who, in a comprehensive study covering nearly the entire Rioja vine-growing region, identified the highest acidity values in the wettest years. These years witnessed substantial water accumulation not only from bloom to veraison but also from veraison to maturity. Furthermore, in those instances, the lower temperatures recorded in wetter years could have potentially impeded the combustion of malic acid during the ripening process.

These changes in grape acidity components were reflected in the wine. Thus, the pH and total acidity of the wine were influenced by the year factor and not by the plot factor. Similarly to the must, 2013 exhibited the highest acidity and the lowest pH of the wines. These results confirm the importance of the climatic conditions of the year on the final acidity of the wines. Another must component related to grape quality is K. Under our conditions, must K (KM) was affected by both plot and year factors. Thus, P21 and P22 soils, with higher AWC, had higher KM content. Increased ASW can raise K levels in vegetative tissues (as above mentioned), leading to higher K levels in berries as they serve as significant sinks for K [11]. While for K content in wine, we found results similar to those found in must, indicating the influence of both the year and plot factors. Additionally, correlations between ASW and KW were observed in the budbreak–bloom and veraison-maturity periods. These are important results since K reduces the free acids in the wine and raises its pH, leading to lower chemical/biological stability of the wine.

Grape anthocyanins, total phenol grape content, and color intensity (CIM) were affected by the plot factor. Thus, soils P53 and P63, with lower AWC and experiencing moderate to severe postveraison water stress, showed higher CIM, TPIM, and AntM. Moreover, in all the periods studied (budbreak–bloom, bloom–fruit set, fruit set–veraison and veraison–maturity), TPIM and AntM were negatively correlated with ASW. These results agree with the observations of other authors [36,67,80–82], although different compounds may be affected differently depending on cultivars and rootstocks [80,83]. Anthocyanin and polyphenol concentrations improved in scenarios where there was an absence of water stress from bloom to fruit set, mild water stress between fruit set and veraison, and moderate to severe water stress post-veraison [84]. This agrees with the positive PLS coefficients for CIM, TPIM, and AntM observed in the weeks W5 to W8, and with the negative PLS coefficients observed after veraison (W10–W12). In this regard, Ferrer et al. [82] noted higher anthocyanin concentrations under mild to moderate water deficits during ripening.

The periods following W9, during which a negative relationship between ASW and phenolic compounds was observed, coincide exactly with the periods in which the ASW increases berry size. Therefore, it could be suggested that the higher concentration of phenolic compounds is influenced by ASW and its effect on berry size. Changes in polyphenol and anthocyanin content observed in the must were also found in the wine. Both the plot factor and the year factor affected CIW, TPIW, and AntW. Additionally, ASW was negatively correlated with these wine properties across all of the studied periods. This result is important as it confirms the impact of AWC on wine quality. A similar result was found in Spain by Ubalde et al. [72], with a higher anthocyanin content in wine from soils with lower AWC.

Between stage H and maturity, there were only two short periods lasting two to three weeks each (except for one lasting five weeks in the case of acidity), during which the impact of water availability diverged, albeit with lesser intensity compared to the overall effect throughout the rest of the vegetative period. The elevated ASW during these specific periods led to a reduction in berry weight and acidity while concurrently boosting the levels of anthocyanins and other phenolic compounds.

These two periods consistently align with the bloom and veraison stages. One plausible explanation may be linked to the general physiological, and especially hormonal, changes observed in the vine during these stages. In both instances, vegetative growth decelerates concurrently with reproductive development—in one instance to form the fruit and in the other to commence the ripening phase [85].

5. Conclusions

From this research, it could be concluded that the evolution of available soil water and the soil and the level of water stress under similar weather conditions (mesoclimate) varied among the study plots due to different soil characteristics (especially the effective soil depth) that determine different available water holding capacities. These differences in available water influence nutritional status, vegetative development, and grape and wine composition. Higher available soil water between the budbreak and ripening phenological stages increased leaf N, P, and K levels. Vegetative development and grape yield per vine also increased, while the probable alcoholic degree of the berries decreased. Higher available soil water from veraison to harvest stages also reduced the berry color and polyphenol and anthocyanin content of the berries. As for acidity components, higher soil water availability around fruit set increased total acidity and malic acid. The impact of soil available water on berry characteristics was transmitted to the wine in the case of color, polyphenols, and anthocyanins, which decreased in plots with higher available soil water, but not in the case of alcohol content, pH, and total acidity. Therefore, the complete description of the soil profile with its rooting depth and the determination of the available water holding capacities can be useful for the selection of the most favorable soils for obtaining quality grapes in current climatic conditions. In addition, in a scenario of climate change where soil water could decrease due to changes in rainfall and increased demand for evapotranspiration, the location of vineyards in soils with greater available water holding capacities should be considered.

Author Contributions: Conceptualization, F.P.; methodology, J.M.M.-V. and M.C.R.; validation, M.C.R.; formal analysis, F.P. and J.M.M.-V.; investigation, F.P., J.M.M.-V. and E.P.P.-Á.; resources, J.M.M.-V.; data curation, J.M.M.-V.; writing—original draft preparation, F.P.; writing—review and editing, F.P., J.M.M.-V. and E.P.P.-Á.; visualization, F.P. and J.M.M.-V.; supervision, E.G.-E.; project administration, J.M.M.-V. and E.G.-E.; funding acquisition, J.M.M.-V. and E.G.-E. All authors have read and agreed to the published version of the manuscript.

Funding: This work was supported by MINECO, Instituto Nacional de Investigación y Tecnología Agraria y Alimentaria INIA, European Social Fund [grant number RTA2009-00101-00-00], and the Government of La Rioja. Eva Pilar Pérez-Álvarez also acknowledges the Spanish Ministry of Science, Innovation and Universities (MCIU) for her "Juan de la Cierva Incorporación" postdoctoral grant (IJC2019-040502-I).

Data Availability Statement: The original contributions presented in the study are included in the article, further inquiries can be directed to the corresponding author.

Acknowledgments: We gratefully acknowledge Zinio Wineries from Uruñuela (La Rioja) for its collaboration with selection of vineyards plots.

Conflicts of Interest: The authors declare no conflicts of interest.

References

1. Cheng, G.; He, Y.N.; Yue, T.X.; Wang, J.; Zhang, Z.W. Effects of climatic conditions and soil properties on cabernet sauvignon berry growth and anthocyanin profiles. *Molecules* **2014**, *19*, 13683–13703. [CrossRef] [PubMed]
2. Zerihun, A.; Mcclymont, L.; Lanyon, D.; Goodwin, I.; Gibberd, M. Deconvoluting effects of vine and soil properties on grape berry composition. *J. Sci. Food Agric.* **2015**, *95*, 193–203. [CrossRef]
3. Lanyon, D.M.; Cass, A.; Hansen, D. *The Effect of Soil Properties on Vine Performance*; Land and Water Technical Report No. 34/04; CSIRO: Canberra, Australia, 2004.
4. Costantini, E.A.C.; Pellegrini, S.; Vignozzi, N.; Barbetti, R. Micromorphological characterization and monitoring of internal drainage in soils of vineyards and olive groves in central Italy. *Geoderma* **2006**, *131*, 388–403. [CrossRef]
5. Ramos, M.C.; Jones, G.V.; Yuste, J. Phenology and grape ripening characteristics of cv Tempranillo within the Ribera del Duero designation of origin (Spain): Influence of soil and plot characteristics. *Eur. J. Agron.* **2015**, *70*, 57–70. [CrossRef]
6. Costantini, E.A.C.; Pellegrini, S.; Bucelli, P.; Barbetti, R.; Campagnolo, S.; Storchi, P.; Magini, S.; Perria, R. Mapping suitability for Sangiovese wine by means of δ13C and geophysical sensors in soils with moderate salinity. *Eur. J. Agron.* **2010**, *33*, 208–217. [CrossRef]
7. Tramontini, S.; van Leeuwen, C.; Domec, J.C.; Destrac-Irvine, A.; Basteau, C.; Vitali, M.; Mosbach-Schulz, O.; Lovisolo, C. Impact of soil texture and water availability on the hydraulic control of plant and grape-berry development. *Plant Soil* **2013**, *368*, 215–230. [CrossRef]
8. Keller, M. Deficit irrigation and vine mineral nutrition. *Am. J. Enol. Vitic.* **2005**, *56*, 267–283. [CrossRef]
9. Conradie, W.J. Partitioning of mineral nutrients and timing of fertilizer applications for optimum efficiency. In Proceedings of the Soil Environment and Vine Mineral Nutrition Symposium, San Diego, CA, USA, 29–30 June 2004; Christensen, P., Smart, D.R., Eds.; American Society for Enology and Viticulture: Davis, CA, USA, 2005; pp. 69–81.
10. Pérez-Álvarez, E.P.; Martínez-Vidaurre, J.M.; Martín, I.; García-Escudero, E.; Peregrina, F. Relationships among Soil Nitrate Nitrogen and Nitrogen Nutritional Status, Yield Components, and Must Quality in Semi-arid Vineyards from Rioja AOC, Spain. *Commun. Soil Sci. Plan.* **2013**, *44*, 232–242. [CrossRef]
11. Mpelasoka, B.S.; Schachtman, D.P.; Michael, M.T.; Thomas, M.R. A review of potassium nutrition in grapevines with special emphasis on berry accumulation. *Aust. J. Grape Wine Res.* **2003**, *9-3*, 154–168. [CrossRef]
12. Matthews, M.A.; Anderson, M.M.; Schultz, H.R. Phenologic and growth-responses to early and late season water deficits in Cabernet franc. *Vitis* **1987**, *26*, 147–160.
13. van Leeuwen, C.; Friant, P.; Chone, X.; Tregoat, O.; Koundouras, S.; Dubourdieu, D. Influence of climate, soil, and cultivar on terroir. *Am. J. Enol. Vitic.* **2004**, *55*, 207–217. [CrossRef]
14. Peacock, W.L. Fertigating drip-irrigated vineyards with macro and micronutrients. In Proceedings of the Soil Environment and Vine Mineral Nutrition Symposium, San Diego, CA, USA, 29–30 June 2004; Christensen, L.P., Smart, D.R., Eds.; American Society for Enology and Viticulture: Davis, CA, USA, 2005; pp. 129–133.
15. Ramos, M.C.; Martínez-Casasnovas, J.A. Soil water balance in rainfed vineyards of the Penedès region (Northeastern Spain) affected by rainfall characteristics and land levelling: Influence on grape yield. *Plant Soil* **2010**, *333*, 375–389. [CrossRef]
16. Esteban, M.A.; Villanueva, M.J.; Lissarrague, J.R. Effect of irrigation on changes in berry composition of Tempranillo during maturation. Sugars, organic acids, and mineral elements. *Am. J. Enol. Vitic.* **1999**, *50*, 418–434. [CrossRef]
17. Deloire, A.; Vaudour, E.; Carey, V.; Bonnardot, V.; Van Leeuwen, C. Grapevine responses to terroir: A global approach. *J. Int. Sci. Vigne Vin* **2005**, *39*, 149–162. [CrossRef]
18. Downey, M.O.; Dokoozlian, N.K.; Krstic, M.P. Cultural Practice and Environmental Impacts on the Flavonoid Composition of Grapes and Wine: A Review of Recent Research. *Am. J. Enol. Vitic.* **2006**, *57*, 257–268. [CrossRef]
19. Gómez-Míguez, M.J.; Gómez-Míguez, M.; Vicario, I.M.; Heredia, F.J. Assessment of colour and aroma in white wines vinifications: Effects of grape maturity and soil type. *J. Food Eng.* **2007**, *79*, 758–764. [CrossRef]
20. van Leeuwen, C.; Tregoat, O.; Choné, X.; Bois, B.; Pernet, D.; Gaudillére, J.P. Vine water status is a key factor in grape ripening and vintage quality for red Bordeaux wine. How can it be assessed for vineyard management purposes? *J. Int. Sci. Vigne Vin* **2009**, *43*, 121–134. [CrossRef]
21. Ojeda, H.; Andary, C.; Kraeva, E.; Carbonneau, A.; Deloire, A. Influence of pre- and postveraison water deficit on synthesis and concentration of skin phenolic compounds during berry growth of *Vitis vinifera* cv. shiraz. *Am. J. Enol. Viticult.* **2002**, *53*, 261–267.
22. Ruby, G.; Harbertson, J.F.; Adams, D.A.; Matthews, M.A. Berry size and vine water deficits as factors in winegrape composition: Anthocyanins and tannins. *Aust. J. Grape Wine Res.* **2004**, *10*, 100–107. [CrossRef]
23. Castellarin, S.D.; Matthews, M.A.; Di Gaspero, G.; Gambetta, G.A. Water deficits accelerate ripening and induce changes in gene expression regulating flavonoid biosynthesis in grape berries. *Planta* **2007**, *227*, 101–112. [CrossRef]
24. Intrigliolo, D.S.; Pérez, D.; Risco, D.; Yeves, A.; Castel, J.R. Yield components and grape composition responses to seasonal water deficits in Tempranillo grapevines. *Irrig. Sci.* **2012**, *30*, 339–349. [CrossRef]
25. Cooley, N.M.; Clingeleffer, P.R.; Walker, R.R. Effect of water deficits and season on berry development and composition of Cabernet Sauvignon (*Vitis vinifera* L.) grown in a hot climate. *Aust. J. Grape Wine Res.* **2017**, *23*, 260–272. [CrossRef]
26. Bellvert, J.; Marsal, J.; Mata, M.; Girona, J. Yield, must composition, and wine quality responses to preveraison water deficits in sparkling base wines of chardonnay. *Am. J. Enol. Vitic.* **2016**, *67*, 1–12. [CrossRef]

27. Torres, N.; Antolín, M.C.; Goicoechea, N. Arbuscular mycorrhizal symbiosis as a promising resource for improving berry quality in grapevines under changing environments. *Front. Plant Sci.* **2018**, *9*, 897. [CrossRef] [PubMed]
28. Balint, G.; Reynolds, A.G. Effect of different irrigation strategies on vine physiology, yield, grape composition and sensory profiles of *Vitis vinifera* L. Cabernet-Sauvignon in a cool climate area. *OENO One* **2014**, *48*, 269–292. [CrossRef]
29. Lizama, V.; Pérez-Álvarez, E.P.; Intrigliolo, D.S.; Chirivella, C.; Álvarez, I.; García-Esparza, M.J. Effects of the irrigation regimes on grapevine cv. Bobal in a Mediterranean climate: II. Wine, skins, seeds, and grape aromatic composition. *Agric. Water Manag.* **2021**, *256*, 107078. [CrossRef]
30. Bucchetti, B.; Matthews, M.A.; Falginella, L.; Peterlunger, E.; Castellarin, S.D. Effect of water deficit on Merlot grape tannins and anthocyanins across four seasons. *Sci. Hortic.* **2011**, *128*, 297–305. [CrossRef]
31. Martínez-Vidaurre, J.M. Influencia del Tipo de Suelo en el Estado Nutricional de la Vid, el Desarrollo Vegetativo, la Producción, la Composición de la Uva y de los Vinos de la Variedad Tempranillo Tinto (*Vitis vinifera* L.) en el Ámbito de la D.O.Ca Rioja. Ph.D. Thesis, Universidad de La Rioja, Logroño, Spain, 2017. Available online: https://digital.csic.es/bitstream/10261/194557/3/Influencia_tipo.pdf (accessed on 20 February 2024).
32. Nelson, O.W.; Sommers, L.E. Total carbon, organic carbon, and organic matter. In *Methods of Soil Analysis Part 2*, 2nd ed.; Page, A.L., Miller, R.H., Keeney, D.R., Eds.; ASA/SSSA: Madison, WI, USA, 1982; pp. 539–577.
33. Orsini, L.; Remy, J.C. Utilisation du chlorure de cobaltihexammine pour la determina-tion simultanée de la capacité d'échange et des bases échangeables des sols. *Bull. L'afes Sci. Sol.* **1976**, *4*, 269–275.
34. Mehlich, A. Mehlich No. 3 soil test extractant, A modification of Mehlich No. 2. *Commun. Soil Sci. Plant Anal.* **1984**, *15*, 1409–1416. [CrossRef]
35. Saxton, K.E.; Rawls, W.J. Soil Water Characteristic Estimates by Texture and Organic Matter for Hydrologic Solutions. *Soil Sci. Soc. Am. J.* **2006**, *70*, 1569–1578. [CrossRef]
36. Ramos, M.C.; Pérez-Álvarez, E.P.; Peregrina, F.; Martínez de Toda, F. Relationships between grape composition of Tempranillo variety and available soil water and water stress under different weather conditions. *Sci. Hortic.* **2020**, *262*, 109063. [CrossRef]
37. Baillod, M.; Baggiolini, M. Les stades repères de la vigne. *Rev. Suisse Vitic. Arboric. Hortic.* **1993**, *38*, 7–9.
38. Pellegrino, A.; Lebon, E.; Voltz, M.; Wery, J. Relationships between plant and soil water status in vine (*Vitis vinifera* L.). *Plant Soil* **2005**, *266*, 129–142. [CrossRef]
39. Romero, I.; García-Escudero, E.; Martín, I. Effects of leaf position on blade and petiole mineral nutrient concentration of Tempranillo grapevine (*Vitis vinifera* L.). *Am. J. Enol. Vitic.* **2010**, *61*, 544–550. [CrossRef]
40. Organization International de la Vigne et du Vine. *Methods of Analysis of Must and Wine. Compendium of International Methods of Wine and Must Analysis*; OIV: Paris, France, 2014.
41. Ribereau-Gayon, J.; Ribereau-Gayon, P.; Peynaud, E.; Sudraud, P. *Traité d'oEnologie-Sciences et Techniques du Vin, Tome 1: Analyse et Contrôle des Vins*; Dunod: Paris, France, 1972; 671p.
42. Ribéreau-Gayon, P.; Stonestreet, E. Le dosage des anthocyanes dans les vins rouges. *Bull. Soc. Chim. Fr.* **1965**, *9*, 2649–2652. [PubMed]
43. EEC. European Commission Commission Regulation (EEC) No 2676/90 of 17 September 1990 determining Community methods for the analysis of wines. *Off. J. Eur. Union* **1990**, *L 272*, 1–272.
44. Sampaio, T.L.; Kennedy, J.A.; Vasconcelos, M.C. Use of microscale fermentations in grape and wine research. *Am. J. Enol. Vitic.* **2007**, *58*, 534–539. [CrossRef]
45. Peregrina, F.; López, D.; Zaballa, O.; Villar, M.T.; González, G.; García-Escudero, E. Soil quality of vineyards in the Origin Denomination Rioja: Index of overcrusting risk (FAO-PNUMA), content of organic carbon and relation with soil fertility. *Rev. Ciências Agrárias* **2010**, *33*, 338–345. [CrossRef]
46. van Leeuwen, C.; Roby, J.P.; de Rességuier, L. Soil-related terroir factors: A review. *OENO One* **2018**, *52*, 173–188. [CrossRef]
47. Ramos, M.C.; Martínez-Casasnovas, J.A. Soil water variability and its influence on transpirable soil water fraction with two grape varieties under different rainfall regimes. *Agric. Ecosyst. Environ.* **2014**, *185*, 253–262. [CrossRef]
48. Marschner, H. *Mineral Nutrition of Higher Plants*; Academic Press: London, UK, 1995.
49. Schreiner, R.P. Mycorrhizae and mineral acquisition in grapevines. In Proceedings of the Soil Environment and Vine Mineral Nutrition Symposium, San Diego, CA, USA, 29–30 June 2004; Christensen, P., Smart, D.R., Eds.; American Society for Enology and Viticulture: Davis, CA, USA, 2005; pp. 49–60.
50. Grossman, A.; Takahashi, H. Macronutrient utilization by photosynthetic eukaryotes and the fabric of interactions. *Ann. Rev. Plant Physiol. Plant Mol. Biol.* **2001**, *52*, 163–210. [CrossRef] [PubMed]
51. Lawlor, D.W.; Cornic, G. Photosynthetic carbon assimilation and associated metabolism in relation to water deficits in higher plants. *Plant Cell Environ.* **2002**, *25*, 275–294. [CrossRef] [PubMed]
52. Bravdo, B.; Hepner, Y. Irrigation management and fertigation to optimise grape composition and vine performance. *Acta Hortic.* **1987**, *206*, 49–67. [CrossRef]
53. Choné, X.; van Leeuwen, C.; Chéry, P.; Ribéreau-Gayon, P. Terroir influence on water status and nitrogen status of non-irrigated Cabernet-Sauvignon (*Vitis vinifera*). Vegetative development, must and wine composition (example of a Medoc top estate vineyard, Saint Julien area, Bordeaux, 1997). *S. Afr. J. Enol. Vitic.* **2001**, *22*, 8–15. [CrossRef]

54. Gaudillère, J.P.; van Leeuwen, C.; Choné, X.; Tregoat, O. The assessment of vine water and nitrogen uptake by means of physiological indicators influence on vine development and berry potential (*Vitis vinifera* L. cv Merlot, 2000, Bordeaux). *J. Int. Sci. Vigne Vin* **2002**, *36*, 133–142. [CrossRef]
55. Pérez-Álvarez, E.P.; García-Escudero, E.; Peregrina, F. Soil Nutrient Availability under Cover Crops: Effects on Vines, Must, and Wine in a Tempranillo Vineyard. *Am. J. Enol. Viticult.* **2015**, *66*, 311–320. [CrossRef]
56. Dundon, C.G.; Smart, R.E.; McCarthy, M.G. The Effect of Potassium Fertilizer on Must and Wine Potassium Levels of Shiraz Grapevines. *Am. J. Enol. Vitic.* **1984**, *35*, 200–205. [CrossRef]
57. King, P.D.; Smart, R.E.; Mcclellan, D.J. Within-vineyard variability in vine vegetative growth, yield, and fruit and wine composition of Cabernet-Sauvignon in Hawke's Bay, New Zealand. *Aust. J. Grape Wine Res.* **2014**, *2*, 234–246. [CrossRef]
58. Zaballa, O.; García-Escudero, E. Ensayos de riego en viñedos de la D.O.Ca Rioja 1984–1994. In *Reuniones del Grupo de Trabajo de Experimentación en Viticultura y Enología: La Rioja, 14, 15 y 16 de Marzo de 1995*; Ministerio de Agricultura, Alimentación y Medio Ambiente: Madrid, España, 1997; pp. 56–94, ISBN 84-491-0286-3.
59. Klein, I.; Strime, M.; Fanberstein, L.; Mani, Y. Irrigation and fertigation effects on phosphorus and potassium nutrition of wine grapes. *Vitis* **2000**, *39*, 55–62.
60. Boyer, J.S. Water transport. *Ann. Rev. Plant Physiol.* **1985**, *36*, 473–516. [CrossRef]
61. Hsiao, T.C.; Xu, L.K. Sensitivity of growth of roots versus leaves to water stress: Biophysical analysis and relation to water transport. *J. Exp. Bot.* **2000**, *51*, 1595–1616. [CrossRef] [PubMed]
62. Williams, L.E.; Dokoozlian, N.K.; Wample, R.L. Grape. In *Handbook of Environmental Physiology of Fruit Crops. Vol. 1. Temperate Crops*; Shaffer, B., Anderson, P.C., Eds.; CRC Press: Orlando, FL, USA, 1994; p. 83133.
63. Stevens, R.M.; Harvey, G.; Aspinall, D. Grapevine growth of shoots and fruit linearly correlate with water stress indices based on root-weighted soil matric potential. *Aust. J. Grape Wine Res.* **1995**, *1*, 58–66. [CrossRef]
64. Tomasi, D.; Gaiotti, F.; Bragato, G.; Mosetti, D.; Missio, A.; Battista, F. Soil Influence on Vine Performance: From Root System To Wine Quality. In Proceedings of the 33rd OIV World Congress of Vine and Wine, 8th General Assembly of the OIV, Tbilissi, Georgia, 20–27 June 2010.
65. Ojeda, H.; Deloire, A.; Carbonneau, A. Influence of water deficits on grape berry growth. *Vitis* **2001**, *40*, 141–146.
66. Reynard, J.S.; Zufferey, V.; Nicol, G.C.; Murisier, F. Vine water status as a parameter of the «terroir» effect under the non-irrigated conditions of the Vaud viticultural area (Switzerland). *OENO One* **2011**, *45*, 139–147. [CrossRef]
67. Wenter, A.; Zanotelli, D.; Montagnani, L.; Tagliavini, M.; Andreotti, C. Effect of different timings and intensities of water stress on yield and berry composition of grapevine (cv. *Sauvignon blanc*) in a mountain environment. *Sci. Hortic.* **2018**, *236*, 137–145. [CrossRef]
68. Ramos, M.C.; Martínez de Toda, F. Variability of Tempranillo grape composition in the Rioja DOCa (Spain) related to soil and climatic characteristics. *J. Sci. Food Agric.* **2019**, *99*, 1153–1165. [CrossRef]
69. Zulini, L.; Rubinigg, M.; Zorer, R.; Bertamini, M. Effects of drought stress on chlorophyll fluorescence and photosynthetic pigments in grapevine leaves (*Vitis vinifera* cv.' White Riesling'). *Acta Hortic.* **2007**, *754*, 289–294. [CrossRef]
70. Lasko, A.N.; Pool, R. Efecto del estrés hídrico en el viñedo y en la calidad de los vino (climas del este). *Rev. Internet Vitic. Enol.* **2006**, *122*, 10.
71. Bucelli, P.; Costantini, E.A.C.; Storchi, P. It is possible to predict Sangiovese wine quality through a limited number of variables measured on the vines? *J. Int. Sci. Vigne Vine* **2010**, *44*, 207–218. [CrossRef]
72. Ubalde, J.M.; Sort, X.; Zayas, A.; Rosa, M.P. Effects of soil and climatic conditions on grape ripening and wine quality of Cabernet Sauvignon. *J. Wine Res.* **2010**, *21*, 1–17. [CrossRef]
73. Jackson, D.I.; Lombard, P.B. Environmental and management practices affecting grape composition and wine quality—A review. *Am. J. Enol. Vitic.* **1993**, *44*, 409–430. [CrossRef]
74. Girona, J.; Marsal, J.; Mata, M.; Del Campo, J.; Basile, B. Phenological sensitivity of berry growth and composition of tempranillo grapevines (*Vitis vinifera* L.) to water stress. *Aust. J. Grape Wine Res.* **2009**, *15*, 268–277. [CrossRef]
75. Reynolds, A.G.; Lowrey, W.D.; Tomek, L.; Hakimi, J.; De Savigny, C. Influence of irrigation on vine performance, fruit composition, and wine quality of Chardonnay in a cool, humid climate. *Am. J. Enol. Vitic.* **2007**, *58*, 217–228. [CrossRef]
76. Lopes, C.M.; Santos, T.P.; Monteiro, A.; Rodrigues, M.L.; Costa, J.M.; Chaves, M.M. Combining cover cropping with deficit irrigation in a Mediterranean low vigor vineyard. *Sci. Hortic.* **2011**, *129*, 603–612. [CrossRef]
77. Peyrot des Gachons, C.P.; Van Leeuwen, C.; Tominaga, T.; Soyer, J.-P.; Gaudillère, J.-P.; Dubourdieu, D. Influence of water and nitrogen deficit on fruit ripening and aroma potential of *Vitis vinifera* L. cv Sauvignon blanc in field conditions. *J. Sci. Food Agric.* **2005**, *85*, 73–85. [CrossRef]
78. Souza, C.R.D.; Maroco, J.; Santos, T.P.; Rodrigues, M.; Lopes, C.M.A.; Pereira, J.S.; Chaves, M.M. Grape berry metabolism in field-grown grapevines exposed to different irrigation strategies. *Vitis* **2005**, *44*, 103–109.
79. dos Santos, T.P.; Lopes, C.M.; Lucília Rodrigues, M.; de Souza, C.R.; Ricardo-da-Silva, J.M.; Maroco, J.P.; Pereira, J.S.; Chaves, M.M. Effects of deficit irrigation strategies on cluster microclimate for improving fruit composition of Moscatel field-grown grapevines. *Sci. Hortic.* **2007**, *112*, 321–330. [CrossRef]
80. Hochberg, U.; Batushansky, A.; Degu, A.; Rachmilevitch, S.; Fait, A. Metabolic and physiological responses of shiraz and cabernet sauvignon (*Vitis vinifera* L.) to near optimal temperatures of 25 and 35 °C. *Int. J. Mol. Sci.* **2015**, *16*, 24276–24294. [CrossRef]

81. Cáceres-Mella, A.; Talaverano, M.I.; Villalobos-González, L.; Ribalta-Pizarro, C.; Pastenes, C. Controlled water deficit during ripening affects proanthocyanidin synthesis, concentration and composition in Cabernet Sauvignon grape skins. *Plant Physiol. Biochem.* **2017**, *117*, 34–41. [CrossRef]
82. Ferrer, M.; Echeverría, G.; Carbonneau, A. Effect of berry weight and its components on the contents of sugars and anthocyanins of three varieties of *Vitis vinifera* L. under different water supply conditions. *S. Afr. J. Enol. Vitic.* **2014**, *35*, 103–113. [CrossRef]
83. Berdeja, M.; Hilbert, G.; Lafontaine, M.; Stoll, M.; Gomes, E.; Renaud, C.; Delrot, S. Effects of drought stress and rootstock genotype on grape berry quality. *Acta Hortic.* **2014**, *1038*, 375–378. [CrossRef]
84. Basile, B.; Marsal, J.; Mata, M.; Vallverdú, X.; Bellvert, J.; Girona, J. Phenological sensitivity of cabernet sauvignon to water stress: Vine physiology and berry composition. *Am. J. Enol. Vitic.* **2011**, *62*, 453–461. [CrossRef]
85. Martínez de Toda, F. *Biología de la Vid. Fundamentos Bilógicos de la Viticultura*; Mundiprensa: Madrid, Spain, 1991; p. 346.

horticulturae

MDPI

Article

Fine-Tuning Grape Phytochemistry: Examining the Distinct Influence of Oak Ash and Potassium Carbonate Pre-Treatments on Essential Components

Ozkan Kaya [1,2,*], Hava Delavar [1], Fadime Ates [3], Turhan Yilmaz [4], Muge Sahin [5] and Nurhan Keskin [6]

1 Department of Plant Sciences, North Dakota State University, Fargo, ND 58102, USA; hava.delavar@ndsu.edu
2 Erzincan Horticultural Research Institute, Republic of Turkey Ministry of Agriculture and Forestry, 24060 Erzincan, Turkey
3 Manisa Viticulture Research Institute, Republic of Turkey Ministry of Agriculture and Forestry, 45125 Manisa, Turkey; fadimeates@gmail.com
4 Faculty of Agriculture, Department of Horticulture, Kahramanmaras Sutcu Imam University, 46040 Kahramanmaras, Turkey; turhany89@gmail.com
5 Aegean Agricultural Research Institute, Republic of Turkey Ministry of Agriculture and Forestry, 35660 Izmir, Turkey; mugesahin67@hotmail.com
6 Faculty of Agriculture, Department of Horticulture, Van Yuzuncu Yıl University, 65090 Van, Turkey; keskin@yyu.edu.tr
* Correspondence: kayaozkan25@hotmail.com or ozkan.kaya@ndsu.edu; Tel.: +1-0-701-2000315

Abstract: Understanding the impact of pre-treatment methods on the phytochemical composition of grapes is essential for optimizing grape quality and producing raisins with desirable characteristics. Therefore, this study meticulously analyzed the impact of two distinct pre-treatment methods, oak ash and potassium carbonate (K_2CO_3), on the composition of essential phytochemical components in grapes. This research encompassed phenolic compounds, anthocyanins, phenolic acids, flavonoids, and phytoalexins. This study investigates the impact of pre-treatment methods, oak ash and K_2CO_3, on the phytochemical composition of grapes. Significant differences were observed in anthocyanins, flavonoids, phytoalexins, and phenolic acids between the treatments. Oak ash exhibited advantages in preserving specific compounds, including higher levels of anthocyanins, flavonols, flavones, flavanones, catechins, resveratrol, pterostilbene, and viniferin, compared to K_2CO_3. Notably, the delphinidin-3-O-glycoside content was significantly higher in the oak ash treatment. An analysis of phenolic compounds revealed distinctions in hydroxycinnamic acids, hydroxybenzoic acids, benzaldehyde, and phenylacetaldehyde. Additionally, gallic acid, vanillic acid, *trans*-caffeic acid, *trans*-p-coumaric acid, and (-)-epicatechin were significantly more prevalent in the K_2CO_3 treatment, while ferulic acid and quercetin were more prevalent in the oak ash treatment. These findings underscore the pivotal role of pre-treatment methods in shaping the phytochemical content of grapes, thus holding critical implications for grape-derived products' quality and potential health benefits.

Keywords: grape pre-treatment; phytochemical composition; raisin production; oak ash; potassium carbonate

1. Introduction

Grapes are a diverse and versatile fruit of considerable importance, and they belong to the *Vitis* genus and come in various colors, including green, red, and purple/black [1]. Grapes are not only enjoyed fresh but are also used to produce a wide range of products, most notably wines [2–4]. Grapes are rich in essential nutrients, such as vitamins (particularly vitamin C and vitamin K), minerals (like potassium and copper), and dietary fiber, contributing to a healthy diet [5]. Raisins have gained recognition for their remarkable antioxidant content, enriched with vital phenolic compounds, namely flavonoids, resveratrol, and quercetin. These antioxidants play a pivotal role in combatting oxidative stress, thus reducing the risk of age-related ailments and promoting overall well-being [6]. Additionally, raisins stand out for their high dietary fiber content, facilitating digestion and offering sustained energy [7]. Notably, they are abundant in polyphenolic compounds such as quercetin and catechins, renowned for their potent antioxidant properties. In contrast, raisins, being dried grapes, not only contribute to robust bone health due to their calcium content but also provide a concentrated source of the same essential nutrients found in fresh grapes, enriched by the drying process [6].

Considering the esteemed status of grapes in our diet, it is noteworthy that they encompass a diverse array of secondary metabolites, referred to as phenolic compounds. These compounds are not exclusive to grapes but are also prevalent in raisins and various plant species, underscoring their significance [7,8]. This protective capacity has far-reaching implications for human health, reducing the risk of chronic diseases and fortifying our overall well-being [9]. Raisins have earned their reputation for harboring potent antioxidants, effectively shielding our cells from the detrimental effects of free radicals. Their well-established antioxidant and anti-inflammatory attributes play a pivotal role in curbing the risk of chronic conditions like diabetes and neurodegenerative disorders [10]. Within this category of phenolic compounds, flavonoids stand out as a prominent subgroup, displaying a substantial presence in both grapes and raisins. Another facet worth mentioning is the role of anthocyanins, a subgroup of flavonoid compounds; in addition to their contribution to coloration, they possess potent antioxidant and anti-inflammatory properties and are responsible for the vibrant red, purple, and blue hues seen in grapes and various other fruits. Beyond their aesthetic value, anthocyanins offer tangible benefits to cardiovascular health by enhancing blood vessel function and diminishing the risk of heart disease [11]. Furthermore, resveratrol, a widely recognized phytoalexin present in grapes, has garnered substantial attention due to its promising health advantages. Its associations with cardiovascular protection, cancer prevention, and potential longevity enhancement have been a subject of considerable research interest [12]. Phytoalexins, which are bioactive compounds produced by plants, including grapes, in response to environmental stressors like pathogens and UV radiation, are equally intriguing. Additionally, phenolic acids constitute a subclass of phenolic compounds found in grapes and raisins, and are celebrated for their anti-inflammatory and antioxidant effects. These attributes have been correlated with the prevention of various chronic diseases, including cancer and cardiovascular disorders [13]. In the realm of viticulture, on the other hand, the application of K_2CO_3 and oak ash for drying grapes introduces an intriguing dimension of sustainability, with potential implications for grape quality and secondary metabolite composition. Scientific investigations into the environmental impact, resource efficiency, and long-term effects on grapevine health resulting from these drying methods are essential for advancing sustainable viticultural practices. Understanding the sustainability of grape drying techniques not only aligns with the broader discourse on responsible agricultural practices but also holds the key to preserving the diverse array of phenolic compounds discussed in this paragraph, thereby ensuring the continued provision of the health benefits associated with grapes in a manner harmonious with ecological considerations.

The grape drying process, which can be accomplished naturally by mechanically or sun drying in specialized drying facilities, concentrates the natural sugars within the grapes, enhancing their flavor and sweetness [7,8]. Drying grapes is a traditional food preservation method that involves the removal of moisture from grapes, ultimately transforming them into the delectable dried fruit known as raisins [14]. As the grapes lose moisture, they shrink and become darker in color, resulting in the recognizable wrinkled appearance of raisins. Raisins are a popular and nutritious snack enjoyed worldwide and are also widely used in baking and cooking, adding natural sweetness and texture to a variety of dishes [15]. In this regard, dipping grapes in various solutions such as K_2CO_3 and oak ash is a technique employed in food science and culinary experimentation. Drying fresh grapes after dipping them in solutions of K_2CO_3 and oak ash is a captivating process that combines traditional and modern culinary techniques [16]. In historical contexts, the preservation of grapes through dipping processes in Asia Minor was characterized by the use of solutions composed of olive oil and wood ash. Subsequently, a shift occurred, with the adoption of specially formulated K_2CO_3 in combination with olive oil replacing wood ash [7,8]. Modern commercial cold dip solutions often comprise ethyl esters of fatty acids and K_2CO_3 as active components, applied using unheated water. Nevertheless, wood ash remains favored due to its organic nature. Notably, the "Hot Dipping Technique" extensively employed in Hadim, Konya, involves immersing grape clusters in solutions consisting of both K_2CO_3–olive oil blends and oak ash solutions, typically reaching temperatures of 70–90 °C for brief durations, preserving grapes under shaded conditions and imparting a distinctive emerald green hue. Moreover, the sustainability of grapes dried through the application of K_2CO_3 and oak ash in viticulture hinges on understanding the environmental impact and resource efficiency associated with these diverse dipping techniques. Scientific scrutiny of the ecological footprint, energy consumption, and long-term effects on grape quality and phenolic compound composition is imperative for advancing sustainable viticultural practices. Aligning with the organic preference for wood ash in cold dip solutions, investigating the sustainability of grape drying methods not only addresses agricultural practices but also underscores the importance of preserving traditional techniques, such as the "Hot Dipping Technique", that contribute to the distinctive qualities of grapes and their environmental compatibility.

While there is a theoretical proposition suggesting that shade drying in raisins could potentially result in superior quality products with enhanced biochemical constituents like phenolic compounds, anthocyanin, phenolic acids, flavonoids, and phytoalexins, empirical evidence supporting this hypothesis specifically in the context of cv. Gök Üzüm raisins is lacking in existing studies. Therefore, the aim of this study is to explore the influence of pretreatment methods, specifically oak ash and K_2CO_3, on some phytochemical components of cv. Gök Üzüm berries through the process of shade drying.

2. Materials and Methods

2.1. Plant Material

The cv. Gök Üzüm was procured from vineyards situated in Konya–Karaman, encompassing both the Karaman and Hadim Bozkir regions (37°16′36″ N 32°11′55″ E 1288 m asl). The grapes were harvested on September 10. This study adopted a randomized complete block design, comprising four replicates, each containing six vines. The experiment was carried out in a single vineyard and the clusters were taken from the vineyard in which the experiment was carried out. The vineyards were meticulously maintained with uniform practices for fertilization, disease control, and pruning. In the context of irrigation, drip irrigation lines were strategically placed along the central axis of the crop rows. These lines were equipped with pressure-compensating emitters, each delivering a consistent 2.5 L/h output, spaced at intervals of 1.0 m (with one emitter per vine positioned between adjacent vines). The irrigation system comprised two lines with emitters spaced 2.0 m apart. Notably, each line served a specific set of vines and operated independently. This deliberate positioning ensured the presence of an emitter between every two adjacent vines, facilitating an alternating supply of water to each side of the root system. This alternating water supply regimen was implemented on a bi-weekly basis. The grapevines, aged between 15 and 20 years, exhibited a planting density ranging from 1100 to 1150 vines per hectare, adhering to a precise vine spacing protocol of 3 m within rows and 3 m between rows. Notably, all the vines were meticulously grafted onto Kober 5BB rootstocks, and their growth was managed through the Goblet training system, ensuring consistency and comparability in the study's vineyard environment.

2.2. Grape Sampling

In this research endeavor, cv. Gök Üzüm grapes, characterized by a ripeness ranging between 20 and 21°Brix, were meticulously hand-harvested from their primary production areas located in Hadim, Turkey, during the year 2020. Following the harvest, clusters underwent careful trimming, involving the removal of damaged, undersized, infected, and immature berries, a process performed with precision using scissors. The sampling methodology remained consistent, involving the selection of 24 vines at random. Approximately 20 kg of fresh grapes were thoughtfully collected from the lower, middle, and upper segments of the clusters, ensuring optimal maturity representation. Each replicate consisted of 5 kg of grape samples from six vines, adding up to a total of 20 kg collected from the 24 vines involved in the study.

2.3. Preparation for the Drying Process

Specific pre-treatments were applied to the fresh grape clusters. To expedite the drying of the grapes, they were immersed in solutions of either oak ash or K_2CO_3 at elevated temperatures, ranging from 70 to 90 °C, for durations of 5 to 10 s. The oak ash solution consisted of an equal volume mixture of oak ash and water, devoid of any additional chemical additives. In contrast, the K_2CO_3 solution was composed of 5 to 6% K_2CO_3, 0.5 to 1% olive oil, and water. Subsequently, the treated grape clusters were arranged in shaded attics, featuring strategically designed walls that facilitated the free circulation of warm breezes within the attics. In this conducive drying environment, the grapes underwent the drying process over a period of approximately 3 to 4 weeks, concluding when their moisture content reached approximately 15% (Figure 1). The study design followed a randomized complete block setup with four replicates, with each replication involving three bags, each containing 300 g of grapes. Consequently, a total of 900 g of raisin samples were extracted from each replication, adding up to 3600 g collected from 24 vines. Following the conclusion of the drying process, the recorded values for the average temperature and relative humidity (RH) were 20 °C and 60%, respectively, and the raisin samples were carefully placed in polyethylene bags, partitioned into twelve separate 300 g portions, securely sealed, and stored at a constant temperature of 20 °C until they were ready for analysis.

Figure 1. Vineyard area and cluster of cv. Gök Üzüm (**a**,**b**), application of oak ash and potassium sulfate solutions (**c**,**d**), grapes dried with exposure to K_2CO_3 (**e**) and wood ash (**f**).

2.4. Analytical Methods for Phenolic Compounds

The mass concentration of the polymer forms of phenolic substances was determined utilizing a colorimetric method, employing the Folin–Ciocalteu reagent by slightly modifying the method developed by Vendramin et al. [17]. The grape berries were subjected to homogenization, with approximately 20 g diluted in 100 mL of deionized water. Subsequently, 5 mL of the homogenized solution was measured into a 20 mL headspace vial, crimped, and maintained at 50 °C for 15 min for the analysis of benzaldehyde. A Solid-Phase Microextraction (SPME) fiber was introduced into the headspace, allowing for the sampling of compounds over a period of 35 min. Following this, the fiber was inserted into the injector port of a gas chromatograph and desorbed for 10 min. The procedure for aliphatic aldehydes followed a similar protocol, with the exception that they were sampled at room temperature (25 °C) for a duration of 15 min. A solution with an initial concentration of 1 g/L, obtained through the dilution of the extract, underwent oxidation by Folin–Ciocalteu reagent (125 mL). Subsequently, the oxidized solution was neutralized using sodium carbonate at a concentration of 20% w/v. The separation of hydroxycinnamic acids was carried out using a C18 Lichrospher column (4 × 250 mm, 5 μm, Agilent Technologies Italia, Milan, Italy) in conjunction with a 1525 Binary Pump (Waters, Milan, Italy) and a 2487 Dual Band Absorbance Detector (Waters, Milan, Italy). The analysis involved freshly prepared verjuice that underwent centrifugation and filtration (0.2 μm) prior to injection (10 μL) and subsequent examination, following the method proposed by Vanzo et al. [17] with certain modifications. The mobile phase composition remained consistent with the original method suggested by the authors, while adjustments were made to the flow rate (0.6 mL/min) and gradient as follows: (A) Milli-Q water and 0.5% of formic acid v/v, and (B) gradient-grade methanol and 2.0% of formic acid v/v. The gradient program unfolded as follows: 0 min, 16% B; 7 min, 50% B; 8 min, 100% B; 8–12 min, 100% B; 13 min, 18% B; and 13–18 min, 18% B. The column temperature was maintained at 40 °C. Detection of hydroxycinnamic acids and esters occurred at a wavelength of 280 nm for purity determination and 330 nm for HCAs quantification. Peak areas were analyzed using Breeze Version 3.3 software (Waters, Milan, Italy).

2.5. Analytical Methods for the Anthocyanin Content

Anthocyanin analysis was conducted using a method that involved slight modifications of Yousef et al.'s [18] approach. The assessment of anthocyanins in grape berries was carried out using a high-performance liquid chromatography (HPLC) system, specifically the Agilent 1100 Series system, manufactured by Agilent in Waldbronn, Germany. This HPLC system was equipped with a diode array detector (DAD) (model G1315B) and coupled to an LC/MSD Trap VL (model G2445C VL) ESI-MS/MS system. Data processing was handled through the Agilent ChemStation (version B.01.03) data processing station. Mass spectral data were further analyzed using the Agilent LC/MS Trap software (version 5.3). For the analysis, anthocyanin extracts were injected at precise volumes of 10 μL onto a reversed-phase column known as Zorbax Eclipse XDB-C18 (2.1 mm × 150 mm; 3.5 μm particle; Agilent, Germany). The column temperature was consistently maintained at 40 °C. The separation method was based on a previously established technique, with minor adjustments made to optimize the separation of the detected anthocyanins. The mobile phase utilized for separation was a mixture of water, acetonitrile, and formic acid, consisting of two solvent compositions: 88.5% water, 3% acetonitrile, and 8.5% formic acid (solvent A); and 41.5% water, 50% acetonitrile, and 8.5% formic acid (solvent B). The separation gradient proceeded as follows: 97% A and 3% B for 8 min; 70% A and 30% B over 20 min; 50% A and 50% B for 6 min; 0% A and 100% B for 4 min, maintained for 2 min, followed by a return to the initial condition over 6 min. An 8 min conditioning period was incorporated between injections. The flow rate was set at 0.19 mL/min. To determine the concentrations of anthocyanin, a comparative approach was adopted using external standards. Calibration standards, including Rutin (Quercetin-rutinoside), Quercetin (Quercetin-rhamnoside), Isoquercitin (Quercetin glucoside), and Cyanidin-3-O-glucoside obtained from Sigma–Aldrich

Fluka in Madrid, Spain, as well as malvidin-3-O-glucoside (Oenin), Isorhamnetin glucoside, and Procyanidin B1 from Extrasynthese in Genay, France, were employed for this purpose.

2.6. Analytical Methods for Phenolic Acids

The determination of phenolic acids involved a methodology with slight adjustments, building upon the method developed by Pantelić et al. [19]. An ultra-high-performance liquid chromatography (UHPLC) system, specifically the Dionex Ultimate 3000 UHPLC, equipped with a diode array detector (DAD) and complemented by a mass spectrometer (TSQ Quantum Access Max triple-quadrupole with heated electrospray ionization (HESI)) from ThermoFisher Scientific in Basel, Switzerland, was employed for this purpose. The chromatographic separation was achieved using a Syncronis C18 column sourced from ThermoFisher Scientific in Bremen, Germany. The mobile phase for gradient elution comprised two components: a 0.1% aqueous formic acid solution (A) and 100% acetonitrile (B). The elution process followed a well-defined gradient pattern: commencing at 5% B for 2.0 min, it gradually increased from 5% to 95% B over the next 10 min (spanning from 2.0 to 12.0 min). Subsequently, there was a rapid transition from 95% to 5% B in just 0.1 min (from 12.0 to 12.1 min), and the mobile phase was maintained at 5% B for the ensuing 3 min. To quantify each individual phenolic compound, a direct comparison was made with commercially available standards. The results of the analysis were expressed in terms of concentrations, measured in milligrams per kilogram (mg/kg), of the berry samples under examination.

2.7. Analytical Methods for Flavonoids

The analysis of flavonoids such as anthocyanidins, flavonols, flavones, flavanones, and catechins extracted from the berries was conducted following a methodology originally outlined by Brossa et al. [20], with specific modifications. In brief, each filtered extract, amounting to 10 μL, was introduced into a liquid chromatography (LC) system, specifically the Acquity UPLC manufactured by Waters in Milford, MA, USA. Chromatographic separation was executed using a C18 column (3.5 μm; 100 × 2.1, Waters, Milford, MA, USA) maintained at a constant temperature of 42 °C, with its flow rate set at 0.3 mL/min. For the detection and quantification of flavonoids, the Electrospray Ionization (ESI) mode was applied, with positive ionization (UPLC-(ESI+)-MS/MS) for anthocyanins and negative ionization (UPLC-(ESI-)-MS/MS) for all other flavonoids. Spectra were recorded within a range spanning from 100 to 900 atomic mass units per charge (amu/z). Additionally, UV/visible light detection was carried out at 365 nm for flavonols and 520 nm for anthocyanins.

2.8. Analytical Methods for the Phytoalexins

Phytoalexin content was assessed with a modified version of the method developed by Jeandet et al. [21]. Initially, samples were finely ground with sand in a mortar, and 30 mL of a methanol–water mixture (8:2, *v*/*v*) was employed for extraction. Following this step, the resulting mixture underwent centrifugation at 10,000× *g* for 15 min, which effectively separated the supernatant for subsequent processing. Subsequently, a purification step was executed on the supernatant using a Sep-Pak C18 cartridge (Waters, Milford, MA). Elution was carried out with a 30 mL methanol–water mixture (8:2, *v*/*v*), and the eluate obtained was evaporated to dryness while ensuring that the temperature did not exceed 40 °C. To prepare the berry extracts for analysis, the dried eluate was reconstituted in methanol at a ratio of 10 mL per gram of fresh weight, and then meticulously filtered. During high-performance liquid chromatography (HPLC) analysis, precisely 10 μL of each sample, equivalent to 1 mg of fresh weight, was injected into the system for examination and quantification.

2.9. Statistical Analysis

Statistical analyses in this study were conducted using the integrated statistical package within R Studio. To comprehensively evaluate the impact of K_2CO_3 and oak ash, and their potential influences on the concentrations of phenolic compounds, anthocyanin content, phenolic acids, flavonoids, and phytoalexins, a rigorous analysis of variance (ANOVA) was performed, making use of the statistical package in R Studio. The statistical model encompassed both main effects and interaction effects, with subsequent assessments to confirm adherence to normality assumptions. Four distinct models were meticulously constructed to assess the primary effects of K_2CO_3 and oak ash on the concentrations of phenolic compounds, anthocyanin content, phenolic acids, flavonoids, and phytoalexins. When statistical significance was identified through the ANOVA, post hoc analysis was conducted using Tukey's test, a well-established method for exploring differences among multiple groups comprehensively [22]. To gain deeper insights and visualize the relationships among various variables, PCAs for phenolic compounds, anthocyanin content, phenolic acids, flavonoids, and phytoalexins were plotted using ggplot2 within R Studio [23]. PCA is a valuable analytical technique for reducing multidimensional data into a more interpretable format, facilitating the identification of underlying patterns and trends within complex datasets. The heatmap was produced utilizing the pheatmap package in R Studio [24].

3. Results

The content of anthocyanidins was significantly higher in the oak ash treatment (215 mg/kg) compared to the K_2CO_3 one (154.0 mg/kg), with a p-value of 0.001. The flavonols' content exhibited a significant difference, with oak ash (68.0 mg/kg) displaying higher levels than K_2CO_3 (50 mg/kg). The flavones' content was significantly higher in the oak ash treatment (92.6 mg/kg) compared to the K_2CO_3 one (66.0 mg/kg), with a p-value of 0.007. The flavanones' content displayed a significant variance, with oak ash (15.1 mg/kg) having higher levels compared to K_2CO_3 (11.0 mg/kg), and a p-value of 0.001. The catechins' content was significantly higher in the oak ash treatment (61.6) compared to the K_2CO_3 one (39.8 mg/kg), with a p-value of 0.005. On the other hand, the resveratrol content showed a significant difference, with oak ash (33.8 mg/kg) having higher levels compared to K_2CO_3 (23.2 mg/kg), with a very low p-value of 1.120. The content of pterostilbene was significantly higher in the oak ash treatment (4.65 mg/kg) compared to the K_2CO_3 one (3.77 mg/kg).

Piceid content exhibited a significant difference, with oak ash (3.04) showing lower levels compared to K_2CO_3 (3.43 mg/kg), at a p-value of 0.003. Viniferin content was significantly higher in the oak ash treatment (29.2 mg/kg) compared to K_2CO_3 (21.4 mg/kg). The comparison of flavonoids and phytoalexins between the oak ash and K_2CO_3 treatments revealed substantial differences in the composition of these compounds (Table 1 and Supplementary Table S1).

The comparative analysis of anthocyanin content and phenolic acids between the oak ash and K_2CO_3 treatments yielded substantial findings. While malvidin-3-O-glycoside content showed no significant difference between the two treatments, other anthocyanins exhibited noteworthy distinctions. The content of delphinidin-3-O-glycoside was significantly higher in the oak ash treatment, indicating its potential as a beneficial pre-treatment choice. The quantitative analysis of phenolic compounds in grape extracts revealed significant variations between different glycoside forms. Malvidin-3-O-glycoside exhibited comparable values of 165.0 and 171.0, with a p-value of 0.514, indicating a non-significant difference (ns). In contrast, Delphinidin-3-glycoside demonstrated a statistically significant decrease from 21.5 to 19.1 (p = 0.047). Similarly, cyanidin-3-O-glycoside and petunidin-3-glycoside both exhibited significant reductions from 8.26 to 6.52 (p = 0.022) and from 32.3 to 24.8 (p = 0.004), respectively. Peonidin-3-O-glycoside showed a substantial decline from 29.9 to 21.9 (p = 0.0016). Furthermore, malvidin-3-O-glucoside experienced a significant reduction from 22.7 to 17.9 (p = 0.003). Additionally, malvidin-3-O-glycoside acetyl demonstrated a noteworthy decrease from 226 to 143 ($p = 0.010$). Conversely, peonidin-3-glycoside acetyl did not exhibit a statistically significant difference, with values of 312.0 and 273.0 ($p = 0.081$). Notably, malvidin-3-O-glycoside-p-coumaryl displayed a reversal in values, with an increase from 273.0 to 299.0 ($p = 0.030$). The analysis of the phenolic acids in the samples yielded notable differences in the concentrations of specific compounds. Hydroxycinnamic acids exhibited a statistically significant decrease from 12.5 to 11.0 ($p = 0.036$). Hydroxybenzoic acids did not demonstrate a statistically significant difference, maintaining values of 14.0 and 13.2 ($p = 0.131$), indicated as ns (non-significant). Similarly, benzaldehyde did not exhibit a significant difference, with concentrations of 148.0 and 103.0 ($p = 0.291$), marked as ns. In contrast, phenylacetaldehyde displayed a significant reduction from 115.7 to 82.3 ($p = 0.048$) (Table 2 and Supplementary Table S2).

Table 1. Flavonoids (mg/kg) and phytoalexins (mg/kg) of the raisin grapes produced from cv. Gök Üzüm that had undergone treatment with the oak ash or K_2CO_3 dipping solutions.

	Oak Ash	K_2CO_3	*p*-Value	Significance Level
Flavonoids (mg/kg)				
Anthocyanidins	215.0 ± 5.621 b	154.0 ± 5.132 a	0.001	**
Flavonols	68.0 ± 2.14 b	50.5 ± 2.72 a	0.010	*
Flavones	92.6 ± 3.31 b	66.0 ± 3.71 a	0.007	**
Flavanones	15.1 ± 2.02 b	11.0 ± 0.35 a	0.001	**
Catechins	61.6 ± 2.12 b	39.8 ± 2.89 a	0.005	**
Phytoalexins (mg/kg)				
Resveratrol	33.8 ± 0.12 b	23.2 ± 0.40 a	0.020	***
Pterostilbene	4.6 ± 0.3 b	3.8 ± 0.12 a	0.016	*
Piceid	3.0 ± 0.1	3.4 ± 0.1	0.003	**
Viniferin	29.2 ± 0.22 b	21.4 ± 0.61 a	0.001	**

Data are stated as averages of the data and their standard deviations. Different letters within a row indicate significant differences (Tukey test: *, significant at p-value < 0.05; **, significant at p-value < 0.01; ***, significant at p-value < 0.001).

Table 2. The anthocyanin content (%) and phenolic acids (peak/IS ratio) of the raisin grapes produced from cv. Gök Üzüm that had undergone treatment with oak ash or K_2CO_3 dipping solutions.

	Oak Ash	K_2CO_3	*p*-Values	Significance Level
Anthocyanin content				
Malvidin-3-O-glycoside	165.0 ± 5.332	171.0 ± 5.834	0.514	ns
Delphinidin-3-O-glycoside	21.5 ± 0.15 b	19.1 ± 0.62 a	0.047	*
Cyanidin-3-O-glycoside	8.3 ± 0.9 b	6.5 ± 0.3 a	0.022	*
Petunidin-3-O-glycoside	32.3 ± 0.74 b	24.8 ± 0.94 a	0.004	**
Peonidin-3-O-glycoside	29.9 ± 0.52 b	21.9 ± 0.71 a	0.001	**
Malvidin-3-O-glucoside	22.7 ± 0.21 b	17.9 ± 0.57 a	0.003	**
Malvidin-3-O-glycoside acetyl	226.0 ± 13.134 b	143.0 ± 11.722 a	0.010	*
Peonidin-3-O-glycoside acetyl	312.0 ± 11.732	273.0± 11.234	0.081	ns
Malvidin-3-O-glycoside-p-coumaryl	273.0 ± 5.543 a	299.0 ± 5.577 b	0.030	*
Phenolic acids				
Hydroxycinnamic acids	12.5 ± 0.31 b	11.0 ± 0.21 a	0.036	*
Hydroxybenzoic acids	14.0 ± 0.12	13.2 ± 0.23	0.131	ns
Benzaldehyde	148.0 ± 12.733	103.0 ± 26.134	0.291	ns
Phenylacetaldehyde	115.7 ± 8.943 b	82.3 ± 8.42 a	0.048	*

Data are stated as averages of the data and their standard deviations. Different letters within a row indicate significant differences (Tukey test: *, significant at p-value <0.05; **, significant at p-value < 0.01; ns, not significant).

The analysis of phenolic compounds revealed noteworthy distinctions between the oak ash and K_2CO_3 treatments. Gallic acid content was significantly higher in the oak ash treatment (12.9 μg/L) compared to the K_2CO_3 treatment (11.0 μg/L), denoted by a p-value of 0.0182. Similarly, vanillic acid content displayed a substantial variance, with the oak ash treatment (19.5 μg/L) exhibiting higher levels in contrast to the K_2CO_3 treatment

(16.5), with a p-value of 0.004. *Trans*-caffeic acid content was notably elevated in the oak ash treatment (14.6 µg/L) compared to the K_2CO_3 treatment (11.7 µg/L), as evidenced by a p-value of 0.008. *Trans*-p-coumaric acid content also demonstrated a significant distinction, with the oak ash treatment (15.4 µg/L) revealing higher levels than the K_2CO_3 treatment (12.1 µg/L), resulting in a p-value of 0.000. Furthermore, the ferulic acid content was markedly different between the two treatments, with the oak ash treatment (2.9 µg/L) displaying lower levels compared to the K_2CO_3 treatment (4.2 µg/L), as indicated by a p-value of 0.000. In general, the contents of gallic acid, vanillic acid, *trans*-caffeic acid, *trans*-p-coumaric acid, and (-)-epicatechin were significantly greater after the K_2CO_3 treatment compared to the oak ash treatment, while ferulic acid and quercetin demonstrated greater contents after the oak ash treatment than the K_2CO_3 one; however, there were no differences in tyrosol, kaftaric acid, rutin catechin, and myricetin phenolic compounds between the applied treatments (Table 3 and Supplementary Materials Table S2).

Table 3. Phenolic compounds (µg/L) of the raisin grapes produced from the Gök Uzüm cultivar with oak ash and K_2CO_3 dipping solution treatments applied.

	Oak Ash	K_2CO_3	p-Values	Significance Level
Phenolic Compounds (µg/L)				
Tyrosol	20.2 ± 0.32	19.1 ± 0.51	0.210	ns
Gallic acid	12.9 ± 0.13 b	11.0 ± 0.32 a	0.018	*
Vanillic acid	19.5 ± 0.41 b	16.5 ± 0.33 a	0.004	**
Trans-caffeic acid	14.6 ± 0.25 b	11.7 ± 0.41 a	0.008	**
Trans-p-coumaric acid	15.4 ± 0.17 b	12.1 ± 0.24 a	0.001	***
Ferulic acid	2.9 ± 0.1 a	4.2 ± 0.1 b	0.001	***
Kaftaric acid	25.4 ± 0.19	23.5 ± 0.65	0.093	ns
Catechin	18.2 ± 0.22	17.8 ± 0.52	0.666	ns
(-)-Epicatechin	16.1 ± 0.24 b	11.7 ± 0.39 a	0.001	**
Quercetin	5.4 ± 0.1 a	7.3 ± 0.1 b	0.001	***
Rutin	9.6 ± 0.3	8.4 ± 0.4	0.101	ns
Myricetin	8.2 ± 0.3	9.2 ± 0.5	0.292	ns

Data are stated as averages of the data and their standard deviations. Different letters within a row indicate significant differences (Tukey test: *, significant at p-value < 0.05; **, significant at p-value < 0.01; ***, significant at p-value < 0.001; ns, not significant).

Principal Component Analyses (PCAs) have been instrumental in unraveling the relationships and alterations within the various components of phenolic compounds, anthocyanins, phenolic acids, flavonoids, and phytoalexins. These PCAs reveal vital insights into the variance within these components. The findings indicate that Dim1 and Dim2 collectively explain 70% and 14.7% of the variance, respectively, shedding light on the distribution and interplay of these compounds. Phenolic compounds, including gallic acid and catechin, are widely distributed in the PCA (Figure 2A). Focusing on anthocyanin content, Dim1 and Dim2 account for 71.5% and 14.9% of the variance, respectively. Anthocyanins like malvidin-3-glycoside and malvidin-3-O-glycoside-p-coumaryl are positioned on the positive side of Dim1, while others reside on the negative side (Figure 2B). Turning to phenolic acids, Dim1 and Dim2 elucidate 61.7% and 24.5% of the variance, respectively. The dominant phenolic acids, hydroxycinnamic acids and hydroxybenzoic acids, exhibit distinct positions in the PCA. The former gravitates toward the positive side of Dim1, while the latter is closer to the positive side of Dim2 (Figure 2C). In the realm of flavonoids, Dim1 captures 93.3% of the variance, whereas Dim2 explains 5.6%. Compounds like anthocyanidin, flavones, flavonols, and catechins are prominently clustered on the negative side of Dim1, signifying the robust relationships among them. This is corroborated by the high cos2 values, further emphasizing their importance in the PCA (Figure 2D). Examining phytoalexins, the primary PC (Dim1) clarifies 93.2% of the variance, while the secondary (Dim2) accounts for 4.1%. Key components, including pterostilbene, viniferin, and resveratrol, gravitate towards the positive end of Dim2, with piceid aligned with the positive end of Dim1. The high cos2 values underpin the robust relationships depicted in this principal component (Figure 2E).

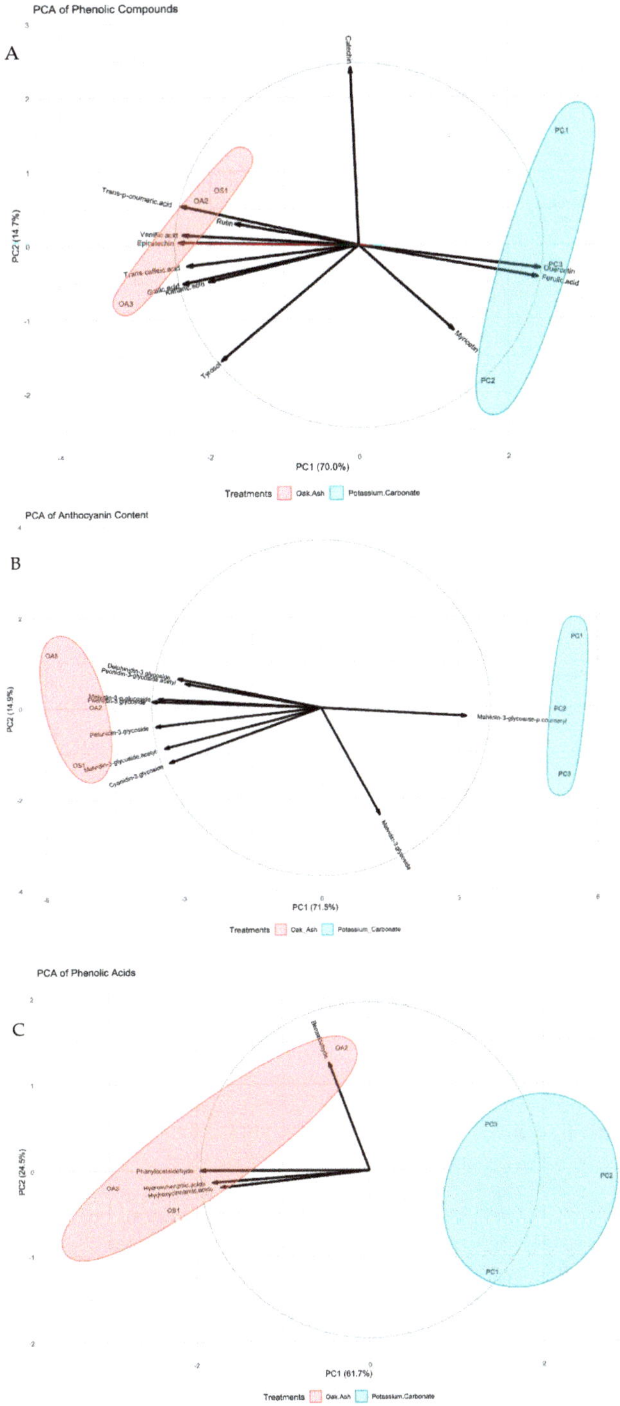

Figure 2. *Cont.*

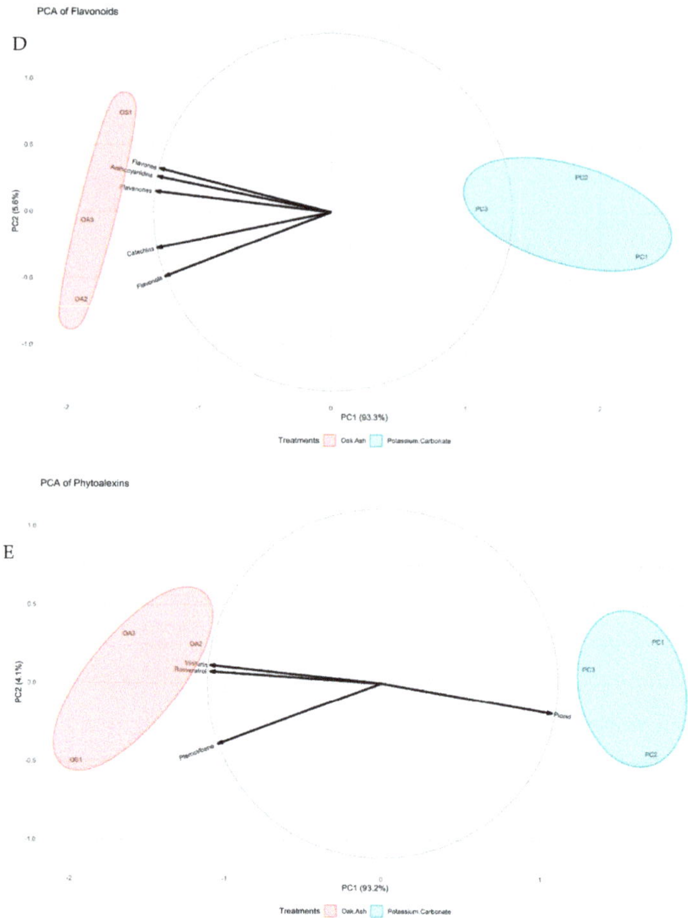

Figure 2. PCA biplots of berries colored by treatments. All phenolic compounds (**A**), anthocyanin (**B**), phenolic acids (**C**), flavonoids (**D**), and phytoalexins (**E**) are displayed. Cos2: the quality of the representation of the variables of the principal components. Data size of each parameter was used to construct the current graphs (n: 9).

Figure 3 presents a hierarchical heatmap that delineates the relative quantities of various phytochemical components within the grape samples, such as phenolic compounds, anthocyanin, phenolic acids, flavonoids, phytoalexins, and phenolic acids. The X axis clusters these components, revealing the similarities and differences among them. Notably, two distinct groups are discernible from right to left, spanning from viniferin to peonidin-3-glycoside acetyl and from rutin to Pterostilbene. It becomes evident that oak ash and K_2CO_3 treatments partition these groups along the Y axis, signifying the treatments' influence on phytochemical compositions, with malvidin-3-O-glycoside and anthocyanidins emerging as deep purple patches in the heatmap, indicative of their high concentration levels (Figure 3).

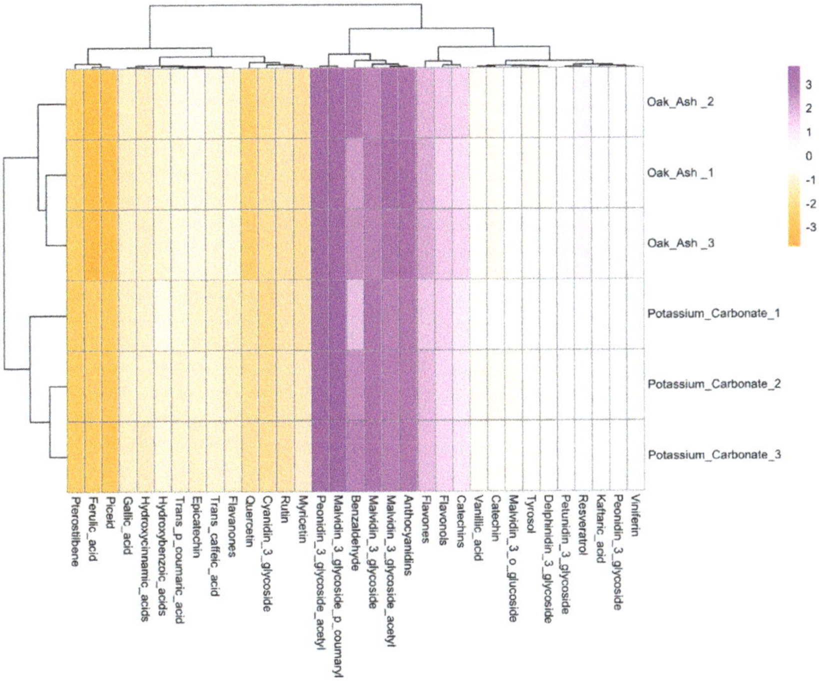

Figure 3. A heatmap analysis that scrutinizes numerous components, from phenolic compounds and anthocyanin to phenolic acids, flavonoids, and phytoalexins. The cos2- meaning of the bar in the top right is the quality of the representation of the variables of the principal components. Data size for each parameter was used to construct the current graph (n: 9).

4. Discussion

The comparative analysis of phenolic compounds in the grape samples revealed significant differences between the oak ash and K_2CO_3 solutions. In general, the results of this comprehensive analysis demonstrate the substantial impact of pre-treatment methods, specifically oak ash and K_2CO_3, on the composition of phenolic compounds, anthocyanins, phenolic acids, flavonoids, and phytoalexins in grapes. These findings underscore the critical role of pre-treatment selection in shaping the phytochemical profile of grapes, with profound implications for the quality and potential health benefits of grape-derived products. The substantial difference in the flavanones' content, with K_2CO_3 surpassing oak ash, indicates the preference for oak ash in preserving these compounds, likely attributed to the unique characteristics of this treatment method. A comparison of phytoalexins and flavonoids between the K_2CO_3 and oak ash solutions also yielded notable insights. The significantly higher anthocyanidins content in the oak ash solution underscores the promise of this pre-treatment method in enhancing these compounds, aligning with the findings of Ünal et al. [25]. The elevated catechins content in the oak ash solution highlights its potential as a preferred choice for enriching these compounds, known for their antioxidant properties [26,27]. Furthermore, the notably higher flavones content in the oak ash solution suggests its efficacy in conserving these compounds, which is significant considering the potential health benefits associated with flavones [28,29]. The significant variance in the flavanols' content, with K_2CO_3 surpassing oak ash, is in line with a previous study indicating the sensitivity of flavanols to different pre-treatment techniques [25]. Similarly, the notably higher pterostilbene content in the oak ash treatment emphasizes its potential positive impact on this phytoalexin. In contrast, the lower piceid content in oak ash compared to K_2CO_3 suggests the sensitivity of this compound to different pre-treatment methods [3,7,14]. Additionally, the significantly higher resveratrol content in the oak ash treatment underscores its efficacy in preserving this compound, highly regarded for its potential health benefits [30]. The greater viniferin content in the oak ash treatment also suggests its potential as a preferred pre-treatment choice for influencing this specific phytoalexin.

The gallic acid content was significantly higher in the oak ash treatment, highlighting its potential health benefits [27]. The K_2CO_3 treatment exhibited higher a ferulic acid content, indicating that different treatments can indeed exert varying effects on specific phenolic acids [26]. Furthermore, the oak ash treatment demonstrated elevated levels of *trans*-caffeic acid and *trans*-p-coumaric acid, suggesting its efficacy in preserving these compounds, possibly attributed to the unique properties of oak ash. The notably higher vanillic acid content in the oak ash treatment may also be a result of oak ash's distinct characteristics, with vanillic acid holding importance due to its antioxidant and antimicrobial properties, rendering it a valuable component of grape products [31]. Collectively, these results highlight the substantial influence of pre-treatment choice on the composition of the phenolic compounds in grapes, aligning with prior research that emphasizes the impact of pre-treatment methods on grape phenolics [3,7,8,14,32].

The substantially higher malvidin-3-O-glycoside acetyl and delphinidin-3-O-glycoside content in the oak ash treatment emphasizes its importance as a pre-treatment choice for influencing this specific anthocyanin, consistent with previous studies underscoring the role of pre-treatment in shaping the anthocyanin profile in grapes [25]. The analysis of the anthocyanin content revealed intriguing disparities between the K_2CO_3 and oak ash treatments, aligning with prior studies highlighting the impact of pre-treatments on anthocyanin content modulation [33]. Similarly, the significantly higher malvidin-3-o-glucoside and peonidin-3-O-glycoside content in the oak ash treatment underscore the potential benefits of this treatment for enhancing the retention of these anthocyanins. Petunidin-3-glycoside and cyanidin-3-O-glycoside contents were notably elevated in the oak ash treatment, reinforcing the idea that different pre-treatment methods can exert varying influences on anthocyanins. These anthocyanins, known for their role in color and their potential health benefits, further underscore the importance of pre-treatment selection [34,35]. On the other hand, the notably higher phenylacetaldehyde content in the oak ash treatment indicates its potential positive impact on this phenolic acid. The higher hydroxycinnamic acids content in the oak ash treatment for phenolic acids' enhancement suggests its potential as a preferred choice for enhancing this specific compound. Taken together, these findings emphasize the substantial influence of pre-treatment selection on the anthocyanin content and phenolic acid composition in grapes, consistent with previous research highlighting the role of pre-treatment methods in shaping these components [28,36]. On the other hand, research on grape harvests, even when conducted over a single year, provides valuable insights, enabling an understanding of the unique characteristics of that specific year. While it highlights the specific environmental factors of that year, it paves the way for acknowledging the need to consider broader variations that may occur annually. For instance, studying the quality parameters of the Gök Üzüm cultivar in a single year can reveal how it responds to specific weather patterns, soil fertility, and other climatic conditions of that year, offering a snapshot of its potential variability. Additionally, recognizing that changes in the characteristics of grapes can influence the raisining process is insightful.

5. Conclusions

Our findings highlight the differential impact of pre-treatment methods, specifically oak ash and K_2CO_3, on the preservation of distinct compounds within grapes. On the one hand, K_2CO_3 excels in maintaining the content of ferulic acid and quercetin within the grapes. The comparative analysis between the oak ash and K_2CO_3 treatments revealed substantial differences in the content of anthocyanins, phenolic compounds, and phenolic acids in the grape extracts. Oak ash demonstrated a significant advantage in preserving specific compounds, including anthocyanins, flavonoids, and phytoalexins. Notably, the higher content of malvidin-3-O-glycoside acetyl, flavonols, and flavones in the oak ash treatment underscores its efficacy as a pre-treatment method. The analysis of phenolic compounds further emphasized their nuanced variations, with oak ash and K_2CO_3 treatments exhibiting distinct impacts on specific phenolic acids. These findings contribute valuable insights into the role of pre-treatment in shaping the phytochemical profile of grapes, offering potential applications for tailored grape-derived products. However, the transformation of grapes into raisins is intricately linked to the grapes' quality and condition. Factors like sugar content, acidity, and moisture levels, which might vary each year, are crucial to determining the quality of the final raisin product. Hence, conducting research over multiple years is vital to thoroughly understand these interactions and to innovate more effective and flexible raisin production techniques. This multi-year approach is instrumental in gaining a comprehensive perspective on how annual variations in grape characteristics can enrich the raisining process and enhance the quality of the end product. Nevertheless, further research is essential to delve into the underlying mechanisms behind these variations and explore their implications for the quality and functionality of grape-based products, considering their diverse applications in the food and pharmaceutical industries.

Supplementary Materials: The following are available online at https://www.mdpi.com/article/10.3390/horticulturae10010095/s1, Table S1: Phenolic compounds (μg/L) of the raisin grapes produced from the Gök Uzüm cultivar applied the treatments Oak ash and K_2CO_3 dipping solutions; Table S2: The anthocyanin content (%) and phenolic acids (peak/IS ratio) of the raisin grapes produced from the Gök Üzüm cultivar applied the treatments Oak ash and K_2CO_3 dipping solutions; Table S3: Flavonoids (mg/kg) and Phytoalexins (mg/kg) of the raisin grapes produced from the Gök Üzüm cultivar applied the treatments Oak ash and K_2CO_3 dipping solutions.

Author Contributions: O.K. conceived and designed the experiments; O.K. and H.D. performed the experiments; T.Y., H.D., O.K., M.S., N.K. and F.A. analyzed the data. O.K. wrote and proofread the final paper. All authors have read and agreed to the published version of the manuscript.

Funding: This research received no external funding.

Institutional Review Board Statement: No ethics approval or consent to participate was required for this study.

Informed Consent Statement: Not applicable.

Data Availability Statement: Data are contained within the article and Supplementary Materials.

Conflicts of Interest: The authors declare no conflicts of interest.

References

1. Keskin, N.; Kunter, B.; Çelik, H.; Kaya, O.; Keskin, S. ANOM approach for the statistical evaluation of organic acid contents of clones of the grape variety 'Kalecik Karası'. *Mitteilungen Klosterneubg.* **2021**, *71*, 126–138.
2. Ates, F.; Kaya, O.; Keskin, N.; Turan, M. Bi ogenic amines in raisins of one vintage year: In fluence of two chemical pre-treatments (dipping in oak ash solution or potassium carbonate solution). *Mitteilungen Klosterneubg.* **2022**, *72*, 51–59.
3. Keskin, N.; Kunter, B.; Celik, H.; Kaya, O.; Keskin, S. Evaluation of Clonal Variability of Berry Phenolics in *Vitis vinifera* L. Cv. Kalecik Karası. *Erwerbs-Obstbau* **2022**, *64* (Suppl. S1), 65–72. [CrossRef]
4. Kalkan, N.N.; Karadoğan, B.; Kadioğlu, Z.; Esmek, İ.; Albayrak, S.; Kaya, O. Response of Karaerik Grape Cultivar (*Vitis vinifera* L.) to Two Training Systems and Three Trunk Heights. *Erwerbs-Obstbau* **2022**, *64* (Suppl. S1), 119–127. [CrossRef]
5. World Health Organization. Fruit and Vegetable Promotion Initiative. 2022. Available online: https://www.who.int/dietphysicalactivity/publications/trs916/intro/en/ (accessed on 20 May 2022).
6. Kaliora, A.C.; Kountouri, A.M.; Karathanos, V.T.; Andrikopoulos, N.K. Vitamin C content in grapes, fresh and dried, and in products made from them. *J. Food Compos. Anal.* **2016**, *46*, 30–37.
7. Karakus, S.; Ates, F.; Keskin, N.; Turan, M.; Kaya, O. Comparison of contents of sugars, organic acids and free amino acids in raisins obtained from Gök Üzüm (*Vitis vinifera* L.). *Mitteilungen Klosterneubg.* **2023**, *73*, 98–113.
8. Karakus, S.; Kaya, O.; Hajizadeh, H.S.; Gutiérrez-Gamboa, G.; Ates, F.; Turan, M.; Araya-Alman, M. Characterization of volatile compounds of Gök Üzüm raisins produced from grapes pre-treated with different dipping solutions. *Chem. Biol. Technol. Agric.* **2023**, *10*, 55. [CrossRef]

9. Nile, S.H.; Park, S.W. Edible Oil. Inhibitory effects of anthocyanins and fruit extracts from Vitis coignetiae on melanin formation in B16F10 cells. *Int. J. Mol. Sci.* **2014**, *15*, 21930–21947.
10. González-Gallego, J.; García-Mediavilla, M.V.; Sánchez-Campos, S.; Tuñón, M.J. Fruit polyphenols, immunity and inflammation. *Br. J. Nutr.* **2010**, *104* (Suppl. S3), S15–S27. [CrossRef]
11. Tsuda, T. Dietary anthocyanin-rich plants: Biochemical basis and recent progress in health benefits studies. *Mol. Nutr. Food Res.* **2012**, *56*, 159–170. [CrossRef]
12. Nunes, T.; Almeida, L.; Rocha, J.F.; Laranjinha, J.; Barbosa, R.M. Inhibition of human in vitro osteoclastogenesis by anthocyanin-rich açai extract. *Inflammopharmacology* **2017**, *25*, 261–268.
13. Chukwumah, Y.; Walker, L.; Verghese, M. Peanut skin color: A biomarker for total polyphenolic content and antioxidative capacities of peanut cultivars. *Int. J. Mol. Sci.* **2009**, *10*, 4941–4952. [CrossRef] [PubMed]
14. Keskin, N.; Kaya, O.; Ates, F.; Turan, M.; Gutiérrez-Gamboa, G. Drying grapes after the application of different dipping solutions: Effects on hormones, minerals, vitamins, and antioxidant enzymes in Gök Üzüm (*Vitis vinifera* L.) raisins. *Plants* **2022**, *11*, 529. [CrossRef]
15. USDA. Statistical Data. 2022. Available online: https://apps.fas.usda.gov/psdonline/circulars/Raisins.pdf (accessed on 20 October 2023).
16. Kaya, O.; Ates, F.; Kara, Z.; Turan, M.; Gutiérrez-Gamboa, G. Study of Primary and Secondary Metabolites of Stenospermocarpic, Parthenocarpic and Seeded Raisin Varieties. *Horticulturae* **2022**, *8*, 1030. [CrossRef]
17. Vendramin, V.; Viel, A.; Vincenzi, S. Caftaric Acid Isolation from Unripe Grape: A "Green" Alternative for Hydroxycinnamic Acids Recovery. *Molecules* **2021**, *26*, 1148. [CrossRef] [PubMed]
18. Yousef, G.G.; Brown, A.F.; Funakoshi, Y.; Mbeunkui, F.; Grace, M.H.; Ballington, J.R.; Loraine, A.; Lila, M.A. Efficient quantification of the health-relevant anthocyanin and phenolic acid profiles in commercial cultivars and breeding selections of blueberries (*Vaccinium* spp.). *J. Agric. Food Chem.* **2013**, *61*, 4806–4815. [CrossRef]
19. Pantelić, M.M.; Zagorac, D.Č.D.; Ćirić, I.Ž.; Pergal, M.V.; Relić, D.J.; Todić, S.R.; Natić, M.M. Phenolic profiles, antioxidant activity and minerals in leaves of different grapevine varieties grown in Serbia. *J. Food Compos. Anal.* **2017**, *62*, 76–83. [CrossRef]
20. Brossa, R.; Casals, I.; Pintó-Marijuan, M.; Fleck, I. Leaf flavonoid content in *Quercus ilex* L. resprouts and its seasonal variation. *Trees* **2009**, *23*, 401–408. [CrossRef]
21. Jeandet, P.; Breuil, A.C.; Adrian, M.; Weston, L.A.; Debord, S.; Meunier, P.; Maume, G.; Bessis, R. HPLC analysis of grapevine phytoalexins coupling photodiode array detection and fluorometry. *Sci. Hortic.* **1997**, *183*, 115–126. [CrossRef]
22. R Core Team. *R: A Language and Environment for Statistical Computing*; R Foundation for Statistical Computing: Vienna, Austria, 2013; Available online: http://www.R-project.org/ (accessed on 8 August 2022).
23. Wickham, H.; Wickham, H. *Data Analysis*; Springer International Publishing: Cham, Switzerland, 2016; pp. 189–201.
24. Kolde, R. Pheatmap: Pretty Heatmaps_.R Package Version 1.0.12. 2019. Available online: https://rdrr.io/cran/pheatmap/ (accessed on 25 February 2022).
25. Ünal, M.S.; Güler, E.; Yaman, M. Changes in Antioxidant and Color Properties of Raisins According to Variety and Drying Method. *Horticulturae* **2023**, *9*, 771. [CrossRef]
26. Teixeira, A.; Eiras-Dias, J.; Castellarin, S.D.; Gerós, H. Berry phenolics of grapevine under challenging environments. *Int. J. Mol. Sci.* **2013**, *14*, 18711–18739. [CrossRef]
27. Badhani, B.; Sharma, N.; Kakkar, R. Gallic acid: A versatile antioxidant with promising therapeutic and industrial applications. *RSC Adv.* **2015**, *5*, 27540–27557. [CrossRef]
28. Williamson, G.; Carughi, A. Polyphenol content and health benefits of raisins. *Nutr. Res.* **2010**, *30*, 511–519. [CrossRef] [PubMed]
29. Olmo-Cunillera, A.; Escobar-Avello, D.; Pérez, A.J.; Marhuenda-Muñoz, M.; Lamuela-Raventós, R.M.; Vallverdú-Queralt, A. Is eating raisins healthy? *Nutrients* **2019**, *12*, 54. [CrossRef] [PubMed]
30. Thiruchelvi, R.; Jayashree, P.; Nayik, G.A.; Gull, A.; Ahad, T.; Thakur, M.; Shah, T.R.; Paray, M.A.; Haq, R.U. Raisin. In *Antioxidants in Vegetables and Nuts-Properties and Health Benefits*; Springer: Berlin/Heidelberg, Germany, 2020; pp. 523–537.
31. Sabel, A.; Bredefeld, S.; Schlander, M.; Claus, H. Wine phenolic compounds: Antimicrobial properties against yeasts, lactic acid and acetic acid bacteria. *Beverages* **2017**, *3*, 29. [CrossRef]
32. Campos, F.; Peixoto, A.F.; Fernandes, P.A.; Coimbra, M.A.; Mateus, N.; de Freitas, V.; Fernandes, I.; Fernandes, A. The antidiabetic effect of grape pomace polysaccharide-polyphenol complexes. *Nutrients* **2021**, *13*, 4495. [CrossRef] [PubMed]
33. Squadrito, M.; Corona, O.; Ansaldi, G.; Di Stefano, R. Evolution of anthocyanin profile from grape to wine. *Oeno One* **2010**, *44*, 167–177. [CrossRef]
34. Karadeniz, F.; Durst, R.W.; Wrolstad, R.E. Polyphenolic composition of raisins. *J. Agric. Food Chem.* **2000**, *48*, 5343–5350. [CrossRef]
35. Kalkan, N.N.; Bozkurt, A.; Gecim, T.; Karadogan, B.; Bahar, E.; Kaya, O. Influence of pruning severity in the table grape variety 'Karaerik' (*Vitis vinifera* L.). *Mitteilungen Klosterneubg.* **2022**, *72*, 137–145.
36. Kelebek, H.; Jourdes, M.; Selli, S.; Teissedre, P.L. Comparative evaluation of the phenolic content and antioxidant capacity of sun-dried raisins. *J. Sci. Food Agric.* **2013**, *93*, 2963–2972. [CrossRef]

horticulturae

Article

Multivariate Analysis and Optimization of the Relationship between Soil Nutrients and Berry Quality of *Vitis vinifera* cv. Cabernet Franc Vineyards in the Eastern Foothills of the Helan Mountains, China

Yashan Li [1,2], Jinnan Xiao [1], Yinfang Yan [1], Weiqiang Liu [1], Ping Cui [3], Chengdong Xu [2], Lijun Nan [2] and Xu Liu [1,*]

1 College of Enology, Northwest A&F University, No. 22 Xi'nong Road, Yangling, Xianyang 712100, China; liyashan@nwafu.edu.cn (Y.L.); 13188355230@163.com (J.X.); y1845737066@163.com (Y.Y.); lwq@nwafu.edu.cn (W.L.)

2 School of Resources, Environment and Chemistry, Chuxiong Normal University, No. 546 Lucheng South Road, Chuxiong 675000, China; chtown@cxtc.edu.cn (C.X.); submit_paper73@126.com (L.N.)

3 Grape Industry Management Committee of the Eastern Foothills of the Helan Mountains in Ningxia Autonomous Region, Yinchuan 750004, China; 13995179902@163.com

* Correspondence: liuxu@nwsuaf.edu.cn

Abstract: The aim of this study is to explore the relationship between soil nutrients and berry quality for the wine grape *Vitis vinifera* cv. Cabernet Franc in the eastern foothills of the Helan Mountains, and subsequently to optimize soil nutrient conditions for optimal berry quality, thus providing guidance for vineyard soil management. Based on the basic data on soil nutrients and berry quality indicators, a partial least squares regression method was used to screen for major soil nutrient factors affecting the grape quality index. Then, the selected soil nutrient factors were taken as independent variables and the corresponding grape quality indicators were taken as dependent variables and a multilinear regression equation was formulated by the method of multivariate linear regression. Finally, the optimal solution for fruit quality and soil nutrients was solved using linear programming equations. The results showed that there was a lack of total nitrogen, organic matter, nitrate nitrogen, ammonium nitrogen, and available phosphorus in the soil nutrients, and an alkaline soil. There is a significant positive correlation between some soil nutrient indices, and there is also a multivariate linearity problem. Among all berry quality indices, titratable acid, tannin, and anthocyanin were negatively correlated with eleven and ten soil indices, respectively, while other berry quality indices were positively correlated with most soil nutrient indices. The optimal parameters for grape quality were determined using the method of linear programming equations, and the corresponding soil nutrient indicators content were defined.

Keywords: *Vitis vinifera* cv. Cabernet Franc; soil nutrients; berry quality; multivariate statistical analysis; optimum proposal

1. Introduction

The grapevine *Vitis vinifera* L. is a perennial plant that is widely distributed around the world due to its environmental adaptability and high economic value. China has a long history of grapevine cultivation and is also one of the origins of the *Vitis* genus, which belongs to the *Vitaceae* family [1]. After years of rapid development, China has become one of the world's largest wine producers and consumers [2]. Based on different terroir characteristics, China has formed eleven wine regions with their own styles [3]. Among them, the eastern foothills of the Helan Mountains are an emerging wine region in China and is one of the most promising [4]. Benefiting from excellent terroir conditions, the use of this wine-producing region has resulted in unique, high-quality wines [5].

As one of the most important components of terroir, soil is also a major source of water and nutrients for plants [6]. Soil nutrients are diverse and complex. This is because, on the one hand, there are various types of soil nutrients and, on the other hand, there are synergistic or antagonistic effects between soil nutrients. Regardless of the intrinsic relationship between soil nutrients, they are bound to have an impact on grapevine and grape quality. Many previous studies have shown that different vineyard soil nutrients are different [7,8], and that different soil nutrients have different effects on grape plant growth and grape quality [9–11].

To better guide production practices, it is necessary to use quantitative research methods to determine the precise relationship between soil nutrients and grape quality. However, previous studies have mostly consisted of qualitative studies of the relationship between soil nutrients and grape quality. To the author's knowledge, there have been few quantitative studies of the relationship between the grape quality and soil nutrients [12]. Nevertheless, there are lessons to be learned from related studies on other fruit deciduous trees. For example, Guo et al. [13] used the typical correlation analysis method to screen for the soil nutrient factors affecting the grape quality index of kiwifruit and establish a regression equation between them. Multivariate statistical analysis was used to determine the relationship between grape quality and soil nutrients in the Osbeck orchard [14]. In addition, canonical correlation analysis and linear equations were used to explore the relationship between soil nutrients and grape quality, and then the corresponding recommendations of soil nutrient content recommendations for the optimal apple quality were obtained [15]. These related research methods not only prove that there is a definite relationship between soil nutrients and grape quality, but also provide ideas for our study.

There is no doubt that there is a complex relationship between grape quality and available nutrients in the soil. Understanding how to quantify the internal relationships between them is essential to guiding viticultural practice. In light of this, six *Vitis vinifera* cv. Cabernet Franc vineyards in the eastern foothills of the Helan Mountains were used as experimental points in this study. Based on the measurements of grape quality and soil nutrient indices in these vineyards, multivariate statistical analysis methods were used to reveal and quantify the complex relationship between grape quality and soil nutrients in the test sites. Finally, the grape quality is optimized by linear programming equations and the corresponding soil nutrient content is obtained, which serves as a reference for soil management in local *Vitis vinifera* cv. Cabernet Franc vineyards.

2. Materials and Methods

2.1. Experimental Sites

In this study, 6 *Vitis vinifera* cv. Cabernet Franc vineyards were selected from 4 subregions (Shizuishan, Yinchuan, Qingtongxia, Hongsipu) in the eastern foothills of Helan Mountains during the 2021 growing season (Figure 1). Cabernet Franc vines grow from self-seeded seedlings.

The eastern foothills of the Helan Mountains in Ningxia are located between the Yellow River alluvial plain and the alluvial fan plain of Helan Mountains in Ningxia, with the Helan Mountains in the west, the upper reaches of the Yellow River in the east, and the ancient city of Yinchuan in the north, with a total area of 200×10^3 hectares. It has a continental monsoon climate with drought and light rainfall, and excellent light and heat resources. The latitude and longitude of the 6 vineyards are 37°27′33″–38°40′33″ N and 105°55′40″–106°18′40″ E (Table 1). The average elevation of the vineyards is about 1.20×10^3 m and there are large temperature differences between day and night. The annual cumulative temperature of the area varies from 3.20×10^3 to 3.35×10^3 °C, with an average annual sunshine of approximately 1.50×10^3 h and annual rainfall of 150–240 mm [16].

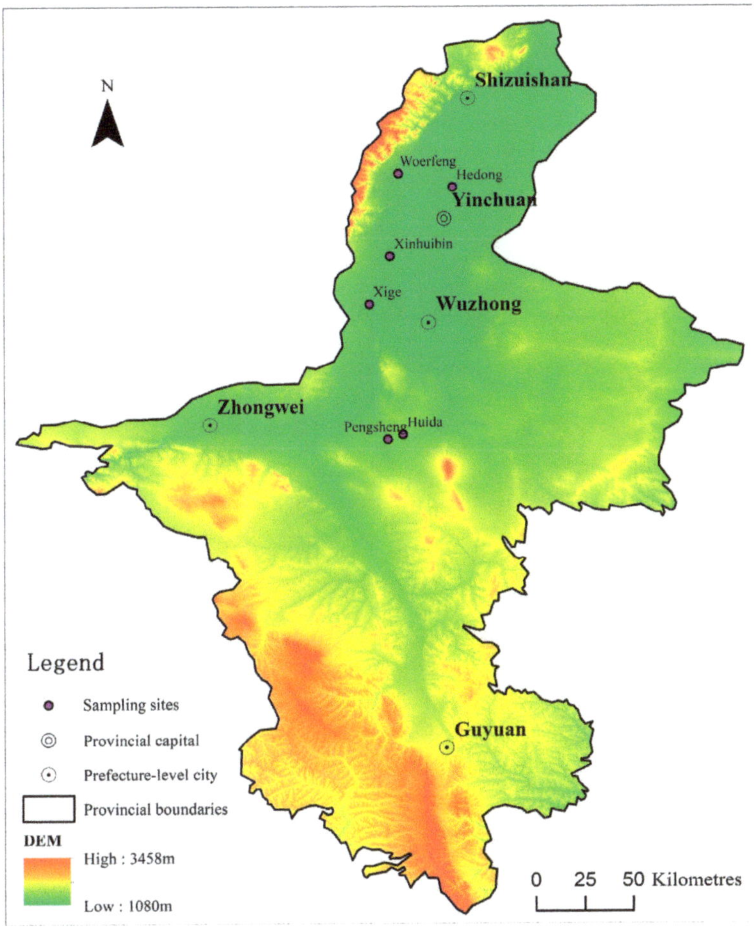

Figure 1. The location of the study sites. Note: The DEM in Figure 1 is an abbreviation of Digital Elevation Model.

Table 1. Basic situation of the sample vineyards of the six representative wineries.

Sub-Regions	Winery	Longitude	Latitude	Altitude (m)	Slope Gradient (°)	Row/within-Row Spaces (m)
Shizuishan	Hedong	106°18′40″ E	38°36′49″ N	1.07×10^3	1.98	3.24×1.51
Yinchuan	Woerfeng	106°03′45″ E	38°40′33″ N	1.16×10^3	2.11	3.47×0.82
	Xinhuibin	106°01′24″ E	38°17′28″ N	1.16×10^3	2.27	3.53×0.78
Qingtongxia	Xige	105°55′40″ E	38°04′08″ N	1.14×10^3	3.06	3.01×1.04
Hongsibu	Pengsheng	106°04′22″ E	37°27′33″ N	1.37×10^3	3.43	3.95×0.80
	Huida	106°04′49″ E	37°27′56″ N	1.32×10^3	3.82	3.45×0.53

2.2. Sampling Methods

In each *Vitis vinifera* cv. Cabernet Franc vineyard of the six wineries, three sampling sites (30.0 m × 30.0 m) were randomly selected. All the vineyard soil samples were taken during the ripening period of the wine grapes, and the soil samples were collected using the S distribution method. Within a single sampling point in the same vineyard, five healthy medium-growth grapevines with medium growth were randomly selected, and each of the two grapevines was more than 5.00 m apart. Three soil collection points were randomly selected at a distance of 30.0 to 40.0 cm from the trunk of each grapevine between rows. To reduce experimental errors, the two trees at the beginning and the end of each row are avoided. A total of 15 soil collection points were selected for each sampling site. After the extraction site was identified, 0.00 to 40.0 cm of soil was drilled with an extraction drill. Plant residues and rocks were manually removed from the borehole soil, and the soil from the 15 sampling points was thoroughly mixed. Approximately 1.00 kg of soil samples were obtained by quartering. Three repetitions were performed. In the end, three soil samples of about 1.00 kg each were taken from each vineyard. The resulting soil samples were wrapped in a self-sealing bag and transported back to the laboratory in an insulated cooler. In these soil samples, a tiny part of the fresh soil (about 100 g for each replicate) was temporarily stored in a refrigerator at 4.0 °C for the determination of the nitrate nitrogen and ammonium nitrogen content, and the rest of the soil samples were stored in a cool place for natural air drying. Subsequently, additional physical and chemical indices of the soil were determined.

Based on the actual condition of each vineyard, mature clusters were collected at commercial maturity for quality analysis. Before harvesting, a professional winemaker tastes the berries to determine the best time to harvest. Harvest conditions can only be achieved when the berry has no green taste, the specific aroma of the variety is established, the flesh and seeds can be easily separated, the tannins of the skin are smooth rather than rough, and the seeds are completely ripe. When the harvest conditions were met, around the soil sampling points, fifteen healthy, disease-free, and insect-free clusters (from both sides of the canopy) were randomly harvested in the single sampling sites. So, in each vineyard, the grape sampling was repeated three times, as was the soil sampling. Finally, a total of 45 clusters of grapes were collected from each sampling site. The grape clusters were immediately packaged and stored in an insulated cooler to be brought back to the laboratory. Then, 6 grape berries from different areas of each cluster were randomly cut with tiny scissors, for a total of 270 grape berries (90 for each replicate) in each winery. These samples were rapidly stored in an ultra-low temperature refrigerator (DW-86L486, Qingdao Haier Biomedical Co., Ltd., Qingdao, China) at −80 °C for subsequent analysis of anthocyanin, flavonol and tannin. A total of 100 grape berries were randomly plucked from the cluster for each replicate to determine the berry mass. After weighing, 100 of the grapes were crushed by a finger press to extract the free-running juice for the subsequent analysis.

2.3. Determination Indices and Methods

The total nitrogen (TN) was determined using the Kjeldahl method [17]; the total phosphorus (TP) and available phosphorus contents were determined by the methods described by Zhang et al. [18]; the total potassium (TK) and available potassium (K) were measured according to Bao [19]; the soil organic matter (OM) was determined following the Walkley–Black method [20]; the nitrate nitrogen (NN) and ammonium nitrogen (AN) were determined using a continuous flow analyzer (AA3, SEAL Analytical GmbH, Norderstedt, Germany) after extraction with 2.00 M Kcl [21]. The contents of available iron (Fe), available manganese (Mn), available copper (Cu) and available zinc (Zn) in the soil were determined using a DTPA-TEA extraction and flame/graphite furnace atomic absorption spectrometer (900T AAS, PerkinElmer Inc., Waltham, MA, USA) [22]. The soil pH was determined using a pH meter (pHS-3C) with a 1.00:2.50 (w/v) soil: water suspension [18]. Three technical replicates were analyzed for each sample.

The single grape berry was weighed on a JY10002 electronic balance (Shanghai Hengping Instrument Co., Ltd., Shanghai, China); yield per plant was weighed using an ES-200 KT electronic balance (Changsha Xiangping Technology Development Co., Ltd., Changsha, China); the soluble solid content was determined using a handheld digital glycometer (PAL-1, Atago, Japan); the reducing sugar content (SSC) was determined using direct titrimetric method [23]; the titratable acid (TA) was measured using an Easyplus

Titration System (Mettler Toledo, Columbus, OH, USA); the pH value was determined with a PHS-3 C pH meter (Precision Science Instruments Co., Ltd., Shanghai, China). Ten grapevines were randomly selected near the soil sampling points in each vineyard, and the total weight of the all grape clusters on each grapevine were weighed. The average value was taken as the yield per plant. Anthocyanin and flavonol from the grape skins were extracted and measured according to a previously published method [24]; extraction and determination of tannins from grape peels following Kennedy and Jones [25] with slight modifications. Briefly, skin samples of skin from 80 berries were hand-separated, dried and crushed into a fine powder. The powder samples were lyophilized in a vacuum freeze dryer (FD5-series; GOLD SIM International Group, Miami, FL, USA). Then, the dry tissues were weighed (0.05 g) and extracted in acetone/water (2.00:1.00, *v*/*v*) for 24 h with agitation. The solids were filtered through Whatman #1 filters and the acetone was removed by rotary evaporation at 38.0 °C. The residues were dissolved with 1 mL methanol, and stored at −80.0 °C to avoid light for further analysis. A Shimadzu LC-20AT HPLC system fitted with 2 Chromolith RP-18e (100 mm × 4.60 mm) columns and a photodiode array detector (Shimadzu Corporation, Kyoto, Japan). The mobile phases include phase A: glacial acetic acid/water, 1.00:99.0, *v*/*v* and phase B: glacial acetic acid/acetonitrile, 1.00:99.0, *v*/*v*. The temperature of column was 30.0 °C, and the column flow rate was 3.00 mL·min^{-1}. The eluting gradient was as follows: 0.00–4.00 min, 3.00% B; 4.00–14.0 min, 18.0% B; and 16.0–18.0 min 3.00% B. Detection was performed at 280 nm, and the injection volume of the sample was 20.0 µL. An external standard method with (−)-epicatechin for quantitative determination of tannins. All analyses were performed in triplicate.

2.4. Method of Screening of Soil Nutrient Factors Affecting the Berry Quality and Establishing the Regression Equation

First, the partial least squares method was used to screen for soil nutrient factors affecting grape quality. This approach is a reference to Wang et al. [26]. Briefly, the partial least squares method extension module is pre-installed in the SPSS 25.0 software, including SPSS Statistics Essentials for Python, the NumPy library, and the SciPy library. Then, the original data were input, and the menu bar "Analysis → Regression → Partial Least Squares" was clicked in turn to enter the interface of partial least squares regression analysis. The default settings were maintained, and after clicking "OK", SPSS would automatically give the analysis results. The results of the latent factor weighting were used to compute standardized regression coefficients for different soil nutrient indicators. Based on the absolute values of these regression coefficients, soil nutrient indicators with large absolute values of the regression coefficients and strong correlations with the corresponding grape quality indicators were screened. Finally, the multivariate linear regression equations were established by the "input" method for the screened soil nutrient indicators, and the significance of differences was tested. The construction of these regression equations will form the basis for the later application of linear programming equations.

2.5. Statistical Analysis of Data

First, TN, TP, TK, OM, NN, AN, available P, available K, available Cu, available Zn, available Fe, available Mn and soil pH were obtained from the soil of the six vineyards; Berry weight, yield per plant, soluble solids content, reducing sugars, titratable acidity, juice pH, tannin, anthocyanin, flavonol were measured from the ripened grape berries. Then, the raw data were then subjected to feature extraction and data normalization. Classical descriptors such as mean, minimum, maximum, coefficient of variation (CV), and standard deviation (SD) were determined by using Microsoft Office Excel 2016. At the same time, we normalized the data by the Z-score method to keep the data in the range from 0 to 1 with the same scale. The equation is as follows:

$$Z = \frac{x - \mu}{\sigma}$$

where Z was the standard score, x was the original measurement data, μ was the mean of all values in corresponding indicator, σ was the standard deviation.

SPSS software 25.0 was used to calculate the simple Pearson correlation coefficient between soil indicators and berry indicators. The formulation of the multiple linear regression model is represented by the following equality:

$$Y_i = \beta_0 + \beta_1 X_1 + \beta_2 X_2 + \beta_3 X_3 + \ldots\ldots + \beta_n X_n + \varepsilon_i$$

where Y_i is the predicted variable (berry quality parameter), $\beta_0 \sim \beta_n$ are the coefficients of regression, $X_1 \sim X_n$ are the input variables (soil nutrient indicators), and ε_i is the error associated with the ith observation.

Multiple linear regression analysis was performed by using SPSS 25.0, the method for variance of analysis tested by F-test, $p < 0.05$ or $p < 0.01$. To perform a valid inference from a multiple linear regression model, assumptions need to be made, such as the independent variables are independent of each other, there is no multicollinearity between the independent variables, and the error terms are normally distributed. Sample independence was checked using Durbin–Watson (DW) statistics. Variance inflation factor (VIF) values were used to test for multicollinearity among the independent variables. The residual normality distribution can be determined from normalized residual error histograms and the normal P–P plots. The SPSS software automatically provides these data and graphs for statistical hypothesis testing.

Microsoft Office Excel 2016 software was used for data statistical analysis, IBM SPSS Statistics 25.0 software (IBM, Chicago, IL, USA) was used for partial least squares regression and multiple linear regression analysis, Origin 2023 software was used for the correlation mapping, and LINGO 18.0 software was used to find the soil nutrient optimization solution.

3. Results

3.1. Soil Nutrition and Berry Quality of *Vitis vinifera* cv. Cabernet Franc Vineyard

3.1.1. Soil Nutrition

The analysis of soil-nutrient related indicators can provide data support for nutrient diagnostics in vineyards. The results of the statistical analysis of soil nutrients in the *Vitis vinifera* cv. Cabernet Franc vineyard in the eastern foothills of Helan Mountains are shown in Table 2, and the appropriate levels of soil nutrients were obtained from the data presented in the nutrient classification standard of the second soil survey in China [27]. In six vineyards, the frequency of deficiencies in TN, OM, NN, and available P in the soil was 83.3%, and the absence of AN reached 100%. Meanwhile, the proportion of soil with a TP content below 0.60 $g \cdot kg^{-1}$ was 66.7%. These data indicate that there TN, OM, NN, available P, AN and TP are deficient in the soil. The optimum frequency of TK, available K, available Cu, available Fe, and available Mn was greater than or equal to 66.7%, and the available Mn reached 100%. Nevertheless, with respect to the contents of the soil OM, NN, AN and pH value, none of the vineyards were within the optimum range. In total, 83.3% of the vineyards had an available Zn content over 1.00 $mg \cdot kg^{-1}$. Notably, all the vineyards surveyed had alkaline soils with pH values above 7.50, with the highest reaching 8.32. This means that it is necessary to lower the pH value of the soil through soil improvement to make it more suitable for grape growth and development.

The coefficients of variation (CV) for the soil nutrition indicators varied considerably (Table 2). The CV of OM and NN were particularly elevated, meaning that the contents of OM and NN varied highly from vineyard to vineyard. However, the CV of TK and soil pH are relatively low. This suggests that there is a slight difference between vineyards in the TK and soil pH.

Table 2. Statistics of soil nutrient status in *Vitis vinifera* cv. Cabernet Franc vineyards.

Indicators	Range	Mean ± SD	CV (%)	Suitable Range	Deficiency Frequency (%)	Optimum Frequency (%)	Excess Frequency (%)
TN ($g \cdot kg^{-1}$)	0.44~1.30	0.71 ± 0.30	42.3	1.00~1.50	83.3	16.7	0.00
TP ($g \cdot kg^{-1}$)	0.25~0.72	0.48 ± 0.17	35.4	0.60~0.80	66.7	33.3	0.00
TK ($g \cdot kg^{-1}$)	14.9~19.0	16.5 ± 1.49	9.05	15.0~20.0	16.7	83.3	0.00
OM ($g \cdot kg^{-1}$)	2.59~36.9	11.5 ± 11.5	101	20.0~30.0	83.3	0.00	16.7
NN ($mg \cdot kg^{-1}$)	3.86~104	41.3 ± 33.8	81.8	67.5~90.0	83.3	0.00	16.7
AN ($mg \cdot kg^{-1}$)	1.48~6.96	3.25 ± 1.91	58.8	22.5~30.0	100	0.00	0.00
Available P ($mg \cdot kg^{-1}$)	3.81~11.5	6.86 ± 2.75	40.1	10.0~20.0	83.3	16.7	0.00
Available K ($mg \cdot kg^{-1}$)	80.1~208	127 ± 41.0	32.4	100~150	16.7	66.7	16.7
Available Cu ($mg \cdot kg^{-1}$)	0.28~1.11	0.65 ± 0.26	40.0	0.20~1.00	0.00	83.3	16.7
Available Zn ($mg \cdot kg^{-1}$)	0.89~3.29	1.73 ± 0.82	47.4	0.50~1.00	0.00	16.7	83.3
Available Fe ($mg \cdot kg^{-1}$)	4.39~8.61	5.67 ± 1.40	24.7	4.50~10.0	16.7	83.3	0.00
Available Mn ($mg \cdot kg^{-1}$)	7.36~14.6	10.3 ± 2.46	24.0	5.00~15.0	0.00	100	0.00
Soil pH	7.87~8.32	8.09 ± 0.16	1.98	6.50~7.50	0.00	0.00	100

Note: TN, TP, TK, OM, NN, AN, P, K, Cu, Zn, Fe and Mn are abbreviations for total nitrogen, total phosphorus, total potassium, organic matter, nitrate nitrogen, ammonium nitrogen, phosphorus, potassium, copper, zinc, iron and manganese, respectively. SD and CV refer to the standard deviation and coefficient of variation, respectively. The same applies in the following.

3.1.2. Berry Quality

Descriptive statistics for the global data set of berry quality parameters for *Vitis vinifera* cv. Cabernet Franc are given in Table 3. From the grape quality measurements, the values of the two yield indicators, i.e., the weight and yield per plant, were 1.40 g and 3.77 kg, respectively. The others are the chemical indices. Among them, the values for the soluble solid content, reducing sugar, titratable acidity and juice pH were 27.8%, 255 $g \cdot L^{-1}$, 3.47 $g \cdot L^{-1}$ and 3.70, respectively. Tannin, anthocyanin, and flavonol belong to the flavonoids, and their average values were 23.3 $g \cdot L^{-1}$, 52.0 $g \cdot L^{-1}$ and 6.76 $g \cdot L^{-1}$, respectively. Meanwhile, there was a great difference between the vineyards regarding the indicators relating to the yield per plant and the content of titratable acidity (TA), tannin, and anthocyanin; meanwhile, the other indicators differ relatively less.

Table 3. Descriptive statistics for the overall data set of berry quality parameters for *Vitis vinifera* cv. Cabernet Franc.

Indicators	Range	Mean ± SD	CV (%)
Berry Weight/g	1.17~1.59	1.40 ± 0.16	11.4
Yield per Plant/kg	1.53~7.15	3.77 ± 1.89	50.1
SSC/%	25.8~30.2	27.8 ± 1.50	5.39
Reducing Sugar/($g \cdot L^{-1}$)	234~278	255 ± 14.9	5.86
TA/($g \cdot L^{-1}$)	2.49~4.47	3.47 ± 0.87	25.1
Juice pH	3.49~3.85	3.70 ± 0.12	3.24
Tannin/($g \cdot L^{-1}$)	12.1~39.7	23.3 ± 9.45	40.5
Anthocyanin/($g \cdot L^{-1}$)	32.5~80.2	52.0 ± 17.3	33.2
Flavonol/($g \cdot L^{-1}$)	5.03~7.96	6.76 ± 0.93	13.8

Note: SSC and TA are short for soluble solids content and titratable acidity, respectively.

3.2. *Internal Correlation of Soil Nutrients and Relationship between Berry Quality in Vineyards*

3.2.1. Correlation Coefficients among Soil Nutrients

The relationships among soil nutrients in vineyards are complex and can be synergistic or antagonistic. The correlation coefficient for soil nutrients is shown in Figure 2. It can be seen from Figure 2 that, except for AN, available K, and soil pH, the TN is positively correlated with other indices, and that the indices with relatively large correlation coefficients are OM (0.94**), NN (0.75), available Fe (0.75), and available Mn (0.92**). In addition to the soil pH, it is clear that TP is positively correlated with other soil nutrient indices, and that the correlation between TP and NN reached a significant level (0.83*). There is a significant correlation between organic matter and TN, NN, available Fe and available Mn, suggesting that increasing the amount of OM increases the content of TN, NN, available Fe and available Mn in the soil. Soil pH is negatively correlated with most soil nutrient indicators, but not to a significant level.

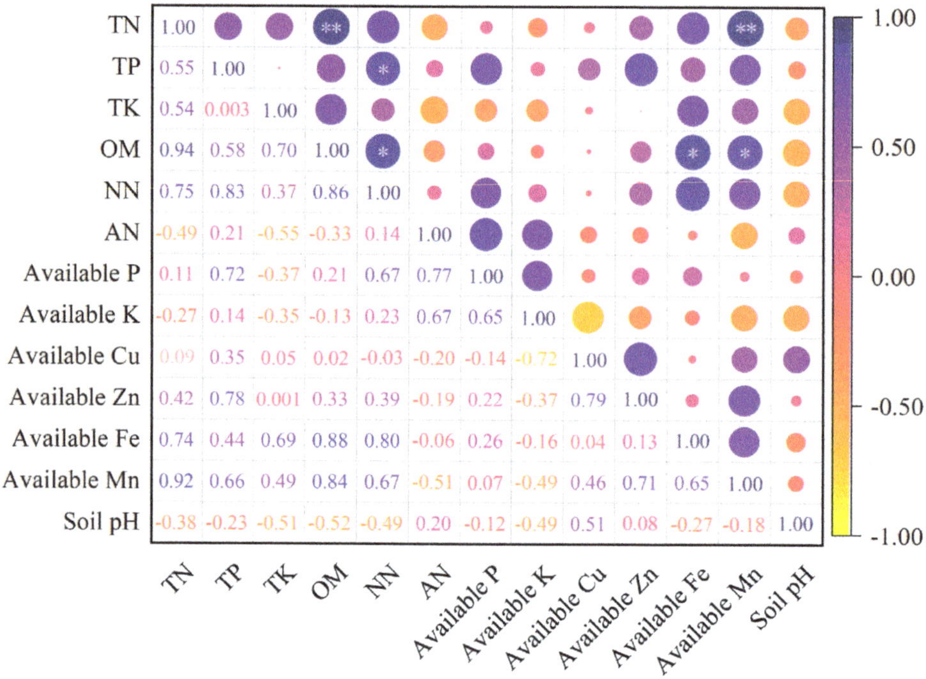

Figure 2. Correlation coefficients among soil nutrients in *Vitis vinifera* cv. Cabernet Franc vineyards. Note: n = 6; ** and * indicate that the correlation reached significant levels of $p < 0.01$ and $p < 0.05$, respectively (Pearson correlation analysis), as shown in the following figure and table; TN, TP, TK, OM, NN, AN, P, K, Cu, Zn, Fe and Mn are abbreviations for total nitrogen, total phosphorus, total potassium, organic matter, nitrate nitrogen, ammonium nitrogen, phosphorus, potassium, copper, zinc, iron, and manganese, respectively.

3.2.2. Correlation between Berry Quality and Soil Nutrients

Soil nutrients are an important material basis for the formation of grape quality, and the type and content of soil nutrients have a crucial effect on grape quality. The effect of different soil nutrient indices on grape quality can be revealed simply by correlation analysis. The correlation coefficients between grape quality and soil nutrients and their significance are shown in Figure 3. As a representative indicator of output, the berry weight and yield per plant showed little correlation with most soil nutrient indicators. However, there was a strong correlation between berry weight and available copper and soil pH, with a significant level of correlation with the latter. The relationship between the soluble solid content (SSC), reducing sugar and soil nutrient indices was very consistent, but the correlation was also not strong. In addition to ammonium nitrogen and available potassium, the titratable acid was negatively correlated with other soil nutrient indicators, with a significant level of correlation with available manganese. The relationship between the juice pH and soil nutrient indicators is opposite to that of titratable acid, which is positively correlated with most soil nutrient indicators, except for available potassium. Tannins, anthocyanins, and flavonols are all flavonoid substances, but there are clear differences between them and the soil nutrient indices. Tannins and anthocyanins were negatively correlated with most soil nutrient indicators, while flavonols were positively correlated with most soil nutrient indicators, but the correlation was not significant. It can be seen that there is a complex correlation between the grape quality indicators and soil nutrient indicators. Overall, the correlation between the two indicator groups is not strong. Simple bivariate correlation analysis does not fully explain the complex relationship between grape quality and soil nutrients; Therefore, in order to obtain a more in-depth and comprehensive understanding of the internal correlation between the two, it is necessary to further explore this issue using multivariate statistical analysis.

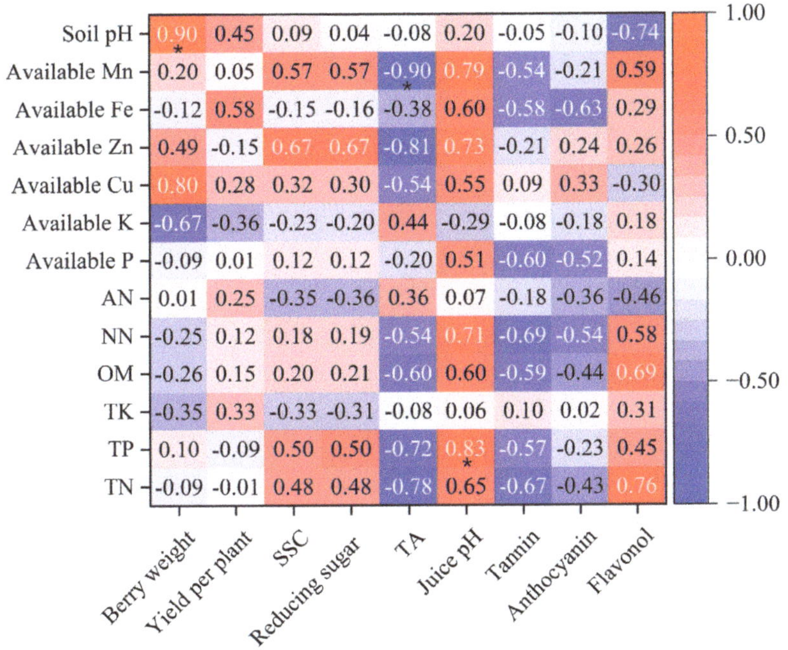

Figure 3. Correlation coefficient between grape quality parameters and soil nutrients. Note: n = 6; TN, TP, TK, OM, NN, AN, P, K, Cu, Zn, Fe, Mn, SSC, and TA are abbreviations for total nitrogen, total phosphorus, total potassium, organic matter, nitrate nitrogen, ammonium nitrogen, phosphorus, potassium, copper, zinc, iron manganese, soluble solid content, and titratable acid, respectively; * indicate that the correlation reached significant levels of $p < 0.05$.

3.3. Screening for Soil Nutrient Factors Affecting Grape Quality and Establishing of the Regression Equations

Grape quality is subject to a combination of soil-nutrient related indicators and is complex and variable. Soil nutrients and grape quality in vineyards are two different indicator systems. They belong to the relation between multivariate independent variables and multivariate dependent variables. Therefore, simple correlations alone do not fully reflect the mechanism of nutrient action and a multivariate analysis is therefore necessary. In this study, there were multiple collinearity relationships among the soil nutrient indices, so it was necessary to employ the partial least squares method for analysis using IBM SPSS Statistics 25 software, combined with professional knowledge, to screen out the main soil nutrient factors that affected a certain grape quality index. Detailed methods was described in Section 2.4 above.

Then, the soil nutrient factors screened by the above method were taken as independent variables and the corresponding grape quality factors were taken as dependent variables to construct a multivariate linear regression equation. The results are shown in Table 4. A significance test of the regression equation was conducted, and the results showed that the established equation reached the level of significant difference, indicating that the established equation was stable and reliable. From Table 4, the berry weight was mainly affected by NN, the available Cu, and soil pH. The yield per plant was more affected by TN, AN, available Fe, and the soil pH. TN, TP, and OM were the main soil nutrient factors affecting SSC. TN, TK, the available Fe, and available Mn mainly affect the content of reducing sugar content. TA and the juice pH are influenced by different soil nutrient factors; the former are mainly affected by TN, TK, and the available Cu, while the latter are mainly affected by TP, the available P, and available K. All three flavonoid components are also affected by different soil nutrient factors; among these, tannin is primarily affected by TN, available K, available Zn, and the soil pH; anthocyanin are mainly affected by TP, the available P, and available Mn; and flavonol is mainly affected by TN, OM, and the soil pH.

Table 4. Screening of soil nutrient factors affecting the berry quality and establishment of the regression equations.

Indicators	Soil Factors	Regression Equations	F Value	*p* Value
Berry weight (Y_1)	X_5, X_9, X_{13}	$Y_1 = -4.69 + (5.58 \times 10^{-4})X_5 + (2.54 \times 10^{-1})X_9 + (7.30 \times 10^{-1})X_{13}$	30.1	0.03
Yield per plant (Y_2)	X_1, X_6, X_{11}, X_{13}	$Y_2 = -50.2 - 6.37X_1 - (2.41 \times 10^{-1})X_6 + 1.94X_{11} + 5.96X_{13}$	7.50×10^3	8.66×10^{-3}
SSC (Y_3)	X_1, X_2, X_4	$Y_3 = 20.3 + 12.6X_1 + 4.67X_2 - (3.19 \times 10^{-1})X_4$	22.5	0.04
Reducing sugar (Y_4)	X_1, X_3, X_{11}, X_{12}	$Y_4 = 316 + 37.9X_1 - 4.81X_3 - 8.48X_{11} + 3.84X_{12}$	267	4.59×10^{-2}
TA (Y_5)	X_1, X_3, X_9	$Y_5 = 1.92 - 2.93X_1 + (2.82 \times 10^{-1})X_3 - 1.56X_9$	109	9.11×10^{-3}
Juice pH (Y_6)	X_2, X_7, X_8	$Y_6 = 3.66 + (2.46 \times 10^{-1})X_2 + (3.65 \times 10^{-2})X_7 - (2.58 \times 10^{-3})X_8$	23.0	0.04
Tannin (Y_7)	X_1, X_8, X_{10}, X_{13}	$Y_7 = 560 - 42.9X_1 - (2.08 \times 10^{-1})X_8 + 1.10X_{10} - 59.6X_{13}$	802	0.03
Anthocyanin (Y_8)	X_2, X_7, X_{12}	$Y_8 = 155.077 + 266X_2 - 14.6X_7 - 12.7X_{12}$	40.8	0.02
Flavonol (Y_9)	X_1, X_4, X_{13}	$Y_9 = 41.6 + 5.56X_1 - (1.12 \times 10^{-1})X_4 - 4.63X_{13}$	33.9	0.03

Note: n = 6; X_1: TN ($g{\cdot}kg^{-1}$), X_2: TP ($g{\cdot}kg^{-1}$), X_3: TK ($g{\cdot}kg^{-1}$), X_4: OM ($g{\cdot}kg^{-1}$), X_5: NN ($mg{\cdot}kg^{-1}$), X_6: AN ($mg{\cdot}kg^{-1}$), X_7: available P ($mg{\cdot}kg^{-1}$), X_8: available K ($mg{\cdot}kg^{-1}$), X_9: available Cu ($mg{\cdot}kg^{-1}$), X_{10}: available Zn ($mg{\cdot}kg^{-1}$), X_{11}: available Fe ($mg{\cdot}kg^{-1}$), X_{12}: available Mn ($mg{\cdot}kg^{-1}$), X_{13}: soil pH.

In summary, there are six and four grape berry quality indicators affected by TN and pH, respectively. However, the two soil nutrient factors had qualitatively different effects on the included grape quality indicators. For example, from Table 4, it can be seen that TN is negatively correlated with the yield per plant, TA and tannin, but positively correlated with the SSC, reducing sugar and flavonol. The soil pH was negatively correlated with tannin and flavonol, but positively correlated with berry weight and the yield per plant. The same is true for other soil indicators. It could be seen that, for the overall quality of wine grapes, it was not the case that the soil nutrient index values were better the larger or smaller they were, but that there may be some appropriate interval (or value) that ensures the optimal overall quality of wine grapes. Of course, determining such an optimal interval (or value) requires the use of linear programming methods.

3.4. Optimization Scheme for Soil Nutrient Content in Vineyards

In order to study the suitable range of soil nutrient contents corresponding to the optimal grape quality of wine grapes, this study took the maximum (or minimum) value of a certain grape quality indicator as the objective function, and the other grape quality indices and soil nutrient factors as the constraint conditions. It then establishes the linear programming equations for determining the optimal grape quality index. In determining the optimal grape quality factor for a given wine grape, it is necessary to ensure that the other grape quality indicators are optimized simultaneously and to give a certain bound on the soil nutrient factor. The upper and lower bounds on the constrained values of the wine grape quality index in this study can be divided into two cases. It is better when the grape quality index is large, and the average value of this index is the lower limit. It is better when the grape quality index is small, and the average value of this index is the upper limit for the vineyards surveyed in this study. The constrained value of the soil nutrient factor was the maximum value for the vineyards surveyed. This was used as the upper limit, and the average value of the vineyards surveyed was used as the lower limit. The range of soil pH suitable for grape production was found to be 6.50–7.50.

Limited by the length of this paper, this study only took berry weight (Y_{1min}) (the smaller the better) as an example in order to demonstrate the establishment and solution of the linear programming equations between the grape berry quality and soil nutrients. The specific equations are as follows.

$$Y_{1min} = -4.69 + (5.58 \times 10^{-4})X_5 + (2.54 \times 10^{-1})X_9 + (7.30 \times 10^{-1})X_{13} <= 1.40, \quad (1)$$

$$-50.2 - 6.37X_1 - (2.41 \times 10^{-1})X_6 + 1.94X_{11} + 5.96X_{13} >= 3.77, \quad (2)$$

$$20.3 + 12.6X_1 + 4.67X_2 - (3.19 \times 10^{-1})X_4 >= 27.8, \quad (3)$$

$$316 + 37.9X_1 - 4.81X_3 - 8.48X_{11} + 3.84X_{12} >= 255, \quad (4)$$

$$1.92 - 2.93X_1 + (2.82 \times 10^{-1})X_3 - 1.56X_9 >= 3.47, \quad (5)$$

$$3.66 + (2.46 \times 10^{-1})X_2 + (3.65 \times 10^{-2})X_7 - (2.58 \times 10^{-3})X_8 <= 3.70, \quad (6)$$

$$560 - 42.9X_1 - (2.08 \times 10^{-1})X_8 + 1.10X_{10} - 59.6X_{13} >= 23.3 \quad (7)$$

$$155 + 266X_2 - 14.6X_7 - 12.7X_{12} >= 52.0, \quad (8)$$

$$41.6 + 5.56X_1 - (1.12 \times 10^{-1})X_4 - 4.63X_{13} >= 6.76, \quad (9)$$

where $0.71 \leq X_1 \leq 1.30$, $0.48 \leq X_2 \leq 0.72$, $16.47 \leq X_3 \leq 19.0$, $11.5 \leq X_4 \leq 36.9$, $41.3 \leq X_5 \leq 104$, $3.25 \leq X_6 \leq 6.96$, $6.86 \leq X_7 \leq 11.5$, $127 \leq X_8 \leq 208$, $0.65 \leq X_9 \leq 1.11$, $1.73 \leq X_{10} \leq 3.29$, $5.67 \leq X_{11} \leq 8.61$, $10.3 \leq X_{12} \leq 14.6$, $6.50 \leq X_{13} \leq 7.50$.

Using the same method, linear programming equations could be established to solve the maximum yield per plant (Y_{2max}), maximum SSC (Y_{3max}), maximum reducing sugar (Y_{4max}), maximum TA (Y_{5max}), minimum juice pH (Y_{6min}), maximum tannin (Y_{7max}), maximum anthocyanin (Y_{8max}), and maximum flavonol (Y_{9max}) of the grape berry quality indices. Then, with the help of LINGO 18.0 software, the optimal solution of the grape berry quality indices and the corresponding range (or value) of soil nutrient factors could be obtained. The results are presented in Table 5. As can be seen in Table 5, the last row showed the optimal value of the grape quality indicators. The last column gave the values or ranges of the soil nutrient factors corresponding to the optimal grape quality indicators conditions.

It should be noted that the linear programming solutions obtained in Table 5 are closely related to the regression equations in Table 4, specifically, the constant terms and the regression coefficients of the regression equations in Table 4. In this study, to ensure consistency between the former and the latter, the constant terms and the regression coefficients of the regression equations are retained with three significant digits on the basis of the exact values. As a result, the linear programming solutions for some of the indicators in Table 5 differ from the exact values. For example, the values of yield per plant (Y2), soluble solids content (Y3), titratable acid (Y5), juice pH (Y6), tannin (Y7), anthocyanin (Y8), TK, available K, available Fe and available Mn in Table 5 differ, but the differences are not large overall. To obtain more accurate optimization results, it is necessary to retain more accurate constant terms and regression coefficients (for example, accurate to three decimal places) for the regression equations in Table 4, and then establish linear programming equations and solve them to obtain more accurate results.

Table 5. The optimum values of the berry quality index and its corresponding range (or value) of soil nutrient factors.

Soil Nutrient Factors	Y_1 Berry Weight (g)	Y_2 Yield per Plant (kg)	Y_3 SSC (°Brix)	Y_4 Reducing Sugar (g·L^{-1})	Y_5 TA (g·L^{-1})	Y_6 Juice pH	Y_7 Tannin (mg·g^{-1})	Y_8 Anthocyanin (mg·g^{-1})	Y_9 Flavonol (mg·g^{-1})	X Optimum Value (or Range)
TN (g·kg^{-1})	0.71	0.71	0.74	0.71	0.71	0.71	0.71	0.71	0.71	0.71~0.74
TP (g·kg^{-1})	0.69	0.69	0.72	0.69	0.69	0.67	0.69	0.72	0.69	0.67~0.72
TK (g·kg^{-1})	16.5	16.5	16.7	16.5	16.7	16.5	16.5	16.5	16.5	16.5~16.7
OM (g·kg^{-1})	11.5	11.5	11.5	11.5	11.5	11.5	11.5	11.5	11.5	11.5
NN (mg·kg^{-1})	41.3	41.3	41.3	41.3	41.3	41.3	41.3	41.3	41.3	41.3
AN (mg·kg^{-1})	3.25	3.25	3.25	3.25	3.25	3.25	3.25	3.25	3.25	3.25
Available P (mg·kg^{-1})	6.86	6.86	6.86	6.86	6.86	6.86	6.86	6.86	6.86	6.86
Available K (mg·kg^{-1})	147	147	150	147	147	208	147	150	147	147~208
Available Cu (mg·kg^{-1})	0.65	0.65	0.65	0.65	0.65	0.65	0.65	0.65	0.65	0.65
Available Zn (mg·kg^{-1})	3.29	3.29	3.29	3.29	3.29	3.29	3.29	3.29	3.29	3.29
Available Fe (mg·kg^{-1})	7.63	7.63	7.60	7.52	7.51	7.52	7.62	7.52	7.63	7.51~7.63
Available Mn (mg·kg^{-1})	14.6	14.6	14.6	14.6	14.6	14.4	14.6	14.4	14.6	14.4~14.6
Soil pH	7.46	7.50	7.50	7.50	7.50	7.50	7.47	7.50	7.46	7.46~7.50
Y optimum value	0.95	3.98	29.3	256	3.53	3.54	57.6	64.0	9.72	——

Note: TN, TP, TK, OM, NN, AN, P, K, Cu, Zn, Fe, Mn, SSC and TA are abbreviations for total nitrogen, total phosphorus, total potassium, organic matter, nitrate nitrogen, ammonium nitrogen, phosphorus, potassium, copper, zinc, iron, manganese, soluble solids content and titratable acidity, respectively.

4. Discussion

From the short study above, it appears that the soil in the eastern foothills of Helan Mountains are comparatively poor, especially in the absence of TN, TP, OM, AN, NN and available P. This may have much to do with the local soil characteristics. Because the wine-grape-producing areas in the eastern foothills of Helan Mountains comprise mostly sandy loam soil and gravelly sandy soil, the soil texture is coarse and rich in gravel, the soil structure is weak, the fertilizer retention capacity is poor, and the nutrient content is lacking [28]. These findings are consistent with those reported by Wang et al. [29] and Qi et al. [5]. In addition, the vineyard soil is alkaline, which is also one of the common characteristics of the local soil [30–33]. Considering the adaptation of grape plants to a wide range of pH levels and the results of this study, it is recommended that local vineyard soil pH be reasonably intervened. Soil is a complex mixture with complex relationships among the nutrients in it; this has not only synergistic but also antagonistic effects. This complex relationship varies according to the study area and plant growth stage [34,35].

Theoretically, the nutrient content of vineyard soil and the quality of the berries are involved in a dynamic shift process. Strictly speaking, the nutrient content of the soil and the quality of berries change throughout the grape growing season. Therefore, determining when to sample in such a dynamic process can undoubtedly be a tricky issue. Fortunately, previous research provides a guide [5,36,37]. Samples were taken during the ripening period, when grape quality tends to be stable and the soil nutrients do not fluctuate much. In this study, only the internal relationship between grape quality and soil nutrients at the grape ripening stage was analyzed, and the related exploration of soil nutrient optimization was carried out. Therefore, in order to explore the relationship between soil nutrients and grape quality formation, it is necessary to evaluate evaluations from a more macroscopic and systematic perspective. Limited by current analytical methods, such a complex problem cannot be solved provisionally; therefore, this study is only a preliminary exploration and more detailed studies are still needed.

Regarding the quality of the berries, as shown by the results of this study, there is a large difference in quality between different wineries. With consistent vineyard management and slight differences in climatic conditions, the vineyard soil nutrients become a key factor affecting in grape quality. Pearson correlation analysis is a common method used to analyze the internal relationship between quality and soil nutrients [38,39]. However, considering the multivariate collinearity of soil nutrients, a more complex multivariate statistical analysis method is needed, such as canonical correlation analysis [12,40], principal component analysis [41], partial least squares regression [14], and so on. In a specific study, the methods that need to be employed must be combined with the actual situation. This method was chosen for critical soil nutrient factors that affect berry quality, as partial least squares regression can better address the problem of multilinearity.

It is important to note that there are some differences between the soil nutrient optimization results obtained through multivariate statistical analysis and the range of soil nutrient issues shown in Table 2. This is because the appropriate ranges in Table 2 are the recommended value for common crops given by the National Soil Census Office, combined with the actual soil conditions in China. Therefore, there is bound to be a difference between different crops. For the results obtained and shown in Table 5, some are numerical values and some are numerical ranges. The reason for this situation is that when solving different soil nutrient indices, the software automatically computes the corresponding independent variable values within the range of independent variables, with the aim of fulfilling the optimal quality index. During this process, the values of some independent variables are more consistent, while the values of some independent variables are somewhat different. In order to consider different berry quality indicators to achieve the optimal conditions, soil nutrient intervention needs to be close to the target value (or within the target range).

On a deeper level, it should be noted that there are numerous external factors affecting the grape quality of wine grapes, in addition to soil nutrients, but also including soil type [42], light [43], temperature [44], irrigation [45], cluster thinning [46], leaf curtain management [47], cover crop [48], and so on. Thus, it can be established that the final formation of the quality of the wine grape is the result of a combination of external influences. The content of this study is only one aspect of grape quality, and it was carried out under conditions in which differences in the external environment were small. Based on this, the best way to get a more accurate picture of the relationship between soil nutrients and grape quality is through a combination of site-specific and quantitative (increment/decrement) studies. In general terms, the results of this study were obtained only from the analysis of available field soil nutrient and grape quality data, which has certain limitations. In the next step, this research team will carry out more detailed studies and improvements in individual nutrient quantification, multi-factor interactions, quality response molecular mechanism, etc.

5. Conclusions

This study showed significant differences in soil nutrients between different *Vitis vinifera* cv. Cabernet Franc vineyards in the eastern foothills of Helan Mountains. In general, the soil is relatively poor, especially lacking TN, OM, NN, AN, and AP, and the soil is typically alkaline. There were both synergistic and antagonistic effects in the soil nutrient index. At the same time, there are multiple collinearity issues. There are also some differences in the berry quality of *Vitis vinifera* cv. Cabernet Franc corresponding to vineyards with different soil nutrient content. In terms of the Pearson correlation coefficient, it was found that titratable acids, tannins, and anthocyanins were negatively correlated with most soil nutrient indicators, while other berry quality indicators were inversely correlated. Based on the results of this study, it is feasible to use multivariate statistical analysis to screen for major soil nutrient deficiencies affecting berry quality. In addition, equations to optimize soil nutrients are expected. The results show that when the soil nutrient indicators are as follows: TN, 0.71~0.74 $g \cdot kg^{-1}$; TP, 0.67~0.72 $g \cdot kg^{-1}$; TK, 16.5~16.8 $g \cdot kg^{-1}$; OM, 11.5 $g \cdot kg^{-1}$; NN, 41.3 $mg \cdot kg^{-1}$; AN, 3.25 $mg \cdot kg^{-1}$; Available P, 6.86 $mg \cdot kg^{-1}$; Available K, 127~208 $mg \cdot kg^{-1}$; Available Cu, 0.65 $mg \cdot kg^{-1}$; Available Zn, 3.29 $mg \cdot kg^{-1}$; Available Fe, 7.49~7.64 $mg \cdot kg^{-1}$; Available Mn, 14.3~14.6 $mg \cdot kg^{-1}$; Soil pH, 7.46~7.50, the grape quality indicators can reach the optimal value.: berry weight, 0.95 g; Yield per plant, 4.04 kg; SSC, 29.4 °Brix; Reducing sugar, 256 $g \cdot L^{-1}$; TA, 3.54 $g \cdot L^{-1}$; Juice pH, 3.45; Tannin, 62.6 $mg \cdot g^{-1}$; Anthocyanin. 64.6 $mg \cdot g^{-1}$; Flavonol, 9.72 $mg \cdot g^{-1}$.

Author Contributions: Conceptualization, Y.L. and X.L.; Software, Y.L. and Y.Y.; Formal analysis, L.N. and C.X.; Investigation, J.X. and Y.Y.; Resources, W.L. and P.C.; Data curation, J.X. and Y.L.; Writing—original draft preparation, Y.L.; Writing—review and editing, Y.L. and X.L.; Visualization, Y.L. and W.L.; Supervision, J.X.; Project administration, X.L.; Funding acquisition, X.L. All authors have read and agreed to the published version of the manuscript.

Funding: This research was funded by the Ningxia Autonomous Region key research and development project (2021BEF02017, 2022BBF02014), Shaanxi Province key research and development plan project (2022NY-118), and the major science and technology project of Shaanxi Agricultural Collaborative Innovation and Promotion Alliance (LMZD202105).

Institutional Review Board Statement: Not applicable.

Informed Consent Statement: Not applicable.

Data Availability Statement: Upon request, the data will be provided.

Conflicts of Interest: The authors declare no conflicts of interest.

References

1. Ding, Q.; Zhang, X.Y.; Li, Q.; Sha, H.M.; Qiao, Y.F.; Hu, Z.H. The best fruit loads for the cold resistance of wine grape (*Vitis Vinifera* L.) in the eastern foothills of Helan Mountain, China. *Bangl. J. Bot.* **2022**, *51*, 263–270. [CrossRef]
2. Yang, C.L.; Song, R.; Ding, Y.T.; Zhang, L.; Wang, H.; Li, H. Review on legal supervision system of the Chinese wine industry. *Horticulturae* **2022**, *8*, 432. [CrossRef]
3. Li, H.; Wang, H. *Overview of Wine in China*; Science Press: Beijing, China, 2020.

4. Wang, L.; Liu, J.M.; Wang, L.E.; Zhu, H.; Lin, J. Tourism resource assessment and spatial analysis of wine tourism development: A case study of the eastern foothills of China's Helan Mountains. *J. Mt. Sci.* **2018**, *15*, 645–656. [CrossRef]
5. Qi, Y.; Wang, R.; Qin, Q.; Sun, Q. Soil affected the variations in grape and wine properties along the eastern foot of Helan Mountain, China. *Acta Agric. Scand. Sect. B-Soil Plant Sci.* **2019**, *69*, 494–502. [CrossRef]
6. Van Leeuwen, C.; Roby, J.P.; De Resseguier, L. Soil-related terroir factors: A review. *Oeno One* **2018**, *52*, 173–188. [CrossRef]
7. Unamunzaga, O.; Besga, G.; Castellón, A.; Usón, M.A.; Chéry, P.; Gallejones, P.; Aizpurua, A. Spatial and vertical analysis of soil properties in a Mediterranean vineyard soil. *Soil Use Manag.* **2014**, *30*, 285–296. [CrossRef]
8. Jiang, T.T.; Yan, P.K.; Ma, T.H.; Wang, R. Nutritional requirements and precise fertilization of wine grapes in the eastern foothills of Helan Mountain. *Int. J. Agric. Biol. Eng.* **2022**, *15*, 147–153. [CrossRef]
9. De Santis, D.; Frangipane, M.T.; Brunori, E.; Cirigliano, P.; Biasi, R. Biochemical markers for oenological potentiality in a grapevine aromatic variety under different soil types. *Am. J. Enol. Vitic.* **2016**, *68*, 15123.
10. Pérez-Álvarez, E.P.; Martínez-Vidaurre, J.M.; Garde-Cerdán, T. Anthocyanin composition of grapes from three different soil types in cv. Tempranillo A.O.C. Rioja vineyards. *J. Sci. Food Agric.* **2019**, *99*, 4833–4841.
11. Zhu, Q.; Xie, X.; Xu, Y. Fertilization regulates grape yield and quality in by altering soil nutrients and the microbial community. *Sustainability* **2022**, *14*, 10857. [CrossRef]
12. Arnó, J.; Rosell, J.R.; Blanco, R.; Ramos, M.C.; Martínez-Casasnovas, J.A. Spatial variability in grape yield and quality influenced by soil and crop nutrition characteristics. *Precis. Agric.* **2012**, *13*, 393–410. [CrossRef]
13. Guo, K.B.; Guo, Z.; Guo, Y.; Qiao, G. The effects of soil nutrient on fruit quality of 'Hayward' kiwifruit (*Actinidia chinensis*) in Northwest China. *Eur. J. Hortic. Sci.* **2020**, *85*, 471–476. [CrossRef]
14. Cao, S.; Zhou, W.; Liu, P.; Tan, J.; Song, B. Multivariate analysis and optimization of relationship between soil nutrients and fruit quality in *C. sinensis* (L.) osbeck orchard. *Soils* **2021**, *53*, 97–104.
15. Zhang, Q.; Wei, Q.P.; Liu, H.P.; Jiang, R.S.; Liu, X.D.; Wang, X.W. Multivariate analysis and optimum proposals of the relationship between soil nutrients and fruit qualities in apple orchard. *Sci. Agric. Sin.* **2011**, *44*, 1654–1661.
16. Ren, Y.Z. *'Terroir' of Wine Region at Eastern Foothills of Helan Mountain Based on Soil and Climate Property*; Northwest A&F University: Xianyang, China, 2021.
17. Bremner, J.M.; Mulvaney, C.S. Nitrogen total. In *Methods of Soil Analysis. Part 2. Chemical and Microbiological Properties*; Page, A.L., Miller, R.H., Keeney, D.R., Eds.; ASA-SSSA: Madison, WI, USA, 1982.
18. Zhang, M.; Dong, L.G.; Wang, Y.; Bai, X.X.; Ma, Z.T.; Yu, X.; Zhao, Z. The response of soil microbial communities to soil erodibility depends on the plant and soil properties in semiarid regions. *Land Degrad. Dev.* **2021**, *32*, 3180–3193. [CrossRef]
19. Bao, S.D. *Agrochemical Analysis of Soil*, 3rd ed.; China Agricultural Publishing: Beijing, China, 2000.
20. Walkley, A.; Black, I.A. An examination of the Degtjareff method for determining soil organic matter, and a proposed modification of the chromic acid titration method. *Soil Sci.* **1934**, *37*, 29–38. [CrossRef]
21. Zhang, C.; Liu, G.B.; Xue, S.; Wang, G.L. Soil bacterial community dynamics reflect changes in plant community and soil properties during the secondary succession of abandoned farmland in the Loess Plateau. *Soil Biol. Biochem.* **2016**, *97*, 40–49. [CrossRef]
22. Lindsay, W.L.; Norvell, W.A. Development of a DTPA soil test for Zn, Fe, Mn and Cu. *Soil Sci. Soc. Am. J.* **1978**, *42*, 421–428. [CrossRef]
23. Wang, H. *Wine Analysis and Inspection*; China Agricultural Publishing: Beijing, China, 2011.
24. Downey, M.O.; Rochfort, S. Simultaneous separation by reversed-phase high-performance liquid chromatography and mass spectral identification of anthocyanins and flavonols in Shiraz grape skin. *J. Chromatogr. A* **2008**, *1201*, 43–47. [CrossRef]
25. Kennedy, J.A.; Jones, G.P. Analysis of proanthocyanidin cleavage products following acid-catalysis in the presence of excess phloroglucinol. *J. Agric. Food Chem.* **2001**, *49*, 1740–1746. [CrossRef]
26. Wang, G.H.; Zhang, H.; Wei, Y.S. Implementation of partial least squares regression in SPSS software. *Stat. Decis.* **2017**, *7*, 67–71.
27. National Soil Census Office. *Soil Census Technology in China*; Agriculture Press: Beijing, China, 1992.
28. Zhang, X.J.; Wu, X.D.; Guo, J.; Sun, Q.; Wang, R.; Wang, Z.P.; Si, H.L. Effect of potassium application rate on wine grape in eastern foot of Helan Mountain. *Acta Agric. Boreali-Occident. Sin.* **2013**, *22*, 144–150.
29. Wang, R.; Sun, Q.; Chang, Q. Soil types effect on grape and wine composition in Helan Mountain Area of Ningxia. *PLoS ONE* **2015**, *10*, e0116690. [CrossRef] [PubMed]
30. Wei, C.F.; Liu, S.T.; Li, Q.; He, J.; Sun, Z.J.; Pan, X.Y. Diversity analysis of vineyards soil bacterial community in different planting years at eastern foot of Helan Mountain, Ningxia. *Rhizosphere* **2023**, *25*, 100650. [CrossRef]
31. Ma, T.H.; Hui, Y.R.; Zhang, L.; Su, B.F.; Wang, R. Foliar application of chelated sugar alcohol calcium fertilizer for regulating the growth and quality of wine grapes. *Int. J. Agric. Biol. Eng.* **2022**, *15*, 153–158. [CrossRef]
32. Li, Z.H.; Liu, B.R.; Chen, Z.F.; Mao, D.C.; Jiang, X.S. Re-vegetation Improves soil quality by decreasing soil conductivity and altering soil microbial communities: A case study of an opencast coal mine in the Helan Mountains. *Front. Microbiol.* **2022**, *13*, 833711. [CrossRef]
33. Wu, M.Y.; Pang, D.B.; Lin, C.; Li, X.B.; Liu, L.Z.; Liu, B.; Li, J.Y.; Wang, J.F.; Ma, L.L. Chemical composition of soil organic carbon and aggregate stability along an elevation gradient in Helan Mountains, northwest China. *Ecol. Indic.* **2021**, *131*, 108228. [CrossRef]

34. Su, B.W.; Zhao, G.X.; Dong, C. Spatiotemporal variability of soil nutrients and the responses of growth during growth stages of winter wheat in northern China. *PLoS ONE* **2018**, *13*, e0203509. [CrossRef]
35. Yan, P.; Lin, K.R.; Yu, C.X.; Tu, X.J. Spatial variability of soil phosphorus and potassium and its influencing factors in the fragile red beds ecosystem in southern China. *Pol. J. Environ. Stud.* **2021**, *30*, 5307–5315. [CrossRef]
36. Li, Y.; Han, M.Q.; Lin, F.; Ten, Y.; Lin, J.; Zhu, D.H.; Guo, P.; Weng, Y.B.; Chen, L.S. Soil chemical properties, 'Guanximiyou' pummelo leaf mineral nutrient status and fruit quality in the southern region of Fujian province, China. *J. Soil Sci. Plant Nutr.* **2015**, *15*, 615–628. [CrossRef]
37. Michopoulos, P.; Solomou, A. Effects of conventional and organic (manure) fertilization on soil, plant tissue nutrients and berry yields in vineyards. The use of the original native soil as a control. *J. Plant Nutr.* **2019**, *42*, 2287–2298. [CrossRef]
38. Wu, W.J.; Lin, Z.; Zhu, X.P.; Li, G.Y.; Zhang, W.J.; Chen, Y.J.; Ren, L.; Luo, S.W.; Lin, H.H.; Zhou, H.K.; et al. Improved tomato yield and quality by altering soil physicochemical properties and nitrification processes in the combined use of organic-inorganic fertilizers. *Eur. J. Soil Biol.* **2022**, *109*, 103384. [CrossRef]
39. Takahashi, H.; Yoshida, C.; Takeda, T. Sugar composition in asparagus spears and its relationship to soil chemical properties. *J. Appl. Glycosci.* **2019**, *66*, 47–50. [CrossRef]
40. Rekik, F.; Van Es, H.; Hernandez-Aguilera, J.N.; Gómez, M.I. Linking coffee to soil: Can soil health increase coffee cup quality in Colombia? *Soil Sci.* **2019**, *184*, 25–33. [CrossRef]
41. Mazzei, P.; Francesca, N.; Moschetti, G.; Piccolo, A. NMR spectroscopy evaluation of direct relationship between soils and molecular composition of red wines from Aglianico grapes. *Anal. Chim. Acta* **2010**, *673*, 167–172. [CrossRef] [PubMed]
42. Gómez-Míguez, M.J.; Gómez-Míguez, M.; Vicario, I.M.; Heredia, F.J. Assessment of colour and aroma in white wines vinifications:: Effects of grape maturity and soil type. *J. Food Eng.* **2007**, *79*, 758–764. [CrossRef]
43. Zhou, L.F.; Yang, Y.X.; Feng, H. Colored plastic film mulching regulates light quality and sucrose mtabolism in wine grape in an arid desert oasis. *J. Plant Growth Regul.* **2023**, *42*, 7705–7714. [CrossRef]
44. Gouot, J.C.; Smith, J.P.; Holzapfel, B.P.; Walker, A.R.; Barril, C. Grape berry flavonoids: A review of their biochemical responses to high and extreme high temperatures. *J. Exp. Bot.* **2019**, *70*, 397–423. [CrossRef]
45. Kang, C.C.; Diverres, G.; Karkee, M.; Zhang, Q.; Keller, M. Decision-support system for precision regulated deficit irrigation management for wine grapes. *Comput. Electron. Agric.* **2023**, *208*, 107777. [CrossRef]
46. Wang, W.; Liang, Y.Y.; Quan, G.R.; Wang, X.F.; Xi, Z.M. Thinning of cluster improves berry composition and sugar accumulation in Syrah grapes. *Sci. Hortic.* **2022**, *297*, 110966. [CrossRef]
47. Drenjančević, M.; Jukić, V.; Zmaić, K.; Kujundžić, T.; Rastija, V. Effects of early leaf removal on grape yield, chemical characteristics, and antioxidant activity of grape variety Cabernet Sauvignon and wine from eastern Croatia. *Acta Agric. Scand. Sect. B-Soil Plant Sci.* **2017**, *67*, 705–711. [CrossRef]
48. Xi, Z.M.; Zhang, Z.W.; Cheng, Y.F.; Li, H. The effect of vineyard cover crop on main monomeric phenols of grape berry and wine in *Vitis viniferal* L. *cv.* Cabernet Sauvignon. *Agric. Sci. China* **2010**, *9*, 440–448. [CrossRef]

horticulturae

MDPI

Article

Grape Yield Prediction Models: Approaching Different Machine Learning Algorithms

Caio Bustani Andrade [1], Jean Michel Moura-Bueno [2,3,*], Jucinei José Comin [1] and Gustavo Brunetto [2]

[1] Department of Rural Engineering, Federal University of Santa Catarina, Florianópolis 88034-000, Brazil; cbagronomo@gmail.com (C.B.A.); j.comin@ufsc.br (J.J.C.)
[2] Department of Soil Science, Federal University of Santa Maria, Santa Maria 97105-900, Brazil; brunetto.gustavo@gmail.com
[3] Health and Agricultural Sciences Center, University of Cruz Alta, Cruz Alta 98020-290, Brazil
* Correspondence: bueno.jean1@gmail.com

Abstract: Efficient marketing of winegrapes involves negotiating with potential buyers long before the harvest, when little is known about the expected vintage. Grapevine physiology is affected by weather conditions as well as by soil properties and such information can be applied to build yield prediction models. In this study, Partial Least Squares Regression (PLSR), Cubist (CUB) and Random Forest (RF) algorithms were used to predict yield from imputed weather station data and soil sample analysis reports. Models using only soil variables had the worst general results (R^2 = 0.15, RMSE = 4.16 Mg ha^{-1}, MAE = 3.20 Mg ha^{-1}), while the use of only weather variables yielded the best performance (R^2 = 0.52, RMSE = 2.99 Mg ha^{-1}, MAE = 2.43 Mg ha^{-1}). Models built with CUB and RF algorithms showed signs of overfitting, yet RF models achieved the best average results (R^2 = 0.58, RMSE = 2.85 Mg ha^{-1}, MAE = 2.24 Mg ha^{-1}) using only weather variables as predictors. Weather data imputation affected RF and CUB models more intensely while PLSR remained fairly insensitive. Plant age, yield level group, vineyard plot, May temperatures, soil pH and exchangeable concentrations of Zn, Cu, K and Mn were identified as important predictors. This exploratory work offers insights for future research on grape yield predictive modeling and grouping strategies to obtain more assertive results, thus contributing to a more efficient grapevine production chain in southern Brazil and worldwide.

Keywords: random forest; cubist; partial least squares regression; grapevine; yield prediction; calibration model

1. Introduction

In 2021, 7.3 million hectares of land were occupied by vineyards worldwide. Total grape production reached 77.8 Mt with over 57% destined to the wine industry. World wine production in 2021 was estimated in 26 GL, excluding juices and musts [1,2]. In Brazil over 75 thousand hectares of vineyards were harvested in 2022 [3] and the socioeconomic importance of viticulture is highlighted by a national wine market of BRL 20.3 billion (~USD 4 billion), which employed around 200 thousand people in 2019 [4]. Despite being the second largest market in South America after Argentina [2], 86 of every 100 bottles consumed by Brazilians in 2019 were imported [4], revealing the enormous growth potential for national wineries.

Efficient marketing of wine grapes involves negotiating with potential buyers long before the harvest, when little is known about the expected vintage. In addition, grape fertilizing practices also rely on yield expectations for the definition of application doses [5]. To anticipate yields, one must inevitably rely on some conceptual or numerical model of how crops respond to surrounding conditions. The use of predictive models in fruit growing can improve the effectiveness of decision-making processes and, as a result, there has been substantial industry and academic interest in building predictive models for grapevines [6–10].

Soil chemical properties and nutrient levels have a decisive impact on vineyards' production [11–13]. Grapevine mineral nutrition is one of the most important factors in fruit production, especially in high-density orchards [14], as minerals are responsible for several functions in plants, such as tissue constitution, enzymatic activation, osmotic regulation of membranes and intermediation of energetic processes [15,16]. Unbalanced grapevine nutrition can compromise yields and the quality of grapes, affecting their appearance, color, flavor, size, aroma, post-harvest storage capacity and tolerance to pests and diseases [17,18], which is later reflected in the characteristics of the resulting must and wines [19–21]. Farmers generally rely on soil analysis to assess properties such as texture and organic matter content, pH and exchangeable nutrient levels in order to guide fertilizing management [22,23] but this type of data can be of limited availability due to cost constraints. Recent advances in soil sensing, however, may lead to more affordable, applicable and shareable soil information that can help circumvent current limitations in the near future [24–28].

Aside from nutrition, plant productivity is strongly driven by environmental factors. Meteorological conditions play a predominant role in grapevine physiology [29], influencing its growth, phenological development, yield and must properties [12,13,30], thus affecting wine quality [31]. Sunlight exposure is a key component of development, as solar radiation is captured by the plants and transformed into biomass [32]. Incoming solar energy is also converted to temperature with direct effects on grapevine production. High temperatures affect photosynthesis, transpiration and grape berry composition [33], whereas low temperatures contribute to the accumulation of chill units and have effects on bud dormancy [34,35]. Water availability is arguably the most important environmental factor limiting crop growth and productivity. Despite the growth of irrigated viticulture worldwide, precipitation remains of paramount importance, making grapevines susceptible to altered water regimes and, while grapevine drought tolerance mechanisms are not fully understood [36], high precipitation is known to increase the risk of downy mildew (*Plasmopora viticola*) occurrence [37]. Fungal spores can be dispersed by wind, spreading infestation and compromising vineyard health [38]. High wind speed, in turn, can reduce grape berry susceptibility to sunburn [39]. Along with humidity and temperature, wind can be used to estimate grapevine canopy evapotranspiration, which reflects water usage by plants [40]. The sheer importance of weather factors to grapevine production raises concerns over climate change impacts on wine-growing regions [31,41–46].

Climate data is now abundant, relatively cheap and can be found in different resolutions, as weather stations across the world continuously record and monitor various parameters for climate classification, planning, modeling and management purposes [47]. However, measuring instruments are subject to recording errors, malfunctioning, maintenance, network transmission and storage failures among other events that can generate data gaps and result in incomplete datasets [48–50]. Data missingness adversely impacts the analysis carried out afterwards and may lead to erroneous findings, false conclusions and inaccurate predictions [51].

The choice of the best strategy to handle missing data is not straightforward but most studies conclude that imputation is more advisable than removing data in order to reduce the risk of introducing bias in datasets and subsequent analyses [50,52,53]. Consequently, data imputation is an important preprocessing task in modeling. The simplest imputation methods consist in replacing the missing values with some measure of central tendency calculated from the non-missing values, such as the mean, mode or median or by randomly selecting values within the entire data or subset (e.g., hot deck imputation) [50,54]. Despite the simplicity, these methods can reduce overall data variance and are bound to the non-missing data range. Climatic data properties such as autocorrelation between time lags, seasonality, periodic trends and cycles can be useful for the development of imputation strategies [47,55]. In addition to leveraging univariate patterns, multivariate approaches can be applied to estimate missing data by means of predictive modeling, where non-missing variables are used as predictors. As a result, different statistical and machine learning methods have long been used to address the data missingness issue in different knowledge domains [51,56,57], including climate data [47,49,50,58,59]. In advanced multiple imputation schemes, the process of generating replacement values for missing data is repeated many times, resulting in m complete datasets that are further analyzed. The literature recommends the number of imputed datasets (m) ought to be between 5 and 10 [60]. These datasets are then used in subsequent analysis and the outcomes are pooled in order to obtain robust results, reducing uncertainty and bias [50,60,61].

Machine learning (ML) is a segment of artificial intelligence that has thrived in the recent context of big data technologies and high-performance computing [62]. ML represents an alternative to statistical models where deterministic processes take precedence over probability or likelihood measures in accomplishing estimation tasks [63]. As such, forecasting models can be built based on nonparametric and semi-parametric structures with validation relying on prediction accuracy. These empirical models are data-driven and do not require deep knowledge of the biophysical mechanisms that produced the data nor a predefined structure of the model, making such techniques inexpensive and relatively easy to apply [10]. Moreover, some ML algorithms can work with a large number of predictors from both categorical and numerical data types all at once, requiring little data preprocessing. These features have led to a prominent adoption of ML methods in agriculture [62] and crop yield prediction [63,64]. Previous studies attempted to model grape yields in wine-growing regions [6,7,10], but to date there have not been such attempts focusing on the well-established viticulture region of southern Brazil. Furthermore, studies have generally not evaluated the effect of imputing missing data on the predictive modeling of grape yields.

The aims of this study were to assess the suitability of different machine learning algorithms in grape yield forecasting; evaluate the effects of data imputations on model performance; and investigate the importance of soil analysis and weather station data in grape yield predictions.

2. Materials and Methods

2.1. Study Region and Grapevine Data

The study area lies in Santana do Livramento (latitude 30°53′27″ S, longitude 55°31′58″ W), located in the Campanha Gaúcha region of Rio Grande do Sul state, in southern Brazil (Figure 1). It is a traditional winegrape region with a humid subtropical climate with hot summers and no dry seasons (Köppen-Geiger classification Cfa) [65]. Data was obtained from a commercial vineyard grown on sandy textured Alisol [66]. Grapevines were grafted onto SO4 (Selection Oppenheim 4) rootstocks and grown in an espalier system.

Yield records (n = 534) ranged from 0.11 to 26.13 ton ha^{-1}, with a mean of 9.10 ± 4.81 Mg ha^{-1}, comprising 14 harvests (1999:2007, 2009, 2011, 2013, 2018, 2019) of 27 cultivars, aging from 2 to 40 years (Alicante Bouschet, Ancelota, Arinarnoa, Cabernet Franc, Cabernet Sauvignon, Chardonnay, Chenin Blanc, Ekigaina, Flora, French Colombard, Gamay Beaujolais, Gewurztraminer, Marselan, Merlot, Moscato Blanco, Moscato d'Amburgo, Napa Gamay, Petite Sirah, Pinot Noir, Pinot Saint Georges, Pinotage, Riesling Italico, Riesling Renano, Saint Emilion, Sauvignon Blanc, Semillon and Tannat).

Yield observations were sorted by harvest. In each harvest, yield distribution was divided into three groups ("high", "medium" and "low" yield) using Jenks Natural breaks optimization [67], implemented via the BAMM tools [68] package in R. Jenks Natural breaks method is borrowed from the field of cartography and seeks to minimize the variance within categories while maximizing the variance between categories. Observations were then assigned to each group based on the yield recorded at a given harvest and the respective defined breaks. After classification, the data was grouped by cultivar and the frequency in which they were classified as "high", "medium" or "low" was calculated. This information was used along with the general yield distribution and the number of observations by cultivar to determine the final yield class of each cultivar. Clustering results (Figure 2) were validated by expert knowledge provided by the agronomist responsible for the studied vineyard.

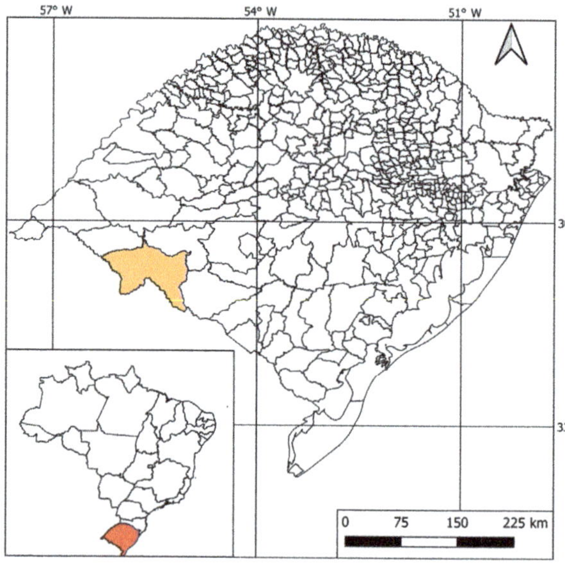

Figure 1. Santana do Livramento location (yellow) in Rio Grande do Sul state (red), southern Brazil.

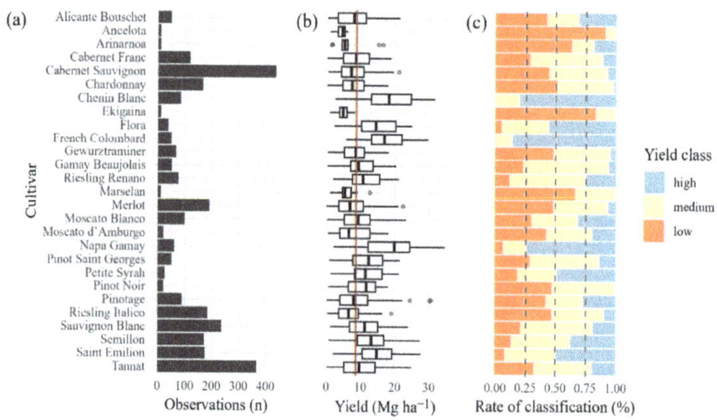

Figure 2. Number of observations (**a**), yield distribution (**b**) and Jenks Natural breaks classification frequency (c) by cultivar. Vertical red line represent cultivar average yield and gray bullets represent outliers.

2.2. Soil Data

Soil data consisted of routine laboratory analysis results obtained from soil top layer samples (0–20 cm) collected under canopy projection and contained information on soil exchange capacity (potential and effective), acidity (potential, pH and SMP index), base and aluminum saturation, clay and organic matter contents, as well as exchangeable concentrations of aluminum (Al), phosphorus (P), potassium (K), calcium (Ca), magnesium (Mg), sulfur (S), copper (Cu), zinc (Zn), boron (B) and manganese (Mn) (Table 1 and Figure 3). Soil analyses followed methodologies recommended by [69,70].

Table 1. Grapevine, soil and weather data used in yield forecasting models, grouped by predictor set.

Dataset		Variable	Unit	Type
Common		Vineyard plot	-	categorical
		Grape berry skin color	-	categorical
		Grape cultivar	-	categorical
		Cultivar yield class	-	categorical
		Plant age	year	numerical
Weather + Soil	Weather	Evaporation	mm	numerical
		Evapotranspiration (potential)	mm	numerical
		Evapotranspiration (real)	mm	numerical
		Sun exposure	hour	numerical
		Cloudiness	tenth	numerical
		Number of rainy days	unit	numerical
		Precipitation (total)	mm	numerical
		Maximum temperature	°C	numerical
		Average temperature	°C	numerical
		Minimum temperature	°C	numerical
		Relative humidity	%	numerical
		Average wind speed	$m\ s^{-1}$	numerical
	Soil	Soil clay content	%	numerical
		Soil pH in water	-	numerical
		Soil SMP index	-	numerical
		Soil exchangeable aluminum	$cmol_c\ dm^{-3}$	numerical
		Soil potential acidity	$cmol_c\ dm^{-3}$	numerical
		Soil cation exchange capacity	$cmol_c\ dm^{-3}$	numerical
		Soil cation exchange capacity (pH 7.0)	$cmol_c\ dm^{-3}$	numerical
		Soil base saturation	%	numerical
		Soil aluminum saturation	%	numerical
		Soil exchangeable calcium	$cmol_c\ dm^{-3}$	numerical
		Soil exchangeable magnesium	$cmol_c\ dm^{-3}$	numerical
		Soil organic matter	%	numerical
		Soil exchangeable phosphorus	$mg\ dm^{-3}$	numerical
		Soil exchangeable potassium	$mg\ dm^{-3}$	numerical
		Soil exchangeable sulfur	$mg\ dm^{-3}$	numerical
		Soil exchangeable copper	$mg\ dm^{-3}$	numerical
		Soil exchangeable zinc	$mg\ dm^{-3}$	numerical
		Soil exchangeable boron	$mg\ dm^{-3}$	numerical
		Soil exchangeable manganese	$mg\ dm^{-3}$	numerical

2.3. Weather Data

Data from the Santana do Livramento ground meteorological station, the closest to the vineyards, was retrieved from the Brazilian National Institute of Meteorology database (https://bdmep.inmet.gov.br/, accessed on 26 July 2023) for the period between 1990 and 2020. Due to the homogeneity of the climate and relief of the study region, data from the nearby weather station of Bagé (~160km apart) was also retrieved in order to provide auxiliary information and aid missing data imputation [60]. Monthly average values of potential and real evaporation, evapotranspiration, sunlight exposure, cloudiness, temperatures (maximum, average and minimum), relative humidity and average wind speed were available, as well as the total number of rainy days and precipitation volumes (Table 1).

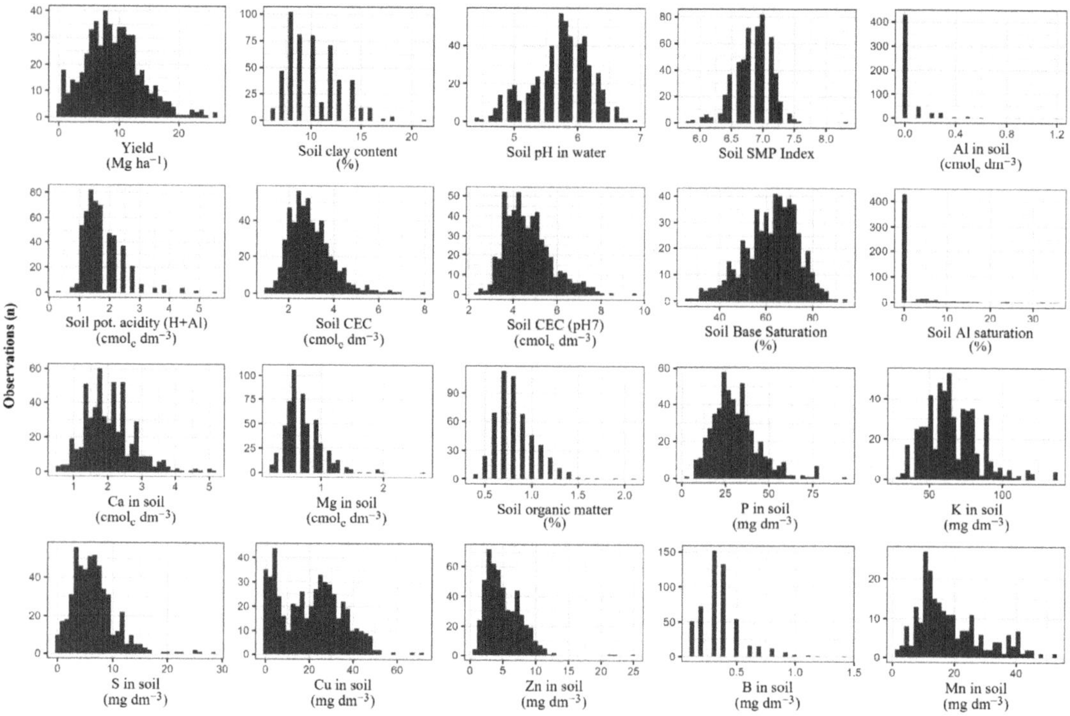

Figure 3. Distribution of yield and soil-related variables used in grapevine prediction models.

2.4. Data Imputation

In this study, missing data from the Santana do Livramento weather station was partially missing within the period covered by soil sampling data (Figure 4) and was imputed using either uni- or multivariate methods, depending on the gap size. Gaps equal to or shorter than nine consecutive observations (i.e., 9 months or 75%) were imputed using a univariate approach, which consisted of splitting the times series into seasons and performing imputation separately for each of the resulting subsets using interpolation. In this study, the univariate imputation of meteorological data was performed using the imputeTS package [71] with a 12-month season.

Univariate methods may fail to provide reasonable imputations for a variable when periods of missing values are large [51], so for gaps larger than 9 months Multivariate Imputation by Chained Equations, or MICE [61] was used. MICE turns the imputation problem into a series of estimations, where each variable has its own imputation model built using the other remaining variables of the dataset. First, only the complete data is used to estimate values for the variable with the smallest number of missing observations. Next, the recently imputed variable is used along with the originally complete data to estimate the variable with the second smallest number of missing observations and so on. The first iteration is finished when all missing values are estimated once. In the second iteration, the order of imputation remains but the imputed values are now updatedconsidering all estimates generated in the previous steps. This process is repeated through a number of iterations to achieve a stable imputation. A single imputed dataset is obtained in the last iteration [72]. To attain robust results, it is recommended that many imputed datasets (m) are used for the desired analysis so that the pooled results can be evaluated [53]. The MICE algorithm was implemented using the mice [61] package to generate five different imputed datasets after 50 iterations each, produced with random forest models containing 500 trees.

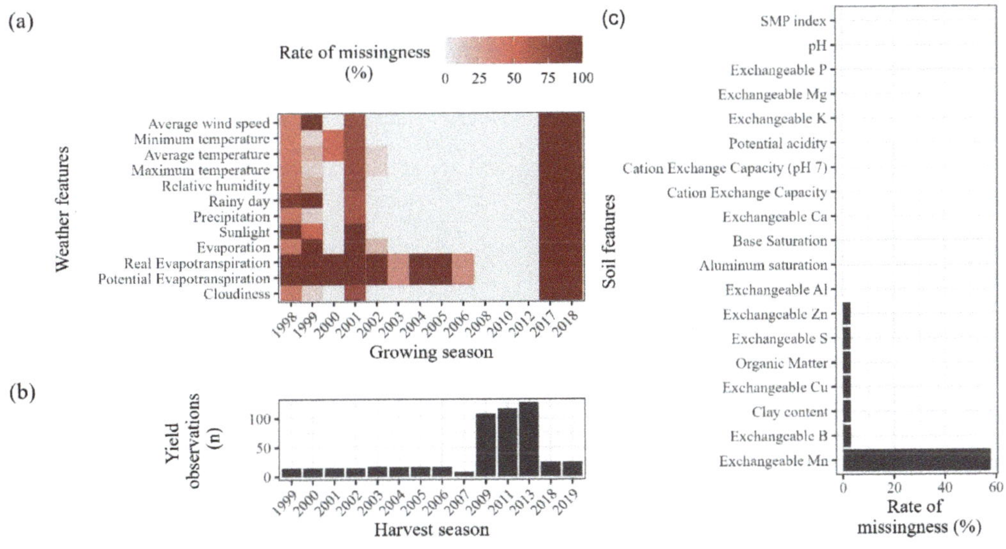

Figure 4. Rate of missingness in Santana do Livramento weather station data, by variable and growing season (**a**). Number of yield observations by harvest (**b**). Rate of missing observations by variable in soil dataset (**c**).

In addition to weather station data, some of the soil analysis reports also hosted missing values, particularly regarding soil micronutrients (Figure 4). Since no seasonality can be attributed to these soil features and given the fact that the missingness mechanism did not seem to be random, a more simple approach was chosen and median imputation was performed to fill soil data gaps.

2.5. Modeling

The underlying assumption of this study is that weather and soil data hold information that is useful to forecast yields and, to investigate its contribution to estimates, models were built using different sets of predictors (p), namely "Soil" (p = 24: four categorical, 20 numeric), "Weather" (p = 149: four categorical, 145 numeric) and "Weather + Soil" (p = 168: four categorical, 164 numeric) (Table 1). In order to handle the high number of predictors and the possible non-linear relationships between variables, three algorithms—Partial Least Squares Regression (PLSR) [73], Cubist (CUB) [74,75] and Random Forest (RF) [76]—were used to predict grape yields. PLSR seeks to produce uncorrelated predictors by resorting to orthogonal projections. As such, predictors are expected to be numeric, therefore, categorical variables in the dataset were one-hot encoded prior to running this algorithm while numeric ones were centered and scaled. Moreover, the "plot" variable, which refers to a vineyard parcel, was removed since it contains 135 possible values and can be considered high cardinality data, yielding too many new variables after one-hot encoding. To evaluate model performance, the coefficient of determination (R^2) (Equation (1)) was used as a measure of variance explained by the predictors while Root Mean Squared Error (RMSE) (Equation (2)) and Mean Absolute Error (MAE) (Equation (3)) were used as a measure of spread in predictions [77].

All models were fit using the caret package [78] ("pls", "ranger" and "cubist" engines) on a training set (n = 399, 75%) with 10-fold cross-validation and five repetitions. Model tuning was performed using grids methods, optimized to reduce Root Mean Squared Error. For PLSR tuning, the number of principal components ranged from 1 to 20. In CUB, 100 committees were used and five different numbers of neighbors were tested (1, 3, 5, 7 and 9). The number of random variables (mtry) used in each RF tree depended on the dataset used for predictions. For "Weather" and "Weather + Soil", four values were tested (50, 65, 80 and 95), whereas the "Soil" dataset was tuned using five values (9, 12, 15, 18 and 21). The same minimum node sizes were tested for all datasets (3, 5, 7 and 10). All RF models were built using 100 trees. Model best fit parameters were applied in the validation model and performance was assessed on a hold-out test set (n = 135, 25%). Data handling, imputing and modeling were performed in R language [79] using RStudio IDE [80].

$$R^2 = \frac{\sum_{i=1}^{N} (\hat{y}_i - \bar{y}_i)^2}{\sum_{i=1}^{N} (y_i - \bar{y}_i)^2} \quad (1)$$

$$RMSE = \sqrt{\frac{1}{N}\sum_{i=1}^{N} (\hat{y}_i - y_i)^2} \quad (2)$$

$$MAE = \frac{1}{N}\sum |\hat{y}_i - y_i| \quad (3)$$

where $\hat{y}$ = predicted value; $\bar{y}$ = mean observed value; y = observed values; N = number of samples with i = 1, 2. . .

3. Results

3.1. Model Performance and Imputation Effects

Model outcomes indicate that the choice of predictors had a considerable impact on grape yield forecasting. Regardless of the algorithm used, models built with the "Soil" dataset had the worst average results (R^2 = 0.15, RMSE = 4.16 Mg ha^{-1}, MAE = 3.20 Mg ha^{-1}), while the use of only the "Weather" dataset yielded the best performance (R^2 = 0.52, RMSE = 2.99 Mg ha^{-1}, MAE = 2.43 Mg ha^{-1}). The combination of soil and weather predictors had slightly worse results than weather variables alone (R^2 = 0.50, RMSE = 3.05 Mg ha^{-1}, MAE = 2.43 Mg ha^{-1}), probably as a consequence of the higher number of variables with low predictive power. On the other hand, as the number of predictors increased, the difference between the three algorithms (i.e., the spread) decreased, suggesting a trade-off between higher precision and lower accuracy (Figure 5 and Table 2).

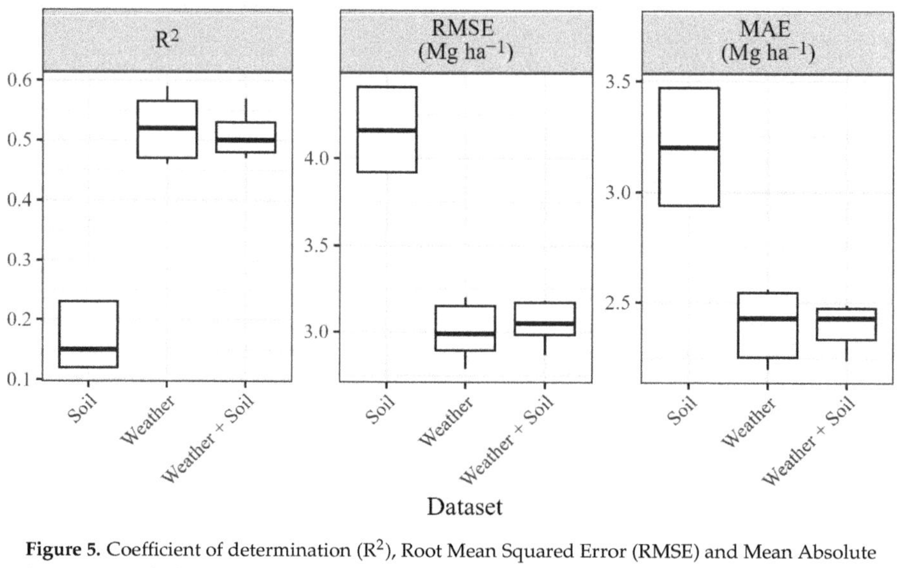

Figure 5. Coefficient of determination (R^2), Root Mean Squared Error (RMSE) and Mean Absolute Error (MAE) of all combined algorithms by dataset.

Table 2. Average coefficient of determination (R^2), Root Mean Squared Error (RMSE) and Mean Absolute Error (MAE) values for Partial Least Squares Regression (PLSR), Cubist (CUB) and Random Forest (RF) algorithms by predictor dataset.

Predictors	Soil			Weather			Weather + Soil		
Algorithm	**PLRS**	**CUB**	**RF**	**PLRS**	**CUB**	**RF**	**PLRS**	**CUB**	**RF**
R^2	0.12	0.15	0.23	0.47	0.52	0.58	0.48	0.51	0.55
RMSE (Mg ha^{-1})	4.41	4.16	3.92	3.16	2.99	2.85	3.18	3.05	2.95
MAE (Mg ha^{-1})	3.47	3.20	2.94	2.55	2.42	2.24	2.48	2.42	2.30

In general, the RF model had the highest R^2 and the lowest RMSE and MAE, followed by CUB and PLSR, regardless of the predictors used (Table 3). However, CUB and RF were the two algorithms most affected by overfitting, presenting very sharp changes in metrics when training and test sets are compared, especially when soil-related predictors were used. In addition, CUB and RF were also the most affected by different imputation sets and, once more, the presence of soil-related predictors contributed to higher variance in model metrics (Figure 6).

Table 3. Grape yield prediction coefficient of determination (R^2), Root Mean Squared Error (RMSE) and Mean Absolute Error (MAE) by algorithm, predictor dataset and imputation set. Best fit refers to optimized tuning parameters for Partial Least Squares Regressions (PLSR: number of principal components), Cubist (CUB: number of neighbors) and Random Forest (RF: random predictors subset/minimum node size).

Imputation Set	Predictors	Algorithm	Train			Best fit	Test		
			R^2	RMSE (Mg ha^{-1})	MAE (Mg ha^{-1})		R^2	RMSE (Mg ha^{-1})	MAE (Mg ha^{-1})
1	Soil	PLSR	0.34	4.00	3.10	8	0.12	4.41	3.47
		CUB	0.90	1.76	1.35	9	0.15	4.16	3.20
		RF	0.94	1.59	1.16	12/3	0.23	3.92	2.94
	Weather	PLSR	0.55	3.33	2.57	10	0.46	3.20	2.54
		CUB	0.89	1.70	1.29	5	0.52	2.99	2.43
		RF	0.89	1.70	1.28	65/5	0.59	2.80	2.23
	Weather + Soil	PLSR	0.60	3.13	2.46	18	0.47	3.18	2.48
		CUB	0.95	1.25	0.95	9	0.50	3.07	2.47
		RF	0.95	1.32	1.00	80/3	0.55	2.94	2.30
2	Soil	PLSR	0.56	3.28	2.54	15	0.47	3.15	2.55
		CUB	0.89	1.69	1.29	5	0.52	3.00	2.44
		RF	0.90	1.70	1.29	65/5	0.56	2.90	2.26
	Weather	PLSR	0.60	3.14	2.46	17	0.48	3.17	2.47
		CUB	0.95	1.23	0.94	9	0.50	3.05	2.43
		RF	0.95	1.33	1.00	95/3	0.52	3.04	2.36
3	Weather	PLSR	0.56	3.29	2.55	14	0.47	3.15	2.55
		CUB	0.89	1.69	1.28	5	0.53	2.98	2.43
		RF	0.92	1.53	1.16	65/3	0.57	2.85	2.24
	Weather + Soil	PLSR	0.60	3.11	2.46	17	0.48	3.17	2.48
		CUB	0.95	1.24	0.95	9	0.52	3.00	2.40
		RF	0.94	1.43	1.08	95/5	0.57	2.87	2.24

Table 3. *Cont.*

Imputation Set	Predictors	Algorithm	Train				Test		
			R^2	RMSE	MAE	Best fit	R^2	RMSE	MAE
				Mg ha^{-1}				Mg ha^{-1}	
4	Weather	PLSR	0.56	3.28	2.54	15	0.47	3.15	2.55
		CUB	0.90	1.63	1.23	5	0.52	2.98	2.41
		RF	0.91	1.54	1.17	65/3	0.57	2.89	2.25
	Weather + Soil	PLSR	0.60	3.13	2.46	18	0.47	3.18	2.49
		CUB	0.95	1.23	0.94	9	0.51	3.03	2.38
		RF	0.95	1.30	0.98	95/3	0.54	2.97	2.31
5	Soil	PLSR	0.56	3.28	2.53	15	0.47	3.17	2.56
		CUB	0.88	1.74	1.33	5	0.52	2.99	2.40
		RF	0.90	1.70	1.29	65/5	0.59	2.79	2.20
	Weather	PLSR	0.60	3.14	2.46	17	0.48	3.18	2.48
		CUB	0.95	1.28	0.98	9	0.50	3.08	2.44
		RF	0.95	1.34	1.00	95/3	0.55	2.92	2.28

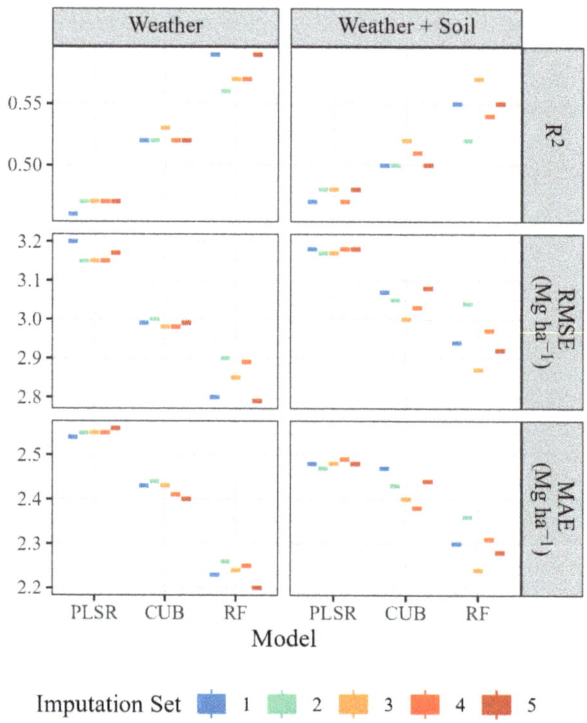

Figure 6. Impact of weather data imputation on coefficient of determination (R^2), Root Mean Squared Error (RMSE) and Mean Absolute Error (MAE) of grapevine yield predictions using Partial Least Squares Regression (PLSR), Cubist (CUB) and Random Forest (RF) algorithms with different sets of predictors.

The impact of imputation sets on model performance arises from changes in predictor distribution and range after filling in the missing values, therefore, the importance of predictors can also be indirectly assessed by investigating the differences between the imputed datasets. Imputation sets were fairly uniform and shaped similarly to the original observed data for most predictors except the mean wind speed and, to some extent, potential evapotranspiration. The number of rainy days, precipitation and relative humidity had at least one imputation set whose distribution appeared different from the others (Figure 7).

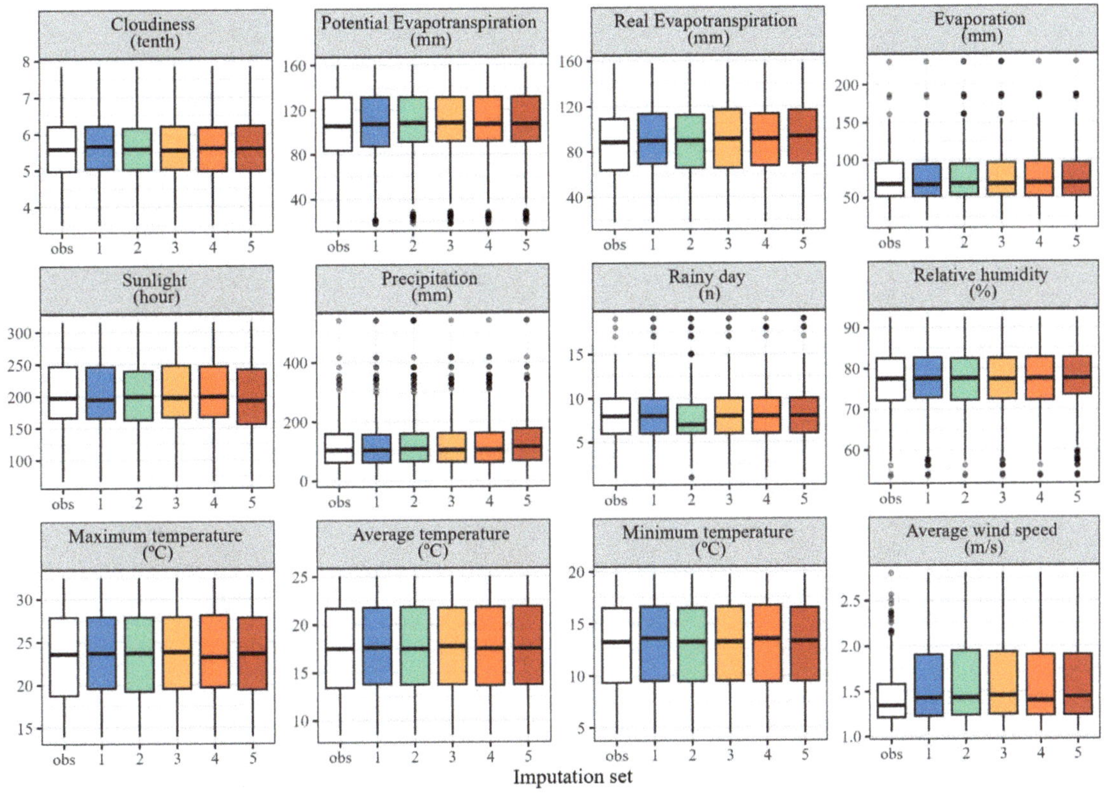

Figure 7. Distributions of weather variables in observed data (obs, in white) and the five different complete datasets produced after multivariate chained equation (MICE) imputations (1–5, coloured). Gray bullets represent outliers.

3.2. Variable Importance

The assessments of variable importance can enable insights about the nature of the most important predictors and point to the direction for data acquisition and prediction improvements. Across all algorithms and all predictor datasets, plant age and yield class were the most important variables. May temperatures, especially the minimum, were the most important weather-related variable. Soil pH and concentrations of Zn and Cu were the most important soil-related predictors, followed by soil K and Mn. In addition, plot and cultivar were important predictors in CUB and RF models. The importance of other predictors varied among dataset x model combinations (Figure 8).

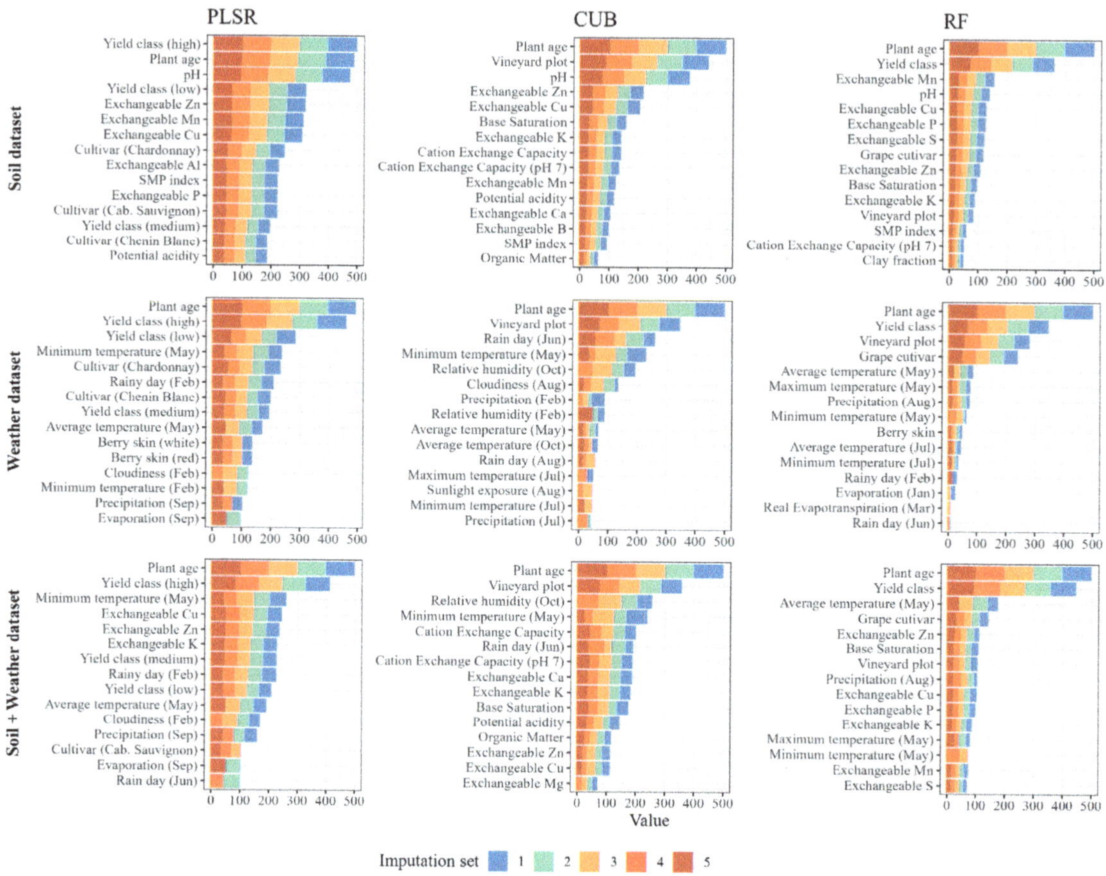

Figure 8. The 15 most important variables by algorithm, predictor and imputation set. For each imputation set, variable importance is given on a 0 to 100 scale and x-axis values are the sum of the importance in m imputation sets. Soil dataset was only imputed using a median method, hence the results are the same for all imputation sets.

4. Discussion

The effects of data imputation on model performance vary according to the algorithm used. The different sensitivities between PLSR, CUB and RF arise due to how each of these algorithms work. The PLSR acts as a supervised Principal Component Regression (PCR) producing new features (PCs) which are linear combinations of the original predictors that maximize the explained variance in the response [73]. Therefore, changes in particular variables due to imputation may have their effects diluted or even ignored given this optimization constraint.

Building on Quinlan's original M5 model [74,75], its modern version called Cubist (CUB) is a rule-based algorithm that makes predictions by building regression models on its terminal nodes [81]. This algorithm works by creating a split in the dataset and fitting a regression with every predictor available. In the next node, data is split again and a new regression is fit using the new subset of predictors. Moreover, CUB includes an ensemble method for predictions called committees and a nearest-neighbor adjustment that occurs after the model predictions [81]. In committee ensembles, the outcome values are modified for each iteration in an effort to reduce over- and under-predictions, whereas the final result is obtained by averaging over the committee model estimates. After model prediction, it is possible to conduct a post-model K-nearest neighbor adjustment where the closest observations in the training set (along with their original predictions) are used to correct prediction errors and trends in the dependent variable [74,81]. In the current study, the number of committees and neighbors used was high and, while effective in improving performance metrics, it is likely that it caused overfitting in the training set, hampering predictive ability during the validation stage.

Random Forest is an ensemble of decision trees that do not build regression models on terminal nodes, but rather use average values and, as a consequence, may be more sensitive to variations caused by imputation, given the changes in imputed data distributions. The number of variables randomly sampled as candidates at each split (mtry) is a tuning parameter that impacts the likelihood of selecting important features that are actually related to the outcome variable for most of the splits that are made. Too large mtry values can reduce randomness in the model, diminishing the benefits of building multiple independent decision trees. Minimum node size, in turn, defines how deep a tree can grow, that is, the minimum number of observations a node requires to proceed with a split. Our results present clear signs of overfitting, suggesting that the combination of large *mtry* values and small node size (Table 3) is detrimental to modeling and should be reconsidered. Nevertheless, the pooled results of RF yielded the best predictive performance, indicating that some of the overfitting issues were addressed by the averaging effect of multiple trees (a.k.a. the law of large numbers) [76], although higher values should also be tested.

Our results corroborate with the well-established importance of meteorological variables in resulting grape yield [12,13,30]. However, our models highlighted end-of-cycle temperatures (May) as being important predictors, as opposed to the most commonly reported temperatures at bud break, flowering, fruit-set and berry development stages [8,10,31,82,83]. Post-harvest temperature might contribute to greater carbohydrate accumulation in plant tissues, which are translocated in the next cycle to the leaves and branches at the beginning of vegetative growth [84].

Relative humidity and precipitation were previously reported as relevant predictors for grape yield models [10,46] but did not show consistent predictive power in the present study. This may arise from the inherit difficulty in capturing rain records. Since rain and humidity can be very heterogeneous across landscape, even nearby stations may not represent the water regime faithfully, let alone in imputed datasets [58], an issue of known concern [58,85]. Imputation of time series and its impact on modeling results is still an active field of research [47,51,53,56] and further studies need to be carried out to avoid erroneous conclusions and inaccurate predictions.

Despite the undeniable importance of plant nutrition to obtain proper grapevine yields, the predictive power of soil variables was considerably low. Grape root systems can grow over a meter deep, therefore, the dynamics of nutrient uptake are not fully captured by top layer soil samples [18]. Moreover, adult grapevines reach nutritional stability and are able to mobilize reserves [86], hindering the relationship between soil nutrient concentrations and yield parameters. This is also reflected in the ubiquitous presence of plant age as the most important yield predictor.

Soil pH plays a key role in plant development due to its effects on nutrient availability and toxicity [87] and such importance is promptly captured by all three models. The importance of soil Cu and Zn as yield predictors can, however, be related to phytosanitary control, since they are components of fungicides largely used against crop diseases, ensuring higher yields but also leading to their accumulation in vineyard soils [88–90]. Grape berries are great sinks of K [91] and clusters can account for over 60% of the total content in the above-ground organs [92], so it is sound that soil K figures as an important yield predictor, which is supported by recent modeling attempts [10]. Grapevines are particularly susceptible to Mn deficiency and shortage of supply can lead to small and poor quality yields [93]. In addition, recent studies suggest that Mn plays a role in enhancing plant resistance to water stress [94], so despite being a micronutrient, Mn soil concentrations seem to play key role in grape production. It is worth noting that Mn was the single most imputed variable in soil-related datasets (Figure 4) and, while having artificially reduced variance, it figured as a relevant predictor when soil variables were considered.

Categorical variables were also important to model estimates. The high importance of vineyard plot information suggests spatial dependency, consistent with recent studies on the effects of intra-vineyard variability and soil heterogeneity on vine performance, dry matter and nutrient partitioning [95], where, based on the Normalized Difference Vegetation Index NDVI, authors divided the vineyard into high, medium and low vigor zones. The availability of georeferenced data can improve yield response estimates by enabling the use of covariates in modeling [96] and imputations [97] but such information was unavailable in the current study. Instead, the cultivars were grouped into low, medium or high yield levels based on numerical clustering, historical yield levels (Figure 2) and expert knowledge. This feature engineering improved model performance by better partitioning the variability while retaining agronomic sense. Similarly, other groupings (or clusterings) could enhance the predictive ability of models. Research has shown the importance of phenological stage grouping for modeling [6,10,98]. Moreover, advances in fruit tree fertilizing indicate different nutrition profiles for each cultivar [12,99,100] or site-specific conditions [101] and, while requiring further elucidation, have the potential to leverage soil data predictive power.

The present study showed that the combined use of databases and robust machine learning methods has the potential to estimate grapevine productivity, helping decision makers to be more assertive not only during commercialization stage but also in fertilizer recommendations and phytosanitary management of vineyards. The herein adopted strategies can be extended for other relevant applications in orchards worldwide, estimating the performance of fruit trees under specific conditions (culture, site, management, etc.), while keeping in mind multiple factors such as increased yield and economic viability, higher visual and bromatological quality of fruits, plant nutrition for proper development, rational use of fertilizers and adaptation within a climate change scenario.

5. Conclusions

In this study, Partial Least Square (PLSR), Cubist (CUB) and Random Forest (RF) algorithms were used to predict grape yields from weather and soil data separately and in conjunction. Overall, the results were best when using only weather predictors, followed by weather and soil data combined and, lastly, soil data alone. Weather data from ground meteorological stations contained observations with missing values and these were imputed by uni- or multivariate methods. The RF and CUB algorithms were the most affected by weather data imputations, while PLSR remained fairly insensitive. Parameter tuning optimized to reduce Root Mean Squared Error led to overfitting in CUB- and RF-based models and (hyper)parameter values should be reconsidered to improve performance. Nevertheless, RF achieved the best metrics, followed closely by CUB, while PLSR yielded the poorest yet most stable results. Plant age, May temperatures, soil pH and concentrations of Zn, Cu, K and Mn were identified as important predictors, even though 57% of soil Mn observations were missing and were imputed using a median method. Yield level groups had high importance in predictions, indicating that clustering can be a useful strategy, whereas the importance of vineyard plots suggests spatial dependencies that can be further explored by using georeferenced data. This exploratory work offers insights for future research on grape yield predictive modeling and highlights the importance of high resolution data and grouping strategies to obtain more assertive results, thus contributing to a more efficient grapevine production chain in southern Brazil and worldwide.

Author Contributions: Conceptualization, G.B., J.J.C., J.M.M.-B. and C.B.A.; methodology, J.M.M.-B. and C.B.A.; software, J.M.M.-B. and C.B.A.; J.M.M.-B. and C.B.A.; formal analysis, C.B.A.; investigation, J.M.M.-B. and C.B.A.; resources, G.B and J.J.C.; data curation, C.B.A.; writing—original draft preparation, J.M.M.-B. and C.B.A.; writing—review and editing, G.B., J.J.C. and J.M.M.-B.; visualization, G.B., J.J.C., J.M.M.-B. and C.B.A.; supervision, G.B., J.J.C. and J.M.M.-B.; project administration, G.B. and J.J.C.; funding acquisition, G.B. and J.J.C. All authors have read and agreed to the published version of the manuscript.

Funding: This research was funded by the National Council for Scientific and Technological Development (CNPq—process: 302023/2019-4) and the Foundation for Support of Research Rio Grande do Sul—Brazil (FAPERGS—process: 21/2551-0002232-9).

Data Availability Statement: The data presented in this study are available on request from the corresponding author. The data are not publicly available due to company data privacy.

Acknowledgments: We would like to thank the professionals involved in data collection, as well as the winery administration that generously shared its data so that the present study could be carried out. We also thank the Coordenação de Aperfeiçoamento de Pessoal de Nível Superior—Brazil (CAPES) and the National Council for Scientific and Technological Development (CNPq).

Conflicts of Interest: The authors declare no conflict of interest.

References

1. OIV. *Statistical Report on World Vitiviniculture*; International Organisation of Vine and Wine: Dijon, France, 2019; p. 23.
2. OIV. *State of the World Vine and Wine Sector*; International Organisation of Vine and Wine: Dijon, France, 2022; p. 20.
3. IBGE. *Levantamento Sistemático da Produção Agropecuária*; IBGE: Rio de Janeiro, Brazil, 2022.
4. De Mello, L.M.R.; Machado, C.A.E. *Vitivinicultura Brasileira: Panorama 2019*; Comunicado Técnico, 214; Embrapa Uva e Vinho: Bento Gonçalves, Brazil, 2020; pp. 1–21.
5. CQFS-RS/SC. *Manual de Calagem e Adubação para os Estados do Rio Grande do Sul e Santa Catarina*; Comissão de Química e Fertilidade do Solo/Núcleo Regional Sul-Sociedade Brasileira de Ciência do Solos: Passo Fundo, Brazil, 2016; ISBN 978-85-66301-80-9.
6. Arab, S.T.; Noguchi, R.; Matsushita, S.; Ahamed, T. Prediction of Grape Yields from Time-Series Vegetation Indices Using Satellite Remote Sensing and a Machine-Learning Approach. *Remote Sens. Appl. Soc. Environ.* **2021**, *22*, 100485. [CrossRef]
7. Barriguinha, A.; Jardim, B.; De Castro Neto, M.; Gil, A. Using NDVI, Climate Data and Machine Learning to Estimate Yield in the Douro Wine Region. *Int. J. Appl. Earth Obs. Geoinf.* **2022**, *114*, 103069. [CrossRef]
8. Cunha, M.; Ribeiro, H.; Abreu, I. Pollen-Based Predictive Modelling of Wine Production: Application to an Arid Region. *Eur. J. Agron.* **2016**, *73*, 42–54. [CrossRef]
9. Kadbhane, S.J.; Manekar, V.L. Development of Agro-Climatic Grape Yield Model with Future Prospective. *Ital. J. Agrometeorol.* **2021**, *2021*, 89–103. [CrossRef]
10. Sirsat, M.S.; Mendes-Moreira, J.; Ferreira, C.; Cunha, M. Machine Learning Predictive Model of Grapevine Yield Based on Agroclimatic Patterns. *Eng. Agric. Environ. Food* **2019**, *12*, 443–450. [CrossRef]
11. Krzyszczak, J.; Brodowska, M.; Bednarek, W.; Dresler, S.; Tkaczyk, P.; Krzyszczak, J.; Baranowski, P. Content of Certain Macro- and Microelements in Orchard Soils in Relation to Agronomic Categories and Reaction of These Soils. *J. Elem.* **2018**, *23*, 1361–1372. [CrossRef]
12. Stefanello, L.; Schwalbert, R.; Schwalbert, R.; Tassinari, A.; Garlet, L.; De Conti, L.; Ciotta, M.; Ceretta, C.; Ciampitti, I.; Brunetto, G. Phosphorus Critical Levels in Soil and Grapevine Leaves for South Brazil Vineyards: A Bayesian Approach. *Eur. J. Agron.* **2023**, *144*, 126752. [CrossRef]
13. Stefanello, L.O.; Schwalbert, R.; Schwalbert, R.A.; Drescher, G.L.; De Conti, L.; Pott, L.P.; Tassinari, A.; Kulmann, M.S.d.S.; da Silva, I.C.B.; Brunetto, G. Ideal Nitrogen Concentration in Leaves for the Production of High-Quality Grapes Cv 'Alicante Bouschet' (*Vitis vinifera* L.) Subjected to Modes of Application and Nitrogen Doses. *Eur. J. Agron.* **2021**, *123*, 126200. [CrossRef]
14. Ebert, G. *Fertilizing for High Yield and Quality: Pome and Stone Fruits of the Temperate Zone*; IPI Bulletin; International Potash Institute: Horgen, Switzerland, 2009; p. 74.
15. Faust, M. *Physiology of Temperate Zone Fruit Trees*; Wiley-Interscience: New York, NY, USA, 1989.
16. Marschner, P. *Marschner's Mineral Nutrition of Higher Plants*; Academic Press: San Diego, CA, USA, 2012; ISBN 978-0-12-384905-2.
17. Brunetto, G.; Melo, G.W.B.D.; Toselli, M.; Quartieri, M.; Tagliavini, M. The Role of Mineral Nutrition on Yields and Fruit Quality in Grapevine, Pear and Apple. *Rev. Bras. Frutic.* **2015**, *37*, 1089–1104. [CrossRef]
18. Rozane, D.E.; Parent, L.E.; Natale, W. Evolution of the Predictive Criteria for the Tropical Fruit Tree Nutritional Status. *Científica* **2015**, *44*, 102. [CrossRef]
19. Brunetto, G.; Ceretta, C.A.; Kaminski, J.; de Melo, G.W.B.; Lourenzi, C.R.; Furlanetto, V.; Moraes, A. Aplicação de Nitrogênio Em Videiras Na Campanha Gaúcha: Produtividade e Características Químicas Do Mosto Da Uva. *Ciência Rural.* **2007**, *37*, 389–393. [CrossRef]
20. Considine, M.J.; Foyer, C.H. Metabolic Responses to Sulfur Dioxide in Grapevine (*Vitis vinifera* L.): Photosynthetic Tissues and Berries. *Front. Plant Sci.* **2015**, *6*, 60. [CrossRef] [PubMed]
21. Skinner, P.W.; Ishii, R.; O'Mahony, M.; Matthews, M.A. Sensory Attributes of Wines Made from Vines of Differing Phosphorus Status. *Oeno One* **2019**, *53*, 205–219. [CrossRef]
22. Brunetto, G.; Ricachenevsky, F.K.; Stefanello, L.O.; de Paula, B.V.; de Souza Kulmann, M.S.; Tassinari, A.; de Melo, G.W.B.; Natale, W.; Rozane, D.E.; Ciotta, M.N.; et al. Chapter 47—Diagnosis and Management of Nutrient Constraints in Grape. In *Fruit Crops*; Srivastava, A.K., Hu, C., Eds.; Elsevier: Amsterdam, The Netherlands, 2020; pp. 693–710. ISBN 978-0-12-818732-6.
23. Rozane, D.E.; de Paula, B.V.; de Melo, G.W.B.; Dos Santos, E.M.H.; Trentin, E.; Marchezan, C.; da Silva, L.O.S.; Tassinari, A.; Dotto, L.; de Oliveira, F.N.; et al. Compositional Nutrient Diagnosis (CND) Applied to Grapevines Grown in Subtropical Climate Region. *Horticulturae* **2020**, *6*, 56. [CrossRef]
24. Tavares, T.R.; Molin, J.P.; Nunes, L.C.; Alves, E.E.N.; Krug, F.J.; De Carvalho, H.W.P. Spectral Data of Tropical Soils Using Dry-Chemistry Techniques (VNIR, XRF, and LIBS): A Dataset for Soil Fertility Prediction. *Data Brief* **2022**, *41*, 108004. [CrossRef] [PubMed]
25. Tavares, T.R.; De Almeida, E.; Junior, C.R.P.; Guerrero, A.; Fiorio, P.R.; De Carvalho, H.W.P. Analysis of Total Soil Nutrient Content with X-ray Fluorescence Spectroscopy (XRF): Assessing Different Predictive Modeling Strategies and Auxiliary Variables. *AgriEngineering* **2023**, *5*, 680–697. [CrossRef]

26. Poppiel, R.R.; Paiva, A.F.D.S.; Demattê, J.A.M. Bridging the Gap between Soil Spectroscopy and Traditional Laboratory: Insights for Routine Implementation. *Geoderma* **2022**, *425*, 116029. [CrossRef]
27. Nadporozhskaya, M.; Kovsh, N.; Paolesse, R.; Lvova, L. Recent Advances in Chemical Sensors for Soil Analysis: A Review. *Chemosensors* **2022**, *10*, 35. [CrossRef]
28. De Mello, D.C.; Barros Souza, A.; Mello, F.A.O.; Marques, K.P.P.; Poppiel, R.R.; Belinasso, H.; Di Raimo, L.D.L.; Francelino, M.R.; Fernandes-Filho, E.I.; Veloso, G.V.; et al. Sensor-Based Field Methods for Pedology and Soil Surveys: Protocol Suggestions for Brazilian Tropical Soils. *Geoderma Reg.* **2023**, *33*, e00651. [CrossRef]
29. Keller, M. Managing Grapevines to Optimise Fruit Development in a Challenging Environment: A Climate Change Primer for Viticulturists. *Aust. J. Grape Wine Res.* **2010**, *16*, 56–69. [CrossRef]
30. Ciotta, M.N.; Ceretta, C.A.; Ferreira, P.A.; Stefanello, L.O.; da Rosa Couto, R.; Tassianri, A.; Marchezan, C.; Girotto, E.; Conti, L.D.; Lourenzi, C.R.; et al. Phosphorus Fertilization for Young Grapevines of Chardonnay and Pinot Noir in Sandy Soil. *IDESIA* **2018**, *36*, 27–34. [CrossRef]
31. Fraga, H.; de Cortázar Atauri, I.G.; Malheiro, A.C.; Santos, J.A. Modelling Climate Change Impacts on Viticultural Yield, Phenology and Stress Conditions in Europe. *Glob. Chang. Biol.* **2016**, *22*, 3774–3788. [CrossRef] [PubMed]
32. Campillo, C.; Fortes, R.; Henar Prieto, M.D. Solar Radiation Effect on Crop Production. In *Solar Radiation*; Babatunde, E.B., Ed.; IntechOpen: London, UK, 2012; ISBN 978-953-51-6163-9.
33. Venios, X.; Korkas, E.; Nisiotou, A.; Banilas, G. Grapevine Responses to Heat Stress and Global Warming. *Plants* **2020**, *9*, 1754. [CrossRef] [PubMed]
34. Anzanello, R. Evolution of the Grapevine Bud Dormancy under Different Thermal Regimes. *Semin. Cienc. Agrar.* **2019**, *40*, 3419. [CrossRef]
35. North, M.; Workmaster, B.A.; Atucha, A. Effects of Chill Unit Accumulation and Temperature on Woody Plant Deacclimation Kinetics. *Physiol. Plant.* **2022**, *174*, e13717. [CrossRef] [PubMed]
36. Gambetta, G.A.; Herrera, J.C.; Dayer, S.; Feng, Q.; Hochberg, U.; Castellarin, S.D. The Physiology of Drought Stress in Grapevine: Towards an Integrative Definition of Drought Tolerance. *J. Exp. Bot.* **2020**, *71*, 4658–4676. [CrossRef]
37. Chen, M.; Brun, F.; Raynal, M.; Makowski, D. Forecasting Severe Grape Downy Mildew Attacks Using Machine Learning. *PLoS ONE* **2020**, *15*, e0230254. [CrossRef]
38. Mezei, I.; Lukić, M.; Berbakov, L.; Pavković, B.; Radovanović, B. Grapevine Downy Mildew Warning System Based on NB-IoT and Energy Harvesting Technology. *Electronics* **2022**, *11*, 356. [CrossRef]
39. Müller, K.; Keller, M.; Stoll, M.; Friedel, M. Wind Speed, Sun Exposure and Water Status Alter Sunburn Susceptibility of Grape Berries. *Front. Plant Sci.* **2023**, *14*, 1145274. [CrossRef]
40. Jenkins, M.; Mannsfeld, A.; Nikzad, S.; Lambert, J.-J.; Miller, K.; Burns, M.; Earles, J.M.; Block, D.E. Novel Algorithms for High-Resolution Prediction of Canopy Evapotranspiration in Grapevine. *OENO One* **2023**, *57*, 315–326. [CrossRef]
41. Bonfante, A.; Alfieri, S.M.; Albrizio, R.; Basile, A.; De Mascellis, R.; Gambuti, A.; Giorio, P.; Langella, G.; Manna, P.; Monaco, E.; et al. Evaluation of the Effects of Future Climate Change on Grape Quality through a Physically Based Model Application: A Case Study for the Aglianico Grapevine in Campania Region, Italy. *Agric. Syst.* **2017**, *152*, 100–109. [CrossRef]
42. Droulia, F.; Charalampopoulos, I. Future Climate Change Impacts on European Viticulture: A Review on Recent Scientific Advances. *Atmosphere* **2021**, *12*, 495. [CrossRef]
43. Fraga, H.; Molitor, D.; Leolini, L.; Santos, J.A. What Is the Impact of Heatwaves on European Viticulture? A Modelling Assessment. *Appl. Sci.* **2020**, *10*, 3030. [CrossRef]
44. Holland, T.; Smit, B. Climate Change and the Wine Industry: Current Research Themes and New Directions. *J. Wine Res.* **2010**, *21*, 125–136. [CrossRef]
45. Moriondo, M.; Ferrise, R.; Trombi, G.; Brilli, L.; Dibari, C.; Bindi, M. Modelling Olive Trees and Grapevines in a Changing Climate. *Environ. Model. Softw.* **2015**, *72*, 387–401. [CrossRef]
46. Fraga, H.; Santos, J.A. Daily Prediction of Seasonal Grapevine Production in the Douro Wine Region Based on Favourable Meteorological Conditions: Predicting Winery Grape Production in the Douro. *Aust. J. Grape Wine Res.* **2017**, *23*, 296–304. [CrossRef]
47. Afrifa-Yamoah, E.; Mueller, U.A.; Taylor, S.M.; Fisher, A.J. Missing Data Imputation of High-Resolution Temporal Climate Time Series Data. *Meteorol. Appl.* **2020**, *27*, 1873. [CrossRef]
48. Yozgatligil, C.; Aslan, S.; Iyigun, C.; Batmaz, I. Comparison of Missing Value Imputation Methods in Time Series: The Case of Turkish Meteorological Data. *Theor. Appl. Climatol.* **2013**, *112*, 143–167. [CrossRef]
49. Parra-Plazas, J.; Gaona-Garcia, P.; Plazas-Nossa, L. Time Series Outlier Removal and Imputing Methods Based on Colombian Weather Stations Data. *Environ. Sci. Pollut. Res.* **2023**, *30*, 72319–72335. [CrossRef] [PubMed]
50. Li, C.; Ren, X.; Zhao, G. Machine-Learning-Based Imputation Method for Filling Missing Values in Ground Meteorological Observation Data. *Algorithms* **2023**, *16*, 422. [CrossRef]
51. Chaudhry, A.; Li, W.; Basri, A.; Patenaude, F. A Method for Improving Imputation and Prediction Accuracy of Highly Seasonal Univariate Data with Large Periods of Missingness. *Wirel. Commun. Mob. Comput.* **2019**, *2019*, 4039758. [CrossRef]
52. Little, R.; Rubin, D. *Statistical Analysis with Missing Data*, 3rd ed.; Wiley Series in Probability and Statistics; Wiley: Hoboken, NJ, USA, 2019; ISBN 978-1-119-48226-0.

53. Mera-Gaona, M.; Neumann, U.; Vargas-Canas, R.; López, D.M. Evaluating the Impact of Multivariate Imputation by MICE in Feature Selection. *PLoS ONE* **2021**, *16*, e0254720. [CrossRef]
54. Andridge, R.R.; Little, R.J.A. A Review of Hot Deck Imputation for Survey Non-Response. *Int. Stat. Rev.* **2010**, *78*, 40–64. [CrossRef] [PubMed]
55. Moritz, S.; Sardá, A.; Bartz-Beielstein, T.; Zaefferer, M.; Stork, J. Comparison of Different Methods for Univariate Time Series Imputation in R. *arXiv* **2015**, arXiv:1510.03924.
56. Ahn, H.; Sun, K.; Kim, K.P. Comparison of Missing Data Imputation Methods in Time Series Forecasting. *Comput. Mater. Contin.* **2021**, *70*, 767–779. [CrossRef]
57. Lin, W.C.; Tsai, C.F. Missing Value Imputation: A Review and Analysis of the Literature (2006–2017). *Artif. Intell. Rev.* **2020**, *53*, 1487–1509. [CrossRef]
58. Kannegowda, N.; Udayar Pillai, S.; Kommireddi, C.V.N.K. Fousiya Comparative Assessment of Univariate and Multivariate Imputation Models for Varying Lengths of Missing Rainfall Data in a Humid Tropical Region: A Case Study of Kozhikode, Kerala, India. *Acta Geophys.* **2023**. [CrossRef]
59. Lara-Estrada, L.; Rasche, L.; Sucar, L.; Schneider, U. Inferring Missing Climate Data for Agricultural Planning Using Bayesian Networks. *Land* **2018**, *7*, 4. [CrossRef]
60. Azur, M.J.; Stuart, E.A.; Frangakis, C.; Leaf, P.J. Multiple Imputation by Chained Equations: What Is It and How Does It Work? Multiple Imputation by Chained Equations. *Int. J. Methods Psychiatr. Res.* **2011**, *20*, 40–49. [CrossRef]
61. Buuren, S.V.; Groothuis-Oudshoorn, K. Mice: Multivariate Imputation by Chained Equations in R. *J. Stat. Soft.* **2011**, *45*, 1–67. [CrossRef]
62. Liakos, K.; Busato, P.; Moshou, D.; Pearson, S.; Bochtis, D. Machine Learning in Agriculture: A Review. *Sensors* **2018**, *18*, 2674. [CrossRef] [PubMed]
63. Elavarasan, D.; Vincent, D.R.; Sharma, V.; Zomaya, A.Y.; Srinivasan, K. Forecasting Yield by Integrating Agrarian Factors and Machine Learning Models: A Survey. *Comput. Electron. Agric.* **2018**, *155*, 257–282. [CrossRef]
64. Chlingaryan, A.; Sukkarieh, S.; Whelan, B. Machine Learning Approaches for Crop Yield Prediction and Nitrogen Status Estimation in Precision Agriculture: A Review. *Comput. Electron. Agric.* **2018**, *151*, 61–69. [CrossRef]
65. Alvares, C.A.; Stape, J.L.; Sentelhas, P.C.; de Moraes Gonçalves, J.L.; Sparovek, G. Köppen's Climate Classification Map for Brazil. *Meteorol. Z.* **2013**, *22*, 711–728. [CrossRef] [PubMed]
66. Soil Survey Staff. *Keys to Soil Taxonomy*, 12th ed.; USDA-Natural Resources Conservation Service: Washington, DC, USA, 2014.
67. Jenks, G.F. The Data Model Concept in Statistical Mapping. *Int. Yearb. Cartogr.* **1967**, *7*, 186–190.
68. Rabosky, D.L.; Grundler, M.; Anderson, C.; Title, P.; Shi, J.J.; Brown, J.W.; Huang, H.; Larson, J.G. BAMM Tools: An R Package for the Analysis of Evolutionary Dynamics on Phylogenetic Trees. *Methods Ecol. Evol.* **2014**, *5*, 701–707. [CrossRef]
69. Tedesco, J.M.; Gianello, C.; Bissani, C.A.; Bohnem, H.; Volkweiss, S.J. *Análise de Solo, Plantas e Outros Materiais*, 2nd ed.; Universidade Federal do Rio Grande do Sul: Porto Alegre, Brazil, 1995.
70. Teixeira, P.C.; Donagemma, G.K.; Fontana, A.; Teixeira, W.G. *Manual de Métodos de Análise de Solo*, 3rd ed.; Embrapa: Brasília, Brazil, 2017; ISBN 978-85-7035-771-7.
71. Moritz, S.; Bartz-Beielstein, T. imputeTS: Time Series Missing Value Imputation in R. *R J.* **2017**, *9*, 207. [CrossRef]
72. Wulff, J.; Ejlskov, L. Multiple Imputation by Chained Equations in Praxis: Guidelines and Review. *Electron. J. Bus. Res. Methods* **2017**, *15*, 2017–2058.
73. Wold, S.; Sjöström, M.; Eriksson, L. PLS-Regression: A Basic Tool of Chemometrics. *Chemom. Intell. Lab. Syst.* **2001**, *58*, 109–130. [CrossRef]
74. Quinlan, J.R. Combining Instance-Based and Model-Based Learning. In Proceedings of the Tenth International Conference on International Conference on Machine Learning, Amherst, MA, USA, 27–29 July 1993; pp. 236–243, ISBN 978-1-55860-307-3.
75. Quinlan, J.R. *Learning with Continuous Classes*; World Scientific: Sinagpore, 1992; Volume 92, pp. 343–348.
76. Breiman, L. Random Forests. *Mach. Learn.* **2001**, *45*, 5–32. [CrossRef]
77. Chai, T.; Draxler, R.R. Root Mean Square Error (RMSE) or Mean Absolute Error (MAE)?—Arguments against Avoiding RMSE in the Literature. *Geosci. Model Dev.* **2014**, *7*, 1247–1250. [CrossRef]
78. Kuhn, M. Building Predictive Models in *R* Using the Caret Package. *J. Stat. Soft.* **2008**, *28*, 1–26. [CrossRef]
79. R Core Team. *R: A Language and Environment for Statistical Computing*; R Foundation for Statistical Computing: Vienna, Austria, 2023.
80. RStudio Team. *RStudio: Integrated Development for R*; PBC: Boston, MA, USA, 2023.
81. Kuhn, M.; Johnson, K. *Applied Predictive Modeling*; Springer: New York, NY, USA, 2013; ISBN 978-1-4614-6848-6.
82. Reis Pereira, M.; Ribeiro, H.; Abreu, I.; Eiras-Dias, J.; Mota, T.; Cunha, M. Predicting the Flowering Date of Portuguese Grapevine Varieties Using Temperature-Based Phenological Models: A Multi-Site Approach. *J. Agric. Sci.* **2018**, *156*, 865–876. [CrossRef]
83. Santos, J.A.; Grätsch, S.D.; Karremann, M.K.; Jones, G.V.; Pinto, J.G. Ensemble Projections for Wine Production in the Douro Valley of Portugal. *Clim. Chang.* **2013**, *117*, 211–225. [CrossRef]
84. Rafael, A.; Souza, P.V.D.D. Conteúdo de Reservas, Vigor Vegetativo e Rendimento de Videiras Submetidas a Duas Safras Por Ciclo Vegetativo. *Sem. Ci. Agr.* **2015**, *36*, 719. [CrossRef]
85. Aieb, A.; Madani, K.; Scarpa, M.; Bonacorso, B.; Lefsih, K. A New Approach for Processing Climate Missing Databases Applied to Daily Rainfall Data in Soummam Watershed, Algeria. *Heliyon* **2019**, *5*, e01247. [CrossRef] [PubMed]

86. Holzapfel, B.P. Seasonal Vine Nutrient Dynamics and Distribution of Shiraz Grapevines. *OENO One* **2019**, *53*, 363–372. [CrossRef]
87. Sparks, D.L. *Environmental Soil Chemistry*, 2nd ed.; Academic Press: San Diego, CA, USA, 2003; ISBN 978-0-12-656446-4.
88. Brunetto, G.; Ferreira, P.A.A.; Melo, G.W.; Ceretta, C.A.; Toselli, M. Heavy Metals in Vineyards and Orchard Soils. *Rev. Bras. De Frutic.* **2017**, *39*, 1–12. [CrossRef]
89. Brunetto, G.; de Melo, G.W.B.; Terzano, R.; Del Buono, D.; Astolfi, S.; Tomasi, N.; Pii, Y.; Mimmo, T.; Cesco, S. Copper Accumulation in Vineyard Soils: Rhizosphere Processes and Agronomic Practices to Limit Its Toxicity. *Chemosphere* **2016**, *162*, 293–307. [CrossRef]
90. Brunetto, G.; Miotto, A.; Ceretta, C.A.; Schmitt, D.E.; Heinzen, J.; de Moraes, M.P.; Canton, L.; Tiecher, T.L.; Comin, J.J.; Girotto, E. Mobility of Copper and Zinc Fractions in Fungicide-Amended Vineyard Sandy Soils. *Arch. Agron. Soil Sci.* **2014**, *60*, 609–624. [CrossRef]
91. Rogiers, S.Y.; Coetzee, Z.A.; Walker, R.R.; Deloire, A.; Tyerman, S.D. Potassium in the Grape (*Vitis vinifera* L.) Berry: Transport and Function. *Front. Plant Sci.* **2017**, *8*, 01629. [CrossRef] [PubMed]
92. Mpelasoka, B.S.; Schachtman, D.P.; Treeby, M.T.; Thomas, M.R. A Review of Potassium Nutrition in Grapevines with Special Emphasis on Berry Accumulation. *Aust. J. Grape Wine Res.* **2003**, *9*, 154–168. [CrossRef]
93. Alloway, B.J. Micronutrients and Crop Production: An Introduction. In *Micronutrient Deficiencies in Global Crop Production*; Alloway, B.J., Ed.; Springer: Dordrecht, The Netherlands, 2008; pp. 1–39. ISBN 978-1-4020-6860-7.
94. Ghorbani, P.; Eshghi, S.; Ershadi, A.; Shekafandeh, A.; Razzaghi, F. The Possible Role of Foliar Application of Manganese Sulfate on Mitigating Adverse Effects of Water Stress in Grapevine. *Commun. Soil Sci. Plant Anal.* **2019**, *50*, 1550–1562. [CrossRef]
95. Gatti, M.; Garavani, A.; Squeri, C.; Diti, I.; De Monte, A.; Scotti, C.; Poni, S. Effects of Intra-Vineyard Variability and Soil Heterogeneity on Vine Performance, Dry Matter and Nutrient Partitioning. *Precis. Agric* **2022**, *23*, 150–177. [CrossRef]
96. Paccioretti, P.; Bruno, C.; Gianinni Kurina, F.; Córdoba, M.; Bullock, D.S.; Balzarini, M. Statistical Models of Yield in On-farm Precision Experimentation. *Agron. J.* **2021**, *113*, 4916–4929. [CrossRef]
97. Grund, S.; Lüdtke, O.; Robitzsch, A. Multiple Imputation of Missing Data in Multilevel Models with the R Package Mdmb: A Flexible Sequential Modeling Approach. *Behav. Res. Methods* **2021**, *53*, 2631–2649. [CrossRef] [PubMed]
98. Gómez-Lagos, J.E.; González-Araya, M.C.; Ortega Blu, R.; Acosta Espejo, L.G. A New Method Based on Machine Learning to Forecast Fruit Yield Using Spectrometric Data: Analysis in a Fruit Supply Chain Context. *Precis. Agric* **2023**, *24*, 326–352. [CrossRef]
99. Melo, G.W.; Rozane, D.E.; Brunetto, G.; Lattuada, D.S.; Brasileira, E.; Agropecua, D.P.; Gonçalves, B.; Grande, R. Discriminant Analysis in the Selection of Groups of Peach Cultivars. *Acta Hortic.* **2018**, *1217*, 335–342. [CrossRef]
100. Parent, S.É.; Parent, L.E.; Rozane, D.E.; Natale, W. Plant Ionome Diagnosis Using Sound Balances: Case Study with Mango (*Mangifera indica*). *Front. Plant Sci.* **2013**, *4*, 00449. [CrossRef]
101. Andrade, C.B.; Comin, J.J.; Moura-Bueno, J.M.; Brunetto, G. Obtaining Reference Values for Nutrients in Vineyard Soils through Boundary Line Approach Using Bayesian Segmented Quantile Regression on Commercial Farm Data. *Eur. J. Agron.* **2023**, *150*, 126928. [CrossRef]

horticulturae

MDPI

Article

Soil Abandonment as a Trigger for Changes in Zn Fractionation in Afforested Former Vineyard Acidic Soils

Paula Pérez-Rodríguez [1,2,*], Juan Carlos Nóvoa-Muñoz [1,2], Manuel Arias-Estévez [1,2] and David Fernández-Calviño [1,2]

[1] Departamento de Bioloxía Vexetal e Ciencia do Solo, Facultade de Ciencias, Universidade de Vigo, As Lagoas s/n, 32004 Ourense, Spain; edjuanca@uvigo.es (J.C.N.-M.); mastevez@uvigo.es (M.A.-E.); davidfc@uvigo.es (D.F.-C.)

[2] Instituto de Agroecoloxía e Alimentación (IAA), Universidade de Vigo, Campus Auga, 32004 Ourense, Spain

* Correspondence: paulaperezr@uvigo.es

Abstract: Zinc is an essential element for plant nutrition, but it may cause toxicity depending on its bioavailability and potential transformation in soil. In vineyard soils, high concentrations of Zn are usually found, mainly due to agricultural practices. However, a great abandonment of vineyards has recently occurred, leading to changes in the total and bioavailable Zn concentrations, as well as Zn fractionation. We analyzed Zn concentrations (total, Zn_T, and bioavailable, Zn_{ED}) and fractionation in the soil of three paired sites (PM, PT, and AR) up to depths of 50 cm in active and adjacent abandoned vineyards that were already transformed into forests. The Zn_T averaged at 210 mg kg^{-1} among all studied vineyards. The results showed changes in the vertical pattern Zn_T concentrations after vineyard abandonment at the PM and PT sites, while at the AR site, no great variation occurred. The Zn_{ED} (mean values = 7 mg kg^{-1}) decreased after abandonment at PM and AR in the uppermost surface layers, while it increased in the top 10 cm at the PT site, reaching up to 60 mg kg^{-1}. Regarding Zn fractionation in active vineyards, the residual fraction (Zn_R) was the most abundant, followed by Zn bound to crystalline Fe and Al oxy-hydroxides (Zn_C) and Zn bound to soil organic matter (Zn_{OM}). After abandonment, the Zn_R slightly increased and the Zn_C slightly decreased at the PM and AR sites at all depths, while the Zn_{OM} showed a noticeable variation in the uppermost 10 cm of the PT site. These results suggest that the soil organic matter that is provided during afforestation may play an important role in Zn fractionation and mobilization, depending on its humification degree and chemical stability. Zn mobilization could result in a positive nutrient supply for plants, but caution must be taken, since an excess of Zn could cause toxicity in long-term abandoned vineyards.

Keywords: land-use change; afforestation; forest soils; zinc bioavailability; zinc mobilization; zinc–organic matter association

1. Introduction

Zinc (Zn) is an essential micronutrient for plants that plays a critical role in various physiological processes, including photosynthesis, hormone synthesis, and enzyme activation. Zinc occurs in natural soils at concentrations ranging from 40 to 120 mg kg^{-1} depending on the soil parental material [1], but Zn concentrations are usually higher in agricultural soils due to the application of fertilizers, liming materials, manure, and, especially, fungicides containing Zn [2,3]. Zinc may accumulate in soils since it is easily adsorbed by minerals and organic compounds [1,4,5]. A zinc excess may result in toxicity for plants, leading to reduced growth and photosynthetic and respiratory rates, as well as imbalanced mineral nutrition and an enhanced generation of reactive oxygen species [6]. It also may affect soil organisms, leading them to develop tolerance when soil concentrations are higher than 125 mg kg^{-1} in acid soils [7]. The availability and mobility of Zn in soil are influenced by its chemical forms or fractions, which can vary depending on soil properties and management practices [8–11]. A threshold of 200 mg kg^{-1} of total Zn in soil was established by the European Union (EU) to consider whether agricultural soil is safe or not for food production [12], while concentrations of available Zn that are higher than 15 mg kg^{-1} are considered to be phytotoxic [1,13]. In vineyards, a Zn excess may decrease pigments and photosynthetic efficiency, and diminish leaves' activity, chlorosis, and even necrosis, among others [6].

The total Zn concentrations in vineyard soils from the NW of the Iberian Peninsula range from 100 to 170 mg kg^{-1} in coastal areas, and from 60 to 149 mg kg^{-1} in vineyards from inland areas, while the values for available Zn (Zn extracted with EDTA or DTPA) varied from 0.8 to 25 mg kg^{-1} [13,14]. In other areas of the Iberian Peninsula, Zn concentrations in vineyards ranged between 16 and 154 mg kg^{-1} [15]. At a global scale, total Zn was found in vineyard soils from Brazil, Slovenia, Greece, and Iran at high values ranging from 58 to 197 mg kg^{-1} [16–19]. On the other hand, low concentrations of total Zn were found in vineyard soils of Brazil, showing severe deficiency, while low to adequate concentrations of available Zn were found in Italy, Brazil, or Syria [20–22].

The availability and mobility of Zn in soils are influenced by its chemical forms and/or fractions, which can vary depending on soil properties and management practices [8]. Organic matter, cation exchange capacity (CEC), and pH are known to play important roles in Zn retention on soil constituents by increasing its adsorption, and thus decreasing Zn mobility [8,23]. However, when Zn concentrations are in exceedance of the soil maximum adsorption capacity, Zn bonded with lower energy may be mobilized through the soil profile and water bodies [20]. In acidic soils, Zn tends to be strongly bound to soil components such as Fe and Al oxyhydroxides, limiting its uptake by plants [1,24]. However, land-use changes may modify the distribution of Zn in the soil's solid phase, influencing its availability and mobility in soil. The abandonment of agricultural soils, and the subsequent natural revegetation recovery, is a widespread phenomenon worldwide, but it is especially common in mountainous regions with steep slopes. Land abandonment occurred in Southern Europe at a proportion of around 25% up to 2011 [25], and about 11% of agricultural land in the EU is at a high potential risk of abandonment in the period of 2015–2030 [26]. It is expected that changes in the management of agricultural soil can affect its physicochemical properties and nutrient dynamics, influencing the availability and mobility of essential and non-essential elements such as Zn, Cu, Cd, and Pb. In the case of Cd and Pb, the time since vineyard abandonment was found to influence their bioavailable content, which increased as the time since the abandonment increased [27]. On the contrary, Cu concentrations were lower in an abandoned vineyard soil compared to an adjacent active vineyard, and even its fractionation changed towards a predominance of less mobile fractions [28]. Moreover, abandoned vineyards may experience changes in soil properties due to the cessation of management practices, potentially affecting Zn availability and speciation in soil [27,29].

Whilst several studies have determined the distribution of Zn in vineyard soils [5,16–18,30,31], none of these studies included abandoned vineyard soils. The effects that land-use changes can promote in the geochemical behavior of Zn deserve a thorough investigation in the case of abandoned vineyard soils, providing knowledge that could minimize potential negative environmental consequences derived from the cease of agricultural activities. Therefore, this study aimed to assess the distribution of Zn fractions in high-resolution sampled active vineyards (ACs) and their adjacent abandoned vineyards (ABs) in which a deciduous forest had grown, and how they changed because of the land-use changes. This will be analyzed from the perspective that Zn fractionation would be different between active and abandoned vineyards, this being a consequence of the changes in the soil properties and its management. The expected findings will provide insights into the effects of land-use changes on Zn cycling, assisting in future management strategies that promote the sustainable use of abandoned agricultural areas, and maintaining soil ecosystem services and biodiversity conservation.

2. Material and Methods

2.1. Study Area and Soil Sampling

Three sites were chosen for this study, named Portotide-O Mato (PM), Portotide (PT), and A Raña (AR). All these sites belong to the vine-growing area, the Ribeira Sacra Designation of Origin, which is partially spread along the valley of the river Miño (Galicia, NW Spain). Vine cultivation in this area has been practiced continuously since the late 19th century, and it is so-called "heroic viticulture" because it is developed in areas with very steep slopes (>30%) which have been skipped by arranging the land in terraces. Due to the hard work that the management of these vineyards demands, as well as the progressive aging of the local population, a great number of vineyards were abandoned, starting at least 35 years ago. Consequently, abandoned vineyards were progressively afforested with species that are characteristic of deciduous Atlantic forests (*Quercus robur* and *Quercus pyrenaica*), as well as some coniferous individuals of the *Pinus* genus (mostly *Pinus pinaster*). Nowadays, a huge number of old vineyards have become young Atlantic forests with the consequent land-use change.

The study area has a characteristic oceanic–Mediterranean transition climate, with 15 °C as the mean annual temperature and a rainfall depth ranging from 800 to 1300 mm year^{-1}. The local lithology is mostly dominated by schist (such as in the PT and PM sites) with small patches of granite and gneissic rocks (as in the AR site). The climatological conditions, also influenced by two water reservoirs located in the surroundings of the study area which provide a high degree of humidity during the vine-growing season, meant that the use of agrochemicals were widely applied for pest management.

The high-resolution soil sampling strategy was described in detail by Vázquez-Blanco et al. (2022) [28]. In brief, in each of the three sites, an active vineyard (AC) and an adjacent afforested abandoned vineyard (AB) were sampled up to a depth of 50 cm. Composite soil samples (made up of 5 subsamples) were obtained for each site and land-use type (active vs. abandoned) every 2 cm in the uppermost 10 cm of the mineral soil, every 5 cm between the 10 and 20 cm depths, and every 10 cm from 20 to 50 cm. This sampling strategy resulted in a total of 10 samples per soil type and land-use type. A composite soil sample, corresponding to the 0–20 cm depth range, was collected per site and land-use type to determine their general physicochemical characteristics. Before analyses, the soil samples were air-dried, sieved with 2 mm sized mesh, and homogenized.

2.2. Soil General Characterization

The general characterization of soil samples from the active and abandoned vineyard soils were previously shown by Vázquez-Blanco et al. (2022) [28] and included the pH in distilled water (pHw) and saline solution (pHk), the total contents of organic C and N, the exchangeable base cations (Na^+, K^+, Ca^{+2}, and Mg^{+2}) displaced with 1M NH_4Cl and exchangeable Al (with 1M KCl). The sum of exchangeable base cations and Al displaced with KCl was considered as an estimation of the effective cation exchange capacity (eCEC). In brief, active vineyard soils (ACs) are slightly acidic (pHw ranged from 5.6 to 6.0), with abandoned vineyards (ABs) being somewhat more acidic (pHw ranged from 4.6 to 5.3). The eCEC is quite similar in active and abandoned vineyards (range of 8.5 to 16.8 $cmol_+$ kg^{-1}), with Ca being the predominant exchangeable cation in all of them. The total organic carbon ranged from 15 to 24 g kg^{-1} in active vineyards (ACs), and ranged from 36 to 71 g kg^{-1} in the abandoned vineyards (ABs), whereas the ranges for the total N were 1.1–2 and 2.2–3.5 g kg^{-1} for ACs and ABs, respectively. The total organic C ($Z = -4.299$, $p = 0.000$, $N = 30$) and N ($Z = -2.251$, $p = 0.024$, $N = 30$), as well as the C/N ratio ($Z = -4.351$, $p = 0.000$, $N = 30$), were significantly higher in abandoned than in active vineyards. The soil textures were sandy clay loam in the PM and PT sites, and sandy loam in the AR site. Detailed information on the soil chemical characteristics is provided as a Supplementary Material (Table S1).

2.3. Total and Potentially Available Zn

Total Zn (Zn_T) was extracted from the soil by digesting 0.5 g of soil with 5 mL of HNO_3, 4 mL of HF, and 1 mL of HCl in a microwave oven at 100 psi. For quality control purposes, two commercially available certified materials (i.e., soil, BCR 142R, and sediment, BCR 227R) were digested in triplicate following the same method. The Zn recoveries for both materials were consistent with the certified values (94.2 ± 2.8 and 183.2 ± 22.6 mg kg^{-1}, respectively). Potentially available Zn (Zn_{ED}) was extracted from the soil by shaking 5 g of soil in 50 mL of a solution containing 0.02M Na_2-EDTA and 0.5M NH_4OAc at pH 4.65 for 1 h [32].

2.4. Zinc Fractionation

Zinc fractions were obtained following a non-sequential procedure [33]. Selective extractants were used to obtain operatively defined metal fractions in which Zn was bounded to different soil components, reporting their potential mobility in soil. So, Zn was measured in a soil extract derived from the application of 1M NH4Ac (pH 7) (Zn_ac), 0.1M Na-pyrophosphate (Zn_p), 0.2M oxalic acid–ammonium oxalate (pH 3) (Zn_o), and 0.2M oxalic acid–ammonium oxalate–ascorbic acid (pH 3.25) (Zn_ao). Consequently, the following operative fractions were obtained according to Zn bioavailability: (i) the exchangeable Zn fraction (Zn_{EX}), which is equivalent to that measured in Zn_ac extraction; (ii) the metal fraction bound to the soil organic matter (Zn_{OM}), resulting from Zn_p–Zn_ac; (iii) the metal bound to non-crystalline Fe and Al oxy-hydroxides (Zn_{IA}) obtained from Zno–Znp; (iv) the metal bound to crystalline Fe and Al oxy-hydroxides (Zn_C) after Zn_ao–Zn_o; and, finally, (v) the residual fraction of metal (Zn_R) obtained from Zn_T–Zn_ao. Zn bioavailability is expected to decrease in the sequence of $Zn_{EX} > Zn_{OM} > Zn_{IA} > Zn_C > Zn_R$.

All exchangeable cations (basic and acidic) as well as the Zn contents in the different extractions were determined via atomic emission spectrometry (Na and K) and atomic absorption spectrometry (Ca, Mg, Al, and Zn) using a Thermo Solaar M Series spectrophotometer (Thermo Fisher Scientific Inc., Franklin, MA, USA). Analyses of the total Zn and in the different extractions of Zn fractionation were performed in duplicate for 10% of samples, showing a variation coefficient below 5%.

2.5. Statistical Analysis

Non-parametric paired tests for 2 related samples were performed after significant results of Levene's test, meaning that the data obtained were non-normal and non-homogeneous in their variances. Tests using all data ($N = 60$) were performed to compare the results from the active and abandoned vineyard soils (factor: land use) with the soil characteristics, while tests per site ($N = 20$) were performed to compare the results on the influence of land-use changes for Zn fractionation parameters, since the total Zn levels were very different at each site.

Spearman's correlation analyses were used to test the correlation between the Zn_{ED} and the total organic C (TOC) from the active and abandoned vineyard soils. All statistical analyses were conducted using IBM SPSS Statistics 25.

3. Results

3.1. Total and Potentially Available Zn in Depth

The total Zn (Zn_T) concentration in the 0–20 cm soil layer was >100 mg kg^{-1} in all sites, independently of land use (Figure 1A,B). The zinc contents under both land-use types were higher in the PM and PT sites (ranging from 250 to 350 mg kg^{-1}) than in the AR site where the Zn_T varied between 120 and 140 mg kg^{-1}. The total Zn concentration was slightly higher in the abandoned vineyards of the PM and PT sites, although differences regarding land use were significant only in the PM site ($Z = -1.988$, $p = 0.047$, $N = 20$). On the contrary, the Zn_T levels in the AR site were slightly higher in the active than in the abandoned vineyard, but without a significant effect of land-use change. The highest Zn_T concentration was found in the uppermost soil layer (0–2 cm) of the PT–AC vineyard, where it reached 353 mg kg^{-1}. In the active vineyards, the Zn_T contents were generally higher in the uppermost 10 cm, and then they gently decreased (as for PT–AC) or remained almost constant between 10 and 50 cm, as occurred in PM–AC and AR–AC (Figure 1A). In the first 10 cm of PT–AC, the Zn_T showed two peaks at the soil layers of 0–2 and 8–10 cm. Regarding the abandoned vineyards, AR–AB showed a very narrow range of Zn_T variation through the 50 cm of analyzed soil (118–130 mg kg^{-1}; Figure 1B), resulting in a vertical pattern that was quite similar to the corresponding active vineyard of that site. The variation in the Zn_T with the soil depth in the PT–AB and PM–AB vineyards was somewhat different compared to their active ones. The PT–AB vineyard also showed two subsurface peaks for the Zn_T, but they occurred at different depths than in active vineyard (at 4–6 and 10–15 cm with 300 and 305 mg kg^{-1}, respectively). Below this secondary peak, the Zn_T decreased progressively with the soil depth until reaching the lowest values of this site in the deepest soil layer (219 mg kg^{-1}). In the abandoned vineyard of the PM site, the lowest concentration of Zn_T (217 mg kg^{-1}) was found at the surface soil layer (0–2 cm), contrary to that in the active vineyard. The vertical pattern of Zn_T in PM–AB was more irregular than in PM–AC, with a double peak of Zn_T in the subsurface layers (4–6 and 15–20 cm), but they did not coincide with those for PT–AB. Below 30 cm, the Zn_T in PM–AB scarcely varied, and it was similar to those for PT–AB (240 mg kg^{-1}).

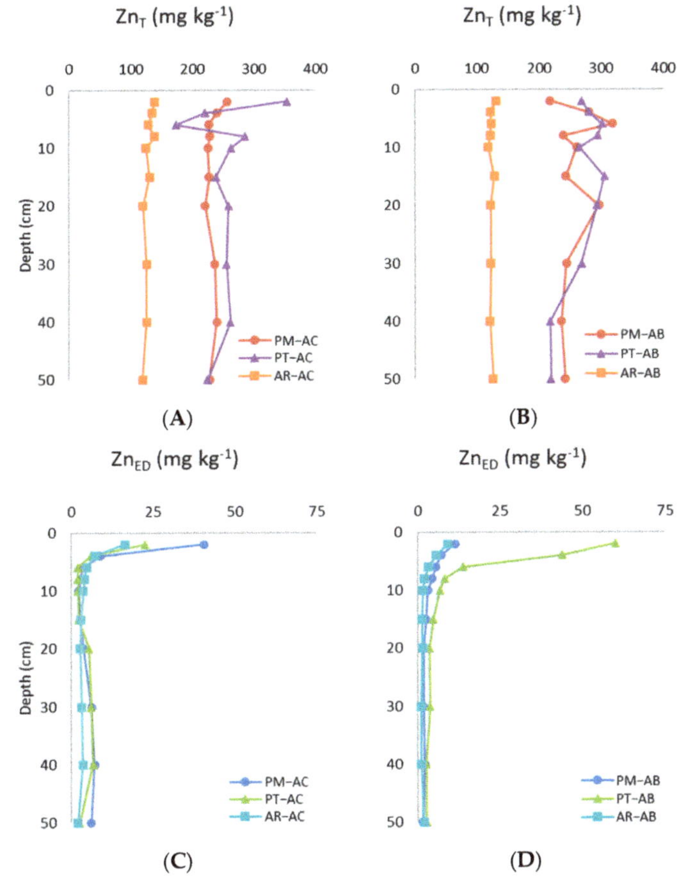

Figure 1. Distribution of total Zn (Zn_T) and potentially available Zn (Zn_{ED}) in active (**A**,**B**) and abandoned (**C**,**D**) vineyard plots.

The potentially bioavailable Zn (Zn_{ED}) in the active vineyards (Figure 1C) ranged from 2 to 40 mg kg^{-1}, following the sequence of PM–AC > PT–AC > AR–AC. It was noticeable that the site with the highest concentration of Zn_T (PT–AC) was not the site with the highest concentration of Zn_{ED}. In the abandoned vineyards (Figure 1D), the Zn_{ED} ranged from 0.6 to 60 mg kg^{-1}, following the order of PT–AB >> PM–AB ≈ AR–AB. The active vineyards showed greater values of Zn_{ED} than the abandoned vineyards in the AR and PM sites, although the land-use change was a significant factor of variation for the Zn_{ED} only for the AR site ($Z = -2.803$, $p = 0.005$, $N = 20$). The contrary occurred in the uppermost 15 cm of the soils from the PT site, where the Zn_{ED} was higher in the abandoned than in the active vineyard. Regarding the vertical variation in the Zn_{ED}, all sites (including active and abandoned vineyards) showed almost the same pattern, and only minor differences were found. In the active vineyards (Figure 1C), the Zn_{ED} was higher in all sites at the uppermost soil layers (0–2 cm), where it ranged between 16 and 40 mg kg^{-1}. Below 2 cm, the Zn_{ED} decreased considerably up to a 6 cm depth, and then it remained relatively stable throughout the soil or even showed a slight increase between the 20 and 50 cm depths, but quite far from the values reached at the surface layers. Below the 6 cm depth, the Zn_{ED} concentrations ranged between 0.9 and 8.0 mg kg^{-1}. In the abandoned vineyards, the highest values of Zn_{ED} were also observed in the uppermost soil layer (0–2 cm), with values between 9 and 60 mg kg^{-1} (Figure 1D). The values of Zn_{ED} also decreased considerably in the upper 10 cm of the soil, but this was much less pronounced than in the active vineyards, except for PT–AB, whose Zn_{ED} concentration decreased up to 10 times in the first 10 cm. As for the active vineyards, the levels of Zn_{ED} in the abandoned vineyards showed few changes at depths below 20 cm (3.3 to 1.3 mg kg^{-1}), but they showed a trend at diminished progressively toward the deepest soil layer.

The percentage of available Zn with regard to the total Zn (Zn_{ED}/Zn_T ratio) gives insights into the amount of potentially bioavailable Zn present in the studied soils. The values of the Zn_{ED}/Zn_T ratio were generally higher in the active than in the abandoned vineyards in the PM and AR sites, particularly in the layers corresponding to the uppermost 6 cm. Below this depth, the proportion of available Zn regarding the total Zn was below 3%. On the contrary, in the PT site, the Zn_{ED}/Zn_T ratio in the uppermost 6 cm of the soil was quite different between the abandoned and active vineyards, being 3–4 times higher in the former (Table 1).

Table 1. Comparison of Zn_{ED}/Zn_T ratios (expressed as percentages) at the different studied sites.

Depth (cm)	PM–AC	PM–AB	PT–AC	PT–AB	AR–AC	AR–AB
0–2	16	5	6	22	12	7
2–4	4	3	3	16	5	5
4–6	2	2	1	5	4	3
6–8	1	2	1	3	3	2
8–10	1	1	1	2	3	1
10–15	1	1	1	1	2	1
15–20	2	1	2	1	2	1
20–30	3	1	2	1	2	1
30–40	3	1	3	1	3	1
40–50	3	1	1	1	2	1

Surnames –AC and –AB indicate active and abandoned vineyards, respectively.

3.2. Variations of Zn Fractionation with Soil Depth and Land Use

The variations in the different Zn fractions with the soil depth is shown in Figure 2. The expected mobility of Zn fractions, from the most to the least, is as follows: $Zn_{EX} > Zn_{OM} > Zn_{IA} > Zn_C > Zn_R$. In the active vineyards, the Zn_{EX} (Figure 2A) ranged from 0.1 to 1.6 mg kg^{-1}, showing a very low variation with soil depth in AR–AC, while in the PM and PT sites, the Zn_{EX} showed an irregular vertical pattern, with high values occurring equally in upper and deeper soil layers. In summary, the exchangeable fraction did not exceed 1% of total Zn in all active vineyards (Figure 3). In the abandoned vineyards, the Zn_{EX} (Figure 2F) ranged from 0.1 to 3.4 mg kg^{-1}, showing the same trend with soil depth as the active vineyards in the PM and AR sites. In PT–AB, the Zn_{EX} was higher compared to the active vineyard, especially in the uppermost 20 cm, reaching up to 3.4 mg kg^{-1} and showing a clear pattern of the Zn_{EX} decreasing with depth (Figure 2F). In general, the Zn_{EX} was higher in the active than in the abandoned vineyards of the PM and AR sites, although the differences according to the land-use changes were only significant in the AR site ($Z = -2.805$, $p = 0.005$, $N = 20$). The percentages of Zn_{EX} regarding the total Zn were low in all former vineyards at any depth (Figure 3D–F), being only slightly above 1% of the total Zn in the uppermost 4 cm of the PT–AB soil (Figure 3E). As well as for the absolute values, the change in land use only brought about significantly different values for the percentage of Zn_{EX} at the AR site ($Z = -2.871$, $p = 0.004$, $N = 20$).

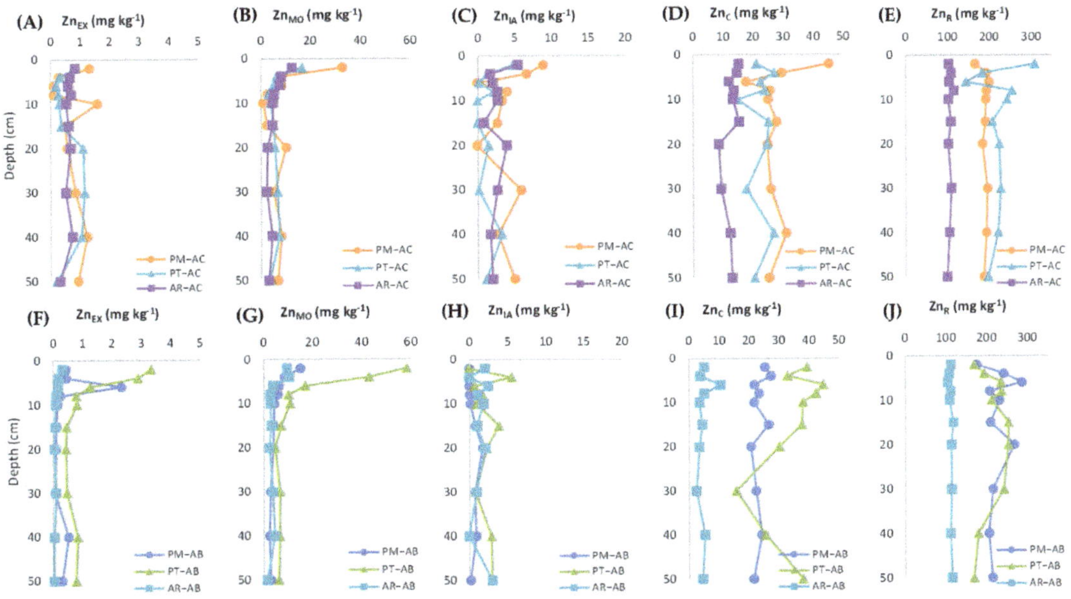

Figure 2. Distribution of exchangeable Zn (Zn_{EX}), Zn associated with soil organic matter (Zn_{OM}), Zn associated with inorganic amorphous oxides (Zn_{IA}), Zn associated with crystalline oxides (Zn_{C}), and residual Zn (Zn_{R}) with depth in active (**A–E**) and abandoned vineyard sites (**F–J**). Note that the *X*-axis has different graduations per fraction.

The Zn associated with the soil organic matter (Zn_{OM}) ranged from 1.2 to 33 mg kg^{-1} in the active vineyards, and from 1.7 to 58.4 mg kg^{-1} in the abandoned vineyards (Figure 2B,G, respectively). The land-use changes only significantly influenced the Zn_{OM} values at the PT site, which increased after abandonment ($Z = -2.312$, $p = 0.021$, $N = 20$). Thus, the values of Zn_{OM} in the upper 10 cm of the PT–AB vineyard were about 3 times higher than those showed in the active vineyard (Figure 2B,G). Contrarily, the Zn_{OM} tended to be lower in the abandoned vineyards compared to the active ones in the PM and AR sites. Regarding its vertical pattern, the Zn_{OM} showed a consistent trend in all sites under both land-use types, being higher in the uppermost surface layers (0–10 cm) and decreasing with the soil depth both in concentration and in percentage (Figure 2B,G; Figure 3A–F). As a percentage of the total Zn, the Zn_{OM} fraction was especially remarkable in the topsoil sample (0–2 cm) of PT–AB, where it became the second largest fraction of Zn accounted for ≈22% of Zn_T after vineyard abandonment and afforestation (Figure 3E). Except for the abovementioned layer and the 0–2 cm layer of the PM–AC soil, where the Zn_{OM} achieves 13% of the total Zn, the Zn fraction bound to the soil organic matter accounted for less than 10% of the total Zn in most of the soil samples (Figure 3).

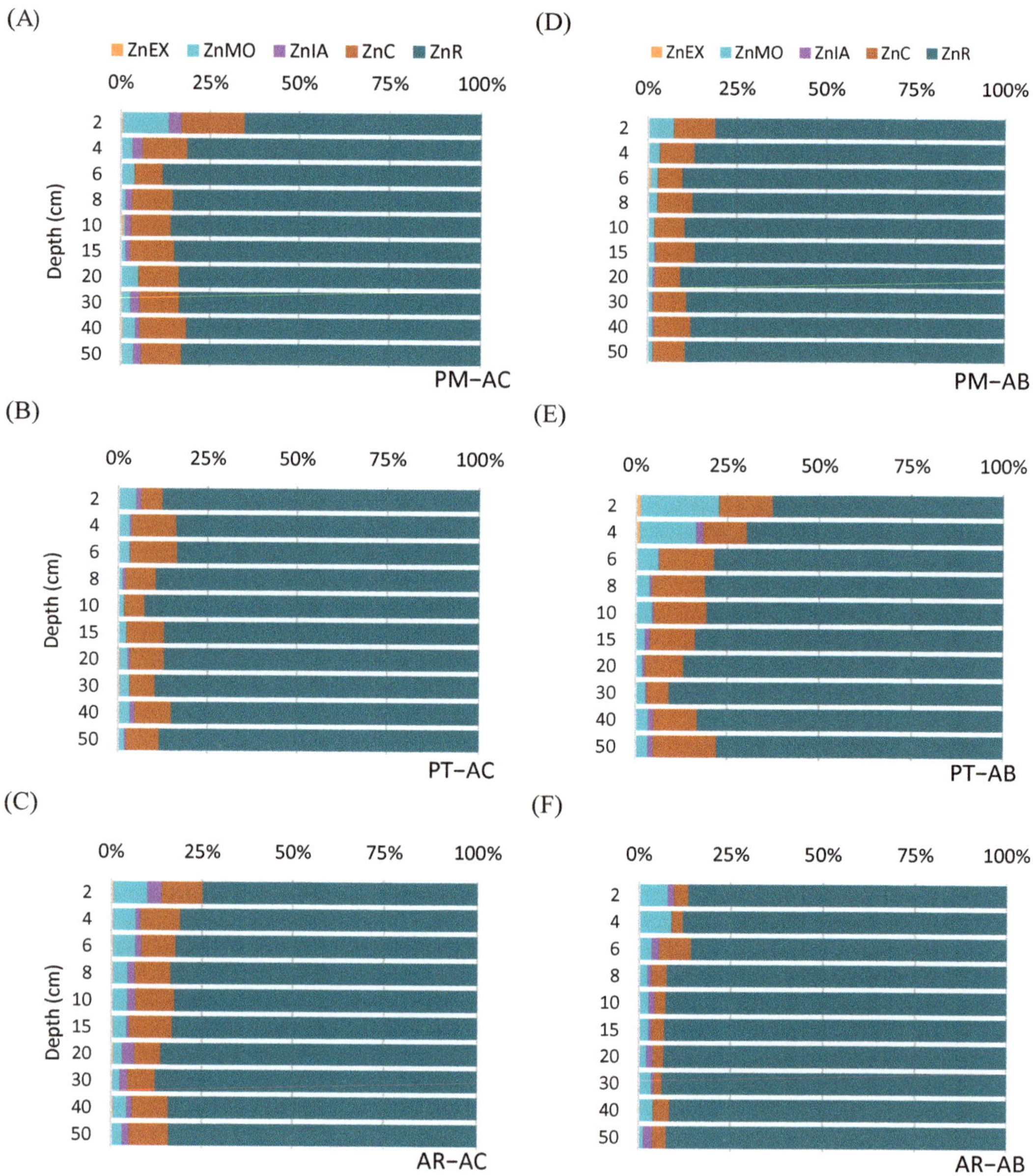

Figure 3. Percentages of exchangeable Zn (Zn_{EX}), Zn associated with soil organic matter (Zn_{OM}), Zn associated with inorganic amorphous oxides (Zn_{IA}), Zn associated with crystalline oxides (Zn_C), and residual Zn (Zn_R) per depth in active (**A–C**) and abandoned vineyard sites (**D–F**).

Regarding the Zn associated with inorganic amorphous oxides (Zn_{IA}), the pattern with the soil depth is quite irregular for the active vineyards in all sites, with values ranging from 0 to 9 mg kg^{-1}, and with the following sequence of abundance: PM–AC > AR–AC ≈ PT–AC (Figure 2C). However, after abandonment, the Zn_{IA} generally decreased throughout the profile, with Zn_{IA} values ranging between 0 and 5.6 mg kg^{-1} (Figure 2H). Vineyard abandonment resulted in statistically significant differences of Zn_{IA} in the PM ($Z = -2.547$, $p = 0.011$, $N = 20$) and AR sites ($Z = -2.091$, $p = 0.037$, $N =$

20), with greater values in the active vineyards (Figure 2C,H). The levels of Zn_{IA} were always lower than those of the Zn_{OM}, being almost 4 and 10 times lesser in the active and abandoned vineyards, respectively. As a percentage of the total Zn, the land-use change resulted in a decrease in the fraction of Zn associated with inorganic amorphous oxides, with values below 2% in the abandoned vineyards and at about 4% in the active ones (Figure 3A–F). The percentages of Zn_{IA} regarding the total Zn were significantly lower in the abandoned than in the active vineyards in the PM ($Z = -2.547$, $p = 0.011$, $N = 20$) and AR ($Z = -2.143$, $p = 0.032$, $N = 20$) sites, a fact that was particularly reflected in the superficial layers (0–4 cm).

In general, the Zn associated with crystalline Al and Fe oxyhydroxides (Zn_C) is the second fraction of Zn in order of abundance, ranging from 3 to 46 mg kg^{-1} in all analyzed samples (Figure 2D,I), except for the uppermost samples (0–4 cm) of the PT–AB vineyard, where this fraction is surpassed by the Zn_{OM} (Figure 2B). As regards its vertical pattern, the Zn_C also showed an irregular trend with the soil depth. For instance, a maximum value (45 mg kg^{-1}) was registered at the uppermost soil layer (0–2 cm) of the PM–AC soil, whilst in the rest of the profile, the Zn_C values remained at about 25 mg kg^{-1} (Figure 2D), whereas in the PT–AC and AR–AC vineyards, the Zn_C showed a saw-toothed pattern ranging from 10 to 15 mg kg^{-1} (Figure 2D). The abandonment of vineyards showed two different behaviors depending on the site. In the PT vineyard, the Zn_C concentrations were significantly higher in the abandoned vineyard compared to the active one ($Z = -2.397$, $p = 0.017$, $N = 20$). Contrarily, the active vineyards showed greater values of Zn_C than the abandoned vineyards in the PM ($Z = -2.803$, $p = 0.005$, $N = 20$) and AR ($Z = -2.805$, $p = 0.005$, $N = 20$) sites (Figure 2D,I). As a percentage of the total Zn, the Zn_C fraction ranged from 2 to 18%, depending on the site and soil depth (Figure 3A–F), being significantly affected by the land-use changes in the three sites ($Z = -2.397$, $p = 0.017$, $N = 20$ in PT; $Z = -2.547$, $p = 0.011$, $N = 20$ in PM; and $Z = -2.143$, $p = 0.032$, $N = 20$ in AR).

The residual Zn (Zn_R), i.e., the least mobile fraction, ranged from 100 to 309 mg kg^{-1} and showed quite a homogeneous vertical pattern in the active vineyard soils in PM–AC and AR–AC, whereas a considerable diminution in the Zn_R values occurred in the PT–AC soil between 2 and 6 cm (Figure 2E). The lowest values of Zn_R were found in the AR–AC soil, with values of around 100 mg kg^{-1}. In the abandoned vineyards, the values of Zn_R as well as its vertical pattern in AR–AB were similar to its equivalent active vineyard. In the other two abandoned vineyards (PT–AB and PM–AB), the Zn_R showed an even more irregular vertical pattern than their corresponding active vineyards. Thus, the Zn_R tended to increase in the uppermost 10 cm in both soils, with some secondary peaks at 15 and 20 cm depth, and finally, a progressive decline towards the deepest soil layers (Figure 2J). The Zn_R fraction represented more than 75% of the Zn_T in the soil of all vineyards, independently of the land-use type (Figure 3A–F). Only some of the uppermost soil layers (0–2 and 2–4 cm) depart from this general trend, as occurred for PM–AC (Figure 3A) and PT–AB (Figure 3E), where the percentage of Zn_R ranged between 50 and 75% of the total Zn.

4. Discussion

4.1. Total and Available Zn Concentrations

The total Zn concentrations found in the abandoned and active vineyard soils of the PM and PT sites were similar to those reported for the active vineyard soils in the same study area by Campillo-Cora et al. (2019) [34]. Our values were somewhat higher than those reported by Fernández-Calviño et al. (2012) [14] and Wightwick et al. (2008) [35] for different wine-growing areas in NW Spain (60–150 mg kg^{-1}) and Australia (14–161 mg kg^{-1}), respectively. On the other hand, the values of total Zn in the AR site were in the range of the mentioned studies. However, all the values of Zn_T that are reported in the present study were several times higher than the mean value (41 mg kg^{-1}) that was recently reported by Van Eynde et al. (2023) [36] in topsoil across Europe, suggesting the occurrence of an excess of Zn.

The difference in the total Zn between the PM and PT sites compared to the AR site may be due, to some extent, to a distinctive contribution of parent materials, as their characteristics may determine Zn levels in topsoil [37]. Thus, the average geochemical background level of Zn in soils developed from granite in NW Spain was stablished to be 51 mg kg^{-1}, slightly below the level that is considered for schist soils, which is 68 mg kg^{-1} [38]. Therefore, the distinct parent material could partially explain the differences observed in the total Zn among the PM and PT sites regarding the AR site. Despite this, it is necessary to assume that a Zn excess above the geochemical background values in all sites results as a consequence of the agricultural activities that were carried out during vineyard management, and that those should be much more intense in the PT and PM sites than in the AR site. Soil pollution with Zn was reported as consequence of agricultural practices related to fungicide application [39], over-fertilization [3,14,40,41], the addition of organic amendments, and the supply of biosolids containing Zn [42,43]. As the abandoned vineyards were also affected by different agricultural practices before afforestation, they can be also considered as soils with a certain Zn pollution degree. Although the historical application of Zn-based fungicides and fertilizers in the studied soils is unknown, these moderate-to-high values of total Zn could be due to the use of mancozeb and Zn phosphates in the study area.

The occurrence of the highest total Zn in the uppermost centimeters of the active vineyards is consistent with a superficial supply of fertilizers and amendments. The studied active vineyards are composed of narrow terraces (sometimes with space for only one vine row) arranged in a steep slope area, which preclude the use of machinery, and the management practices are restricted to shallow tilling by hand ("heroic viticulture"). Moreover, organic matter is mainly accumulated in the few uppermost centimeters of the vineyard soils [28], and its strong affinity to heavy metals in the surface soil layers of active vineyards creates a favorable geochemical environment for Zn enrichment. Something similar occurred in abandoned plots where, in addition to the legacy of their past management as vineyards, Zn could also reach the uppermost soil layer through litterfall, as it is an essential micronutrient for plants [44]. The additional contribution of Zn from decaying aboveground biomass would explain the higher values of total Zn found in some of the abandoned vineyards than in the active ones (as in the PM and PT sites). This accumulation of Zn in the soil surfaces of abandoned vineyards is consistent with the role of the uppermost organic-matter-rich layers of forest soils as a temporal sink for Zn [45]. The trend in Zn accumulation in the uppermost soil layers observed in the present study agrees with previous research on vineyard soils worldwide [20,46].

The general decreasing trend of the total Zn with the soil depth is expected, considering the strong association of Zn and organic C, and the vertical variation of the latter. Despite this, the observed peaks in the total Zn at subsurface levels in the PT and PM sites can be attributed to the different soil depths that were reached with the agricultural practices, as well as a natural variability in the lithogenic Zn levels. However, this does not preclude that Zn could be mobilized to deeper soil layers, bound to low-molecular-weight organic acids [47,48]. This potential mobility of Zn along the soil profile could help to explain the homogeneous vertical pattern of the total Zn in the vineyards of the AR site, a fact that could be facilitated via the sandy texture of these soils, derived from granitic rocks.

Regarding the potentially bioavailable Zn (Zn_{ED}), the range of values in the present study is wider than that (0.8–11 mg kg^{-1}) reported for vineyards soils in NW Spain by Fernández-Calviño et al. (2012) [14]. Our values are also slightly higher than those reported by Brunetto et al. (2018) [40] in Brazilian vineyard soils, which reached around 40 mg kg^{-1} in the uppermost soil layers, or those recently published for vineyard soils in wine-growing areas of Hungary, which ranged from 1 to 6 mg kg^{-1} [49]. It is important to note that the highest Zn_{ED} concentrations tended to be found where the values of TOC were higher, particularly in the uppermost soil layers of the active and abandoned vineyards (Figure 4). In fact, the TOC and Zn_{ED} were significantly correlated when all samples were considered (rho = 0.276; P = 0.033, n = 60). This relationship is expected because organic matter has a strong affinity for metals and provides available sites for Zn binding in active vineyards, whereas in abandoned vineyards, the accumulation of organic matter derived from litterfall in the uppermost soil layers would also contribute to Zn complexation. When the vertical pattern of the Zn_{ED} is examined in the active and abandoned vineyard soils, the C/N ratio is the parameter that showed a better correlation (rho = 0.485; P = 0.000, N = 60). This suggests that, rather than the quantity of total organic matter, the Zn_{ED} distribution in the active and abandoned vineyards is the most dependent on the humification degree of the soil organic matter. So, despite the total Zn concentration being lower in the abandoned vineyards, it may result in more bioavailability than the active vineyards if the Zn is complexed by organic matter, resulting in Zn mobilization being dependent on soil organic matter (SOM) mineralization. Mineralization is mainly dependent on edaphoclimatic factors, such as temperature, humidity, the abundance of decomposers, and the type of organic matter formed [50]. Considering that fresh litters (in abandoned vineyards) usually contain little N compared to amount that the decomposers need (C/N ratios $\geq$ 20), the mineralization process is slow [51,52] and, therefore, eventually less concerning.

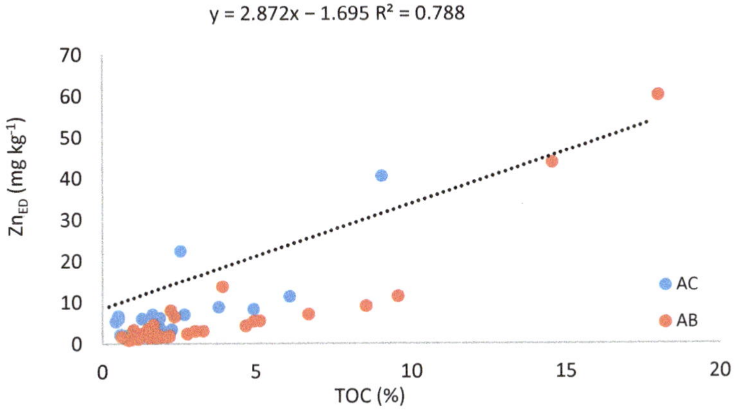

Figure 4. Relationship between the concentration of available Zn (Zn_{ED}) and TOC in active (blue dots) and abandoned (red dots) vineyards.

Additionally, our values of Zn_{ED}/Zn_T, even those of the abandoned vineyards, were lower than those obtained by Brunetto et al. (2018) and Brunetto et al. (2014) [17,40] in Brazilian acid vineyard soils, which ranged from 30 to 52% in the uppermost soil layers. However, our results were higher than those found in neutral to alkaline vineyard soils from the Tokaj region (Hungary) and from alkaline vineyards from NE Spain, which obtained ratios below 7 and 10% in the upper 20 cm [49,53]. Such values mean that a small percentage of the total Zn may be obtained by plants without inducing toxicity, and that the soil's pH is a key parameter in driving Zn bioavailability.

Nevertheless, values of Zn_{ED} deserve special attention and must be examined, especially in the uppermost soil layers of most of the vineyard soils (active and abandoned) that have been studied, since, depending on concentrations, they can exceed the phytotoxic threshold for agricultural crops [1] and may result in environmental concern. Thus, caution must be taken when new agricultural crops might be planted into those sites, as although high concentrations of Zn_{ED} have been previously found in active vineyards exceeding the limits of phytotoxicity [13], they have also found in long-term abandoned vineyards. However, the relatively high levels of Zn_{ED} found in the studied soil could favor the natural instauration of native trees after abandonment, as trees usually require a good supply of Zn for optimal nutrition [54].

4.2. Changes in Zn Fractionation after Abandonment

Generally, >75% of Zn belongs to the residual fraction at all studied depths, except for PT–AB up to the 4 cm depth. This was also found by Beygi and Jalali (2019) [16] in vineyard soils from Iran. Similar results were found in vineyards from France [9], Brazil [5], and from other designations of origin in NW Spain [55]. Zahedifar (2017) and Korchagin et al. (2020) [5,56] found a predominance of the Zn residual fraction in soils under different land uses, including agricultural (cereals and vineyards, respectively) and forest soils, which indicates that, generally, Zn predominates as being geochemically non-mobile, independently of land use, and indicating that its origin is mainly natural. In the present study, the more abundant Zn fractions followed the sequence of $Zn_R > Zn_C > Zn_{OM}$. Generally, a level of $\geq$90% corresponded to the least mobile fractions, i.e., Zn bound to Fe and Al non-crystalline (Zn_{IA}), crystalline oxy-hydroxides (Zn_C), and the residual fraction (Zn_R). The exceptions were found in PM–AC at the upper 2 cm, where the Zn mobile fraction (Zn_{EX} + Zn_{OM}) reached 13%, and in the uppermost layers (0–2 and 2–4 cm) of PT–AB, where the Zn associated with mobile fractions reached 23 and 16%, respectively. The higher concentration of SOM in the uppermost soil layers may explain the differences in Zn fractionation in the surface layers, since SOM can mobilize Zn through the complexation or chelation processes via organic functional groups, improving the humic substances–metal interactions [5].

In the active vineyards of the PM and AR sites, two out of the three most abundant fractions (Zn_R and Zn_C) were barely constant with depth, especially below 10 cm, suggesting that this homogeneity could be consistent with lithology as the main Zn source in these sites and, therefore, showing a lesser influence of the land-use type. Our results hardly differ from those obtained by Korchagin et al. (2020) [5], in which they found a smooth decrease in Zn_R with a pronounced decrease in Zn linked to minerals. However, the Zn fractionation changed after abandonment, since the Zn_C decreased, while the Zn_R slightly increased. This could be explained by the ageing of Zn_C forms and their progressive transformation to more recalcitrant Zn forms that become part of Zn_R, a process that may be more favorable to occur when no Zn addition due to anthropogenic activities takes place [57]. At these sites, the Zn_{OM}, which resulted in being the third most abundant fraction, varied with depth, tending to be higher at the soil's uppermost 4 cm, possibly due to a greater occurrence of organic matter supply from aboveground tree biomass deposition. Similar results were also found by Korchagin et al. (2020) [5] regarding the abundance of this fraction and its importance on surface layers, both in vineyard and forest soils. Despite the Zn bounded to SOM showing a trend to slightly decrease with soil depth, it did not vary significantly after the land-use changes (Figure 3A,C,D,F). At the PM and AR sites, such a diminishing trend in the Zn_{OM} could be due to a less abundant litterfall input or its occurrence during a shorter period, resulting in a modification of this Zn fraction in only the uppermost cm of soil (Figure 3D,F).

The PT site showed a different behavior regarding Zn fractionation. Firstly, the OM fraction increased after abandonment, being quite different between the active and abandoned vineyards at the top 10 cm as a result of the continuous contribution of the senescent biomass of the vegetation growth with the progress of the abandonment. Secondly, the Zn bounded to OM was the second most abundant fraction at the top 4 cm (Figure 3E), which could be expected considering the increase in organic matter and its affinity to metals. The proportion of zinc bounded to OM increased from 5 and 3% (Figure 3B) to 22 and 15% at 2 and 4 cm depths (Figure 3E), respectively. It should be noted that the Zn_{EX} also increased from values near to 0 (Figure 3B) to being higher than 1% (Figure 3E). Consequently, the most mobile Zn fractions (Zn_{OM} + Zn_{EX}) reached nearly 25% of the total Zn after abandonment, while in the active vineyards, these values were lower than 5%. The previous hypothesis of a longer period of abandonment at the PT site could explain the increase in the most mobile fractions after transformation due to a probable higher degree of SOM humification. These results must be considered, since Khoshgoftarmanesh et al. (2018) [58] found a positive correlation of Zn_{EX} and Zn_{OM} with the Zn uptake in the aboveground biomass of wheat cultivars, meaning in the present study that high concentrations of SOM may mobilize Zn to be used by trees or wild plants that are naturally grown in the new afforested sites. Additionally, the Zn_C generally increased after abandonment at all depths, resulting in an overall diminution of Zn_R. Consequently, Zn fractionation becomes modified (especially in the topsoil) after abandonment in favor of the most mobile fractions, increasing its geochemical mobility in the uppermost soil layers. This could enhance the Zn toxicity risk for plants and soil microorganisms, although this concerning scenario would also depend on the SOM composition as well as its humification degree. The results of the present study agree with those reported by Korchagin et al. (2020) [5] in vineyard and in forest soils, who observed that Zn_R was the most abundant fraction and that Zn_{OM} is the most variable fraction, with nutritional and environmental consequences. Interestingly, Vázquez-Blanco et al. (2022) [28] found opposite results for copper in the same vineyard sites, evidencing that the Cu concentrations declined after abandonment, and they tended to be associated with the less mobile fractions.

5. Conclusions

Zn is an essential nutrient that may be bounded to different geochemical fractions, and may provoke phytotoxicity depending on its mobility. The fractionation of Zn in soil may provide information about its potential transformation into soil. Land-use changes after vineyard abandonment into forests may vary Zn fractionation, resulting in a change from Zn bounded to the most immobile fractions in active vineyards towards an increase in Zn associated with SOM fractions in abandoned vineyards, especially in the uppermost soil layers. This study showed that a change in the land-use type resulted in an increase of nearly 25% in the geochemical mobility of Zn in abandoned vineyards compared to active ones. This means that a land-use change may trigger the potential toxicity of Zn in the soil environment, whose intensity would depend on the degree of SOM humification, stabilization, and mineralization. Therefore, caution must be taken on behalf of stakeholder authorities when abandonment occurs and/or when a land-use change is authorized.

Supplementary Materials: The following supporting information can be downloaded at https://www.mdpi.com/article/10.3390/horticulturae9101121/s1: Table S1: Soil chemical properties of all studied vineyards at 10 depths classified by site and land use.

Author Contributions: Conceptualization, P.P.-R., J.C.N.-M., M.A.-E. and D.F.-C.; methodology, P.P.-R., J.C.N.-M. and M.A.-E.; formal analysis, P.P.-R., J.C.N.-M. and D.F.-C.; investigation, P.P.-R. and J.C.N.-M.; resources, P.P.-R. and J.C.N.-M.; data curation, P.P.-R., J.C.N.-M., M.A.-E. and D.F.-C.; writing—original draft preparation, P.P.-R.; writing—review and editing, P.P.-R., J.C.N.-M., M.A.-E. and D.F.-C.; visualization, P.P.-R., J.C.N.-M., M.A.-E. and D.F.-C.; supervision, P.P.-R., J.C.N.-M., M.A.-E. and D.F.-C.; project administration, J.C.N.-M. and M.A.-E.; funding acquisition, J.C.N.-M. and M.A.-E. All authors have read and agreed to the published version of the manuscript.

Funding: This research was funded by Consellería de Cultura, Educación e Universidade (Xunta de Galicia), grant number ED431C2021/46-GRC. Paula Pérez-Rodríguez was funded by Juan de la Cierva contract, grant number JC2020-044426-I, from the Spanish Ministry of Science and Innovation and NextGenerationEU. The APC was partially funded by University of Vigo.

Acknowledgments: Paula Pérez-Rodríguez acknowledges her Juan de la Cierva contract from the Spanish Ministry of Science and Innovation and NextGenerationEU. The financial support of the Consellería de Cultura, Educación e Universidade (Xunta de Galicia), granted to the research group BV1 of the University of Vigo is also acknowledged.

Conflicts of Interest: The authors declare no conflict of interest.

References

1. Kabata-Pendias, A. *Trace Elements in Soils and Plants*; CRC Press: Boca Raton, FL, USA; Taylor and Francis Group LLC: Boca Raton, FL, USA, 2011; 520p.
2. Komárek, M.; Čadková, E.; Chrastný, V.; Bordas, F.; Bollinger, J.-C. Contamination of vineyard soils with fungicides: A review of environmental and toxicological aspects. *Environ. Int.* **2010**, *36*, 138–151. [CrossRef] [PubMed]
3. Senesil, G.S.; Baldassarre, G.; Senesi, N.; Radina, B. Trace element inputs into soils by anthropogenic activities and implications for human health. *Chemosphere* **1999**, *39*, 343–377. [CrossRef] [PubMed]
4. Kabala, C.; Singh, B.R. Fractionation and mobility of copper, lead, and zinc in soil profiles in the vicinity of a copper smelter; 11285909. *J. Environ. Qual.* **2001**, *30*, 485–492. [CrossRef] [PubMed]
5. Korchagin, J.; Moterle, D.F.; Escosteguy, P.A.V.; Bortoluzzi, E.C. Distribution of copper and zinc fractions in a Regosol profile under centenary vineyard. *Environ. Earth Sci.* **2020**, *79*, 1–13. [CrossRef]
6. Kaur, H.; Garg, N. Zinc toxicity in plants: A review. *Planta* **2021**, *253*, 1–28. [CrossRef]
7. Santás-Miguel, V.; Rodríguez-González, L.; Núñez-Delgado, A.; Álvarez-Rodríguez, E.; Díaz-Raviña, M.; Arias-Estévez, M.; Fernández-Calviño, D. Soil Bacterial Community Tolerance to Three Tetracycline Antibiotics Induced by Ni and Zn. *Span. J. Soil. Sci.* **2023**, *13*, 10799. [CrossRef]
8. Alloway, B.J. *Heavy Metals in Soils: Trace Metals and Metalloids in Soils and Their Bioavailability*; Springer Science & Business Media: Berlin/Heidelberg, Germany, 2013.
9. Duplay, J.; Semhi, K.; Errais, E.; Imfeld, G.; Babcsanyi, I.; Perrone, T. Copper, zinc, lead and cadmium bioavailability and retention in vineyard soils (Rouffach, France): The impact of cultural practices. *Geoderma* **2014**, *230–231*, 318–328. [CrossRef]
10. Wieczorek, J.; Baran, A.; Bubak, A. Mobility, bioaccumulation in plants, and risk assessment of metals in soils; 37084910. *Sci. Total Environ.* **2023**, *882*, 163574. [CrossRef]

11. Meite, F.; Granet, M.; Imfeld, G. Ageing of copper, zinc and synthetic pesticides in particle-size and chemical fractions of agricultural soils. *Sci. Total Environ.* **2022**, *824*, 153860. [CrossRef]
12. Tóth, G.; Hermann, T.; Da Silva, M.R.; Montanarella, L. Heavy metals in agricultural soils of the European Union with implications for food safety. *Environ. Int.* **2016**, *88*, 299–309. [CrossRef]
13. Vázquez-Blanco, R.; González-Feijoo, R.; Campillo-Cora, C.; Fernández-Calviño, D.; Arenas-Lago, D. Risk Assessment and Limiting Soil Factors for Vine Production—Cu and Zn Contents in Vineyard Soils in Galicia (Rías Baixas D.O.). *Agronomy* **2023**, *13*, 309. [CrossRef]
14. Fernández-Calviño, D.; Pateiro-Moure, M.; Nóvoa-Muñoz, J.C.; Garrido-Rodríguez, B.; Arias-Estévez, M. Zinc distribution and acid-base mobilisation in vineyard soils and sediments. *Sci. Total Environ.* **2012**, *414*, 470–479. [CrossRef] [PubMed]
15. García-Navarro, F.J.; Jiménez-Ballesta, R.; Garcia-Pradas, J.; Amoros, J.A.; de los Reyes, C.P.; Bravo, S. Zinc concentration and distribution in vineyard soils and grapevine leaves from valdepeñas designation of origin (Central spain). *Sustainability* **2021**, *13*, 7390. [CrossRef]
16. Beygi, M.; Jalali, M. Assessment of trace elements (Cd, Cu, Ni, Zn) fractionation and bioavailability in vineyard soils from the Hamedan, Iran. *Geoderma* **2019**, *337*, 1009–1020. [CrossRef]
17. Brunetto, G.; Schmitt, D.E.; Comin, J.J.; Miotto, A.; De Moraes, M.P.; Heinzen, J. Copper and zinc fractions in vineyard soils in the middle western of Santa Catarina. *Rev. Bras. Eng. Agric. Ambient.* **2014**, *18*, 805–810. [CrossRef]
18. Kelepertzis, E.; Botsou, F.; Patinha, C.; Argyraki, A.; Massas, I. Agricultural geochemistry in viticulture: An example of Cu accumulation and geochemical fractionation in Mediterranean calcareous soils (Nemea region, Greece). *Appl. Geochem.* **2018**, *88*, 23–39. [CrossRef]
19. Vrščaj, B.; Česnik, H.B.; Velikonja Bolta, Š.; Radeka, S.; Lisjak, K. Pesticide Residues and Heavy Metals in Vineyard Soils of the Karst and Istria. *Land* **2022**, *11*, 2332. [CrossRef]
20. Brunetto, G.; Miotto, A.; Ceretta, C.A.; Schmitt, D.E.; Heinzen, J.; de Moraes, M.P.; Canton, L.; Tiecher, T.L.; Comin, J.J.; Girotto, E. Mobility of copper and zinc fractions in fungicide-amended vineyard sandy soils. *Arch. Agron. Soil. Sci.* **2014**, *60*, 609–624. [CrossRef]
21. Genova, G.; Della Chiesa, S.; Mimmo, T.; Borruso, L.; Cesco, S.; Tasser, E.; Matteazzi, A.; Niedrist, G. Copper and zinc as a window to past agricultural land-use. *J. Hazard. Mater.* **2022**, *424*, 126631. [CrossRef]
22. Mohammed, S.; Alsafadi, K.; Enaruvbe, G.O.; Harsányi, E. Assessment of soil micronutrient level for vineyard production in southern Syria. *Model. Earth Syst. Environ.* **2022**, *8*, 407–416. [CrossRef]
23. Bradl, H.B. Adsorption of heavy metal ions on soils and soils constituents. *J. Colloid. Interface Sci.* **2004**, *277*, 1–18. [CrossRef] [PubMed]
24. Voegelin, A.; Tokpa, G.; Jacquat, O.; Barmettler, K.; Kretzschmar, R. Zinc fractionation in contaminated soils by sequential and single extractions: Influence of soil properties and zinc content. *J. Environ. Qual.* **2008**, *37*, 1190–1200. [CrossRef] [PubMed]
25. Gabarrón-Galeote, M.A.; Trigalet, S.; van Wesemael, B. Effect of land abandonment on soil organic carbon fractions along a Mediterranean precipitation gradient. *Geoderma* **2015**, *249–250*, 69–78. [CrossRef]
26. Perpiña Castillo, C.; Kavalov, B.; Diogo, V.; Jacobs, C.; Batista e Silva, F.; Lavalle, C. *Agricultural Land Abandonment in the EU within 2015–2030*; No. JRC113718; Joint Research Centre (Seville site): Ispra, Italy, 2018.
27. De Santiago-Martín, A.; Vaquero-Perea, C.; Valverde-Asenjo, I.; Quintana Nieto, J.R.; González-Huecas, C.; Lafuente, A.L.; Vázquez de la Cueva, A. Impact of vineyard abandonment and natural recolonization on metal content and availability in Mediterranean soils. *Sci. Total Environ.* **2016**, *551–552*, 57–65. [CrossRef] [PubMed]
28. Vázquez-Blanco, R.; Nóvoa-Muñoz, J.C.; Arias-Estévez, M.; Fernández-Calviño, D.; Pérez-Rodríguez, P. Changes in Cu accumulation and fractionation along soil depth in acid soils of vineyards and abandoned vineyards (now forests). *Agric. Ecosyst. Environ.* **2022**, *339*, 108146. [CrossRef]
29. Sciubba, L.; Mazzon, M.; Cavani, L.; Baldi, E.; Toselli, M.; Ciavatta, C.; Marzadori, C. Soil response to agricultural land abandonment: A case study of a vineyard in Northern Italy. *Agronomy* **2021**, *11*, 1841. [CrossRef]
30. Iñigo, V.; Marín, A.; Andrades, M.; Jiménez-Ballesta, R. Evaluation of the copper and zinc contents of soils in the vineyards of la rioja (Spain). *Environments* **2020**, *7*, 55. [CrossRef]
31. Sacchi, E.; Brenna, S.; Genot, S.F.; Leoni, A.; Sale, V.M.; Setti, M. Potentially toxic elements (Ptes) in cultivated soils from lombardy (northern Italy): Spatial distribution, origin, and management implications. *Minerals* **2020**, *10*, 298. [CrossRef]
32. Lakanen, E.; Erviö, R. A comparison of eight extractants for the determination of plant available micronutrients in soils. *Acta Agric. Fenn.* **1971**, *123*, 223–232.
33. Arias, M.; López, E.; Fernández, D.; Soto, B. Copper distribution and dynamics in acid vineyard soils treated with copper-based fungicides. *Soil Sci.* **2004**, *169*, 796–805. [CrossRef]
34. Campillo-Cora, C.; Fernández-Calviño, D.; Pérez-Rodríguez, P.; Fernández-Sanjurjo, M.J.; Núñez-Delgado, A.; Álvarez-Rodríguez, E.; Arias-Estévez, M.; Nóvoa-Muñoz, J.C. Copper and zinc in rhizospheric soil of wild plants growing in long-term acid vineyard soils. Insights on availability and metal remediation. *Sci. Total Environ.* **2019**, *672*, 389–399. [CrossRef] [PubMed]
35. Wightwick, A.M.; Mollah, M.R.; Partington, D.L.; Allinson, G. Copper fungicide residues in Australian vineyard soils. *J. Agric. Food Chem.* **2008**, *56*, 2457–2464. [CrossRef] [PubMed]
36. Van Eynde, E.; Fendrich, A.N.; Ballabio, C.; Panagos, P. Spatial assessment of topsoil zinc concentrations in Europe. *Sci. Total Environ.* **2023**, *892*, 164512. [CrossRef] [PubMed]

37. Cicchella, D.; Giaccio, L.; Dinelli, E.; Albanese, S.; Lima, A.; Zuzolo, D.; Valera, P.; De Vivo, B. GEMAS: Spatial distribution of chemical elements in agricultural and grazing land soil of Italy. *J. Geochem. Explor.* **2015**, *154*, 129–142. [CrossRef]
38. Macías-Vázquez, F.; De Anta, R.C. *Niveles Genéricos de Referencia de Metales Pesados y Otros Elementos Traza en Suelos de Galicia*; Conselleria de Medio Ambiente e Desenvolvemento Sostible: Santiago de Compostela, Spain, 2009.
39. Bortoluzzi, E.C.; Korchagin, J.; Moterle, D.F.; dos Santos, D.R.; Caner, L. Accumulation and Precipitation of Cu and Zn in a Centenarian Vineyard. *Soil Sci. Soc. Am. J.* **2019**, *83*, 492–502. [CrossRef]
40. Brunetto, G.; Comin, J.J.; Miotto, A.; de Moraes, M.P.; Sete, P.B.; Schmitt, D.E.; Gatiboni, L.C.; de Melo, G.W.B.; Morais, G.P. Copper and zinc accumulation, fractionation and migration in vineyard soils from Santa Catarina State, Brazil. *Bragantia* **2018**, *77*, 141–151. [CrossRef]
41. Nziguheba, G.; Smolders, E. Inputs of trace elements in agricultural soils via phosphate fertilizers in European countries. *Sci. Total Environ.* **2008**, *390*, 53–57. [CrossRef]
42. Hua, K.; Wang, T.; Guo, Z.; Zhan, L.; He, C.; Wang, D. Accumulation of potentially toxic elements under long-term application of different organic amendments. *Nutr. Cycl. Agroecosyst.* **2023**, *126*, 293–309. [CrossRef]
43. Kumar, R.; Whelan, A.; Cannon, P.; Sheehan, M.; Reeves, L.; Antunes, E. Occurrence of emerging contaminants in biosolids in northern Queensland, Australia. *Environ. Pollut.* **2023**, *330*, 121786. [CrossRef]
44. Zeng, H.; Wu, H.; Yan, F.; Yi, K.; Zhu, Y. Molecular regulation of zinc deficiency responses in plants. *J. Plant Physiol.* **2021**, *261*, 153419. [CrossRef]
45. Ots, K.; Mandre, M. Monitoring of heavy metals uptake and allocation in Pinus sylvestris organs in alkalised soil. *Environ. Monit. Assess.* **2012**, *184*, 4105–4117. [CrossRef] [PubMed]
46. Pietrzak, U.; Uren, N.C. Remedial options for copper-contaminated vineyard soils. *Soil. Res.* **2011**, *49*, 44–55. [CrossRef]
47. Barbosa, F.L.A.; da Silva, M.G.; Cavalcante, R.M.; de Freitas Sousa, H.H.; dos Santos Dias, C.T.; Escobar, M.E.O. Fe, Zn, and Cu in the Organic Fraction of Soils with Different Textures Under Intense Use of Agrochemicals in Flower Cultivation. *Water Air Soil. Pollut.* **2023**, *234*, 66. [CrossRef]
48. Girotto, E.; Ceretta, C.A.; Brunetto, G.; Rheinheimer dos Santos, D.; Souza da Silva, L.; Lourenzi, C.R.; Lorensini, F.; Beber Vieira, R.C.; Schmatz, R. Copper and zinc forms and accumulation in soil after successive pig slurry applications. *Rev. Bras. De Ciência Do Solo* **2010**, *34*, 955–965. [CrossRef]
49. Pham, N.T.H.; Babcsányi, I.; Balling, P.; Farsang, A. Accumulation patterns and health risk assessment of potentially toxic elements in the topsoil of two sloping vineyards (Tokaj-Hegyalja, Hungary). *J. Soils Sediments* **2022**, *22*, 2671–2689. [CrossRef]
50. Giweta, M. Role of litter production and its decomposition, and factors affecting the processes in a tropical forest ecosystem: A review. *J. Ecol. Environ.* **2020**, *44*, 11. [CrossRef]
51. Berg, B.; Laskowski, R. Litter decomposition: A guide to carbon and nutrient turnover. *Adv. Ecol. Res.* **2006**, *38*, 20–71.
52. Swift, M.J.; Heal, O.W.; Anderson, J.M.; Anderson, J.M. *Decomposition in Terrestrial Ecosystems*; University of California Press: Berkeley, CA, USA, 1979.
53. Ramos, M.C.; López-Acevedo, M. Zinc levels in vineyard soils from the Alt Penedès-Anoia region (NE Spain) after compost application. *Adv. Environ. Res.* **2004**, *8*, 687–696. [CrossRef]
54. Masullo, L.S.; Ferraz, A.d.V.; Rocha, J.H.T.; Valdo, G.d.C.; Ávila, P.A.d.; Florentino, A.L.; de Toledo, F.H.S.F.; Alleoni, L.R.F.; de Moraes Gonçalves, J.L. Soil Cu and Zn pools and their availability in response to forest residue management and micronutrient fertilization in a Eucalyptus grandis plantation. *For. Ecol. Manag.* **2022**, *506*, 119954. [CrossRef]
55. Gómez-Armesto, A.; Carballeira-Díaz, J.; Pérez-Rodríguez, P.; Fernández-Calviño, D.; Arias-Estévez, M.; Nóvoa Muñoz, J.C.; Álvarez-Rodríguez, E.; Fernández-Sanjurjo, M.J.; Núñez-Delgado, A. Copper content and distribution in vineyard soils from Betanzos (A Coruña, Spain). *Span. J. Soil Sci.* **2015**, *5*, 60–71. [CrossRef]
56. Zahedifar, M. Sequential extraction of zinc in the soils of different land use types as influenced by wheat straw derived biochar. *J. Geochem. Explor.* **2017**, *182*, 22–31. [CrossRef]
57. Zhang, X.; Jiang, B.; Ma, Y. Aging of zinc added to soils with a wide range of different properties: Factors and modeling. *Environ. Toxicol. Chem.* **2017**, *36*, 2925–2933. [CrossRef] [PubMed]
58. Khoshgoftarmanesh, A.H.; Afyuni, M.; Norouzi, M.; Ghiasi, S.; Schulin, R. Fractionation and bioavailability of zinc (Zn) in the rhizosphere of two wheat cultivars with different Zn deficiency tolerance. *Geoderma* **2018**, *309*, 1–6. [CrossRef]

horticulturae

MDPI

Article

Early Growth Assessment of *Lolium perenne* L. as a Cover Crop for Management of Copper Accumulation in Galician Vineyard Soils

Raquel Vázquez-Blanco [1,2], Manuel Arias-Estévez [1,2], David Fernández-Calviño [1,2] and Daniel Arenas-Lago [1,2,*]

1 Departamento de Bioloxía Vexetal e Ciencias do Solo, Area de Edafoloxía e Química Agrícola, Facultade de Ciencias, Universidade de Vigo, As Lagoas s/n, 32004 Ourense, Spain; raquel.vazquez.blanco@uvigo.gal (R.V.-B.); mastevez@uvigo.es (M.A.-E.); davidfc@uvigo.gal (D.F.-C.)
2 Instituto de Agroecoloxía e Alimentación (IAA), Campus Auga, Universidade de Vigo, 32004 Ourense, Spain
* Correspondence: darenas@uvigo.es

Abstract: This study investigates the potential use of *Lolium perenne* L. as a cover crop to improve vineyard soils with varying levels of copper (Cu). Cu-based fungicides are commonly used to control fungal diseases in vineyards, but their accumulation in soils poses environmental risks. This study aims to address this issue by evaluating the influence of soil properties on Cu availability and *L. perenne* growth. A total of 42 vineyard soils from different Designations of Origin (D.O.s) in Galicia were sampled and their physicochemical properties were analyzed. The results showed most soils exceeded recommended Cu limits due to fungicide applications. Pot experiments were conducted to assess *L. perenne* growth and Cu accumulation. *L. perenne* biomass did not vary significantly with total soil Cu content, indicating that other factors such as organic matter and cation exchange capacity were more important for plant growth. While *L. perenne* showed Cu tolerance, its aerial Cu accumulation was inversely correlated with available Cu. This study provides insight into the potential of *L. perenne* as a cover crop for sustainable vineyard management and soil improvement and emphasizes the importance of considering Cu accumulation from fungicide applications.

Keywords: copper; fungicide; phytosanitary; cover crop; vineyard; organic viticulture; sustainable agriculture

1. Introduction

In recent decades, the viticulture sector in Galicia (Spain) has gained considerable importance. According to global data from the Designations of Origin (D.O.s) in Galicia, the hectares dedicated to vine monoculture, as well as grape and wine production, have increased significantly in recent years, reaching an area of nearly 10,000 hectares dedicated to viticulture, with an annual grape yield of nearly 50,000 tons [1]. Viticulture in Galicia is organized into five Designations of Origin (D.O.s), which encompass regions where vineyards hold greater importance. Each D.O. in Galicia is characterized by the cultivation of different grape varieties, adapted to the specific edaphoclimatic conditions of each region [2–6]

Galicia's vineyard soils display significant edaphological variation depending on the viticultural area and geological features of each D.O. Nevertheless, they are mostly recognized for being acidic, shallow, and nutrient-deficient as a result of leaching caused by precipitation. The climate in Galicia's viticultural areas is generally categorized as oceanic, producing mild winters and cool summers. Rainfall is distributed throughout the year [7]. More specifically, the Rías Baixas D.O. experiences an Atlantic climate with mild temperatures and high humidity. Similarly, the Ribeiro D.O. also has an Atlantic climate with Mediterranean influence and experiences higher temperatures during hot and dry summers. The precipitation concentrated in winter and spring stimulates the vegetative cycle of the vines. The Ribeira Sacra D.O. features an Atlantic climate with some continental influence. The region experiences extreme temperatures including cold winters and hot summers. However, rainfall is infrequent, leading to potential water stress for the vines. The Valdeorras D.O. has a continental climate with Atlantic influence, marked by low precipitation, cold winters, and hot summers. The region has a marked thermal oscillation between day and night. The Monterrei D.O. has a Mediterranean climate with Atlantic influence, presenting variations depending on the altitude and orientation of the slopes. Overall, the Monterrei D.O. has mild winters and hot, dry summers. However, the region experiences scarce precipitation. The edaphoclimatic characteristics of these regions favor the occurrence of fungal diseases [8]. As a result, there is intensive application of phytosanitary treatments, primarily copper-based fungicides, to maintain the productivity of the sector [9,10].

Copper (Cu) is an essential element for plant growth as it is a necessary micronutrient in plant metabolism. Even though the concentration of Cu in the dry tissue of plants is typically low (<0.01%), it is vital for essential functions, including photosynthesis, protein synthesis, and enzyme activation [11]. Nevertheless, elevated concentrations of Cu in the soil can prove harmful to plants. Soil Cu levels exceeding 100 mg kg^{-1} could affect plant growth and development and pose a threat to soil quality [12]. Copper in the soil mainly exists in the form of divalent cations in the soil solution or interacts with organic matter to form organometallic complexes, along with soil mineral components, such as aluminum, iron, and manganese oxihydroxides, and clays [13]. Copper mobility in the soil depends on the pH; it tends to form insoluble and less available compounds in alkaline soils while being more available to plants in acidic or neutral soils, potentially leading to toxicity issues [14,15].

Copper toxicity in plants causes various symptoms, including leaf darkening, chlorosis, and root malformation. Moreover, different harmful effects have been observed as a result of copper toxicity in plants. These effects include damage to plant tissues, abnormal elongation of root cells, alterations in membrane permeability, lipid peroxidation in chloroplasts, inhibition of electron transport in photosynthesis, immobilization of Cu in cell walls, and damage to genetic material. Despite this, certain plant species have demonstrated a greater capacity for resistance and adaptation to the presence of copper, while others are more sensitive and can experience negative impacts even at lower concentrations [15]. As a result, it is essential to assess the availability and toxicity of Cu in the soil to ensure proper crop management, as well as prevent negative impacts on plant health and soil quality [16].

The presence of Cu in soil may have natural origins, but contamination and toxicity of Cu in soils and plants are primarily due to anthropogenic activities such as mining, the timber industry, metallurgy, and excessive use of fertilizers and pesticides in the agricultural sector [17]. Although Cu is an essential nutrient for plant growth, it is also used in agriculture as a component of fungicides and fertilizers to control plant diseases and enhance productivity, particularly in crops such as vineyards [18].

Cu-based phytosanitary products have been widely used since the mid-nineteenth century to protect vineyards due to their high susceptibility to fungal diseases, such as mildew (*Plasmopara vitícola*) and powdery mildew (*Uncinula necátor*) [19]. Chemical substances with antifungal properties, containing Cu, have been used to combat these diseases. The most commonly used compounds include Bordeaux mixture ($Ca(OH)_2 + CuSO_4$), copper oxychloride ($Cu_2(OH)_3Cl$), and commercial brand compounds. These substances provide an effective defense against vine diseases while maintaining plant health and vineyard productivity. These compounds are frequently applied, leading to Cu accumulation that often exceeds the 100 mg kg^{-1} threshold in vineyard soil [12,13]. This jeopardizes soil and environmental health by creating a potential risk of pollution.

The application of fungicides containing Cu in vineyard protection can significantly affect the soil in which they are applied. Studies have shown that Cu present in such fungicides can enter the soil by directly depositing from treated leaves or by washing due to precipitation [20]. When Cu enters the soil, it tends to accumulate in the top layer of the soil due to its high affinity for organic matter [21]. The accumulation of Cu in the soil can have harmful effects on future crops established in the same area or on crops established in nearby areas. The high copper content in the soil can be toxic to plants, including wild plants or cover crops grown in vineyards [22]. This can potentially affect the plants due to the presence of high levels of Cu in the soil [23]. The concurrent presence of copper and cationic herbicides in the soil can intensify the negative effects on soil quality and ecosystem health. This interaction can cause copper retention and increase the risk of environmental contamination, particularly through the leaching of these compounds [24]. Several studies have reported that excessive or improper use of Cu products in agriculture can increase Cu concentrations in the soil and potentially harm plants and the environment [25–27]. Regarding soils, where Cu-based fungicides are commonly used, the long-term cumulative effects of this metal on the soil are related to: (i) problems with the uptake of essential nutrients due to changes in the ionic balance in roots [28]; (ii) reduction in biodiversity and microbial activity, which can affect soil biological processes, including organic matter decomposition, nitrogen fixation, and nutrient availability to plants [21]; (iii) alteration of soil structure, causing compaction and reduced porosity, which can affect water infiltration, soil aeration, and nutrient retention capacity [29]; (iv) contamination of groundwater as a result of Cu leaching into deeper soil layers and eventually reaching groundwater [30]; (v) and accumulation in fruits and wine products, which can lead to the presence of Cu residues in wine products, raising concerns for human health and wine quality standards [26]. Therefore, proper management during fungicide application and the adoption of sustainable agricultural practices are required to minimize the impact of Cu on soils and ensure the long-term health of vineyards and the quality of wine products.

Over the years, vineyard management techniques have undergone significant changes. Initially, economic and production criteria were the main drivers of changes in vineyard management systems, before a holistic assessment of the environmental and agricultural impacts associated with vineyard operations began [31]. In the context of the Green Revolution that began in the mid-20th century, there was an increase in global agricultural productivity due to the introduction of new techniques, technologies, and inputs into agro-ecosystems. This led to the establishment of intensive agriculture as we know it today. In vineyards, the use of phytosanitary products contributed significantly to the stabilization of productive yields and grape quality over time. However, this intensification has not been without negative environmental and human health impacts, including soil degradation and loss of fertility, groundwater contamination, pest proliferation, excessive use of agricultural inputs, and loss of biodiversity [32,33]. In response to these challenges, alternatives based on the principles of agroecology have emerged in recent years, which are based on ecological principles and biological interactions in agrosystems, with the aim of making agricultural processes sustainable. This includes pest control, appropriate fertilization, and environmental protection in all dimensions [34,35].

In recent years, the implementation and advancement of organic viticulture have been fundamental steps to improve soil health and vineyard quality [36,37], mainly through sustainable management practices such as (i) minimum tillage to preserve soil structure, reduce erosion, and promote microbiological activity; (ii) nutrient application and management to determine the nutritional needs of the vine and balanced fertilizer use; (iii) moisture management to optimize water use and prevent soil compaction; (iv) integrated pest and disease control: implementation of control strategies and regular monitoring of pest populations and use of environmentally friendly biological control methods; and (v) use of cover crops to improve soil quality, maintain biodiversity, and reduce erosion.

Specifically, the use of cover crops in vineyards is an emerging agricultural strategy aimed at improving the sustainability of the viticultural system. Cover crops are the planting of selected plant species between rows of vines to protect the soil, increase its fertility, and regulate weed competition [38]. Cover crops can be categorized into natural cover and planted cover. Natural covers consist of native plant species that grow spontaneously in the vineyard environment, are generally well adapted to local conditions, contribute to the soil characteristics and biodiversity of the vineyard ecosystem, and provide valuable ecosystem services [39]. On the other hand, planted cover involves the selection and planting of specific plant species for agronomic purposes. Planted covers may include legumes such as clover or vetch, grasses such as rye or barley, or mixtures of species that provide complementary benefits [40]. Both types of cover crops offer benefits, and their selection will depend on the specific viticultural management objectives and environmental conditions of each vineyard. Regardless of their type, cover crops offer several characteristics that make them beneficial to the viticultural system. Cover crops protect the soil from climate-induced erosion, as both the roots and above-ground parts act as a physical barrier to prevent soil and nutrient loss; they help improve soil structure by promoting the formation of aggregates that increase porosity and water-holding capacity; they add plant material to the soil, increasing organic matter and soil biodiversity; they compete with weeds, reducing their growth and minimizing the need for herbicide application; they can promote soil nitrogen fixation; they contribute to biological pest and disease control, reducing the need for pesticides; and they influence the vineyard microclimate, reducing temperature fluctuations and maintaining soil moisture, which can promote optimal vine development and grape quality [41]. However, the use of cover crops also presents challenges and inconveniences that must be properly addressed to maximize their effectiveness and benefits. Cover crops can compete with vines for resources such as water and nutrients, especially in the early stages of growth. Therefore, the selection of cover crop species is a critical factor to ensure they do not become invasive weeds in the vineyard. In addition, managing the height and density of cover crops is essential to balance their interaction with vines and maximize their agronomic performance [42].

Among the species with potential for use as cover crops in vineyards is *Lolium perenne* L. This species, commonly known as perennial ryegrass or English ryegrass, is an herbaceous plant belonging to the grass family (Poaceae). It has a fibrous root system and erect stems that can reach heights of 30 to 100 cm. It has high germination and vigor, which facilitate its expansion. *Lolium perenne* is widely used due to its various applications, including its use as a high-quality forage species used in livestock feed; for lawns due to its resistance and rapid germination; in seed mixtures used for re-vegetation of eroded or degraded areas; for erosion control on slopes and areas with excessive steepness; as a phytoremediation species in soils contaminated with potentially toxic elements, thanks to its ability to accumulate metals in its tissues; and as a cover crop to improve soil conditions and control erosion [43–48].

In the present study, the use of *L. perenne* as a cover crop in vineyard soils from different Denominations of Origin (D.O.s) in Galicia was evaluated from an agroecological perspective. This species can provide soil protection against environmental factors that cause erosion, improve soil structure, and sequester part of the Cu present in the soil as a result of the continuous use of fungicides. Therefore, the main objectives of this research are: (i) to study the influence of physicochemical soil characteristics in the availability of Cu accumulated in vineyard soils as a consequence of fungicide application; (ii) to know the possible influence of available Cu in the vineyard soils on the early growth of *L. perenne*, and (iii) to assess the capacity of *L. perenne* to be used as a cover crop and to improve the conditions of Galician vineyard soils under stress caused by high Cu contents.

2. Materials and Methods

2.1. Study Area and Selection and Sampling of Vineyard Soils

In this study, vineyard soils were selected from the five Designations of Origin (D.O.s) in Galicia: Rías Baixas, Ribeiro, Ribeira Sacra, Monterrei, and Valdeorras. The vineyards were selected based on different criteria: ecological, socio-economic, topographical, and climatic, to ensure a wide variability within the study area. All the vineyards studied suffer from high humidity, which requires intensive application of Cu-based fungicides, mainly in the form of a Bordeaux mixture and copper oxychloride, among others.

A total of 42 vineyard soils were sampled from 34 different vineyards in the five D.O.s mentioned above (Rías Baixas: 5 samples; Ribeiro: 12 samples; Ribeira Sacra: 10 samples; Valdeorras: 11 samples; and Monterrei: 4 samples) (Table 1). The numbering of the soils was based on the total Cu content in each soil, with soil 1 having the lowest Cu content and soil 42 having the highest. In addition, the soils were distinguished between those sampled under the vine strain (VS) (20 samples) and those sampled in the alleys between the vine rows (VR) (22 samples).

Table 1. Location and origin of the sampled vineyard soils.

Soil	Sampling Area	D.O.	Municipality	Soil	Sampling Area	D.O.	Municipality
1 M	VR	Monterrei	Vilardevós	**22** RS	VR	Ribeira Sacra	Sober
2 RB	VS	Rías Baixas	Salvaterra de Miño	**23** M	VR	Monterrei	Verín
3 RS	VR	Ribeira Sacra	A Pobra de Trives	**24** RS	VR	Ribeira Sacra	Parada de Sil
4 M	VR	Monterrei	Monterrei	**25** RS	VR	Ribeira Sacra	A Teixeira
5 M	VS	Monterrei	Oímbra	**26** V	VS	Valdeorras	Vilamartín de Valdeorras
6 V	VR	Valdeorras	O Barco de Valdeorras	**27** RS	VS	Ribeira Sacra	Parada de Sil
7 RS	VS	Ribeira Sacra	A Pobra de Trives	**28** RB	VR	Rías Baixas	Arbo
8 RB	VR	Rías Baixas	Tomiño	**29** R	VR	Ribeiro	Sober
9 RB	VS	Rías Baixas	Meis	**30** RS	VS	Ribeira Sacra	Sober
10 V	VR	Valdeorras	Vilamartín de Valdeorras	**31** RS	VS	Ribeira Sacra	Sober
11 V	VS	Valdeorras	Petín	**32** R	VS	Ribeiro	Beade
12 R	VS	Ribeiro	Ribadavia	**33** RS	VS	Ribeira Sacra	Parada de Sil
13 R	VR	Ribeiro	Ribadavia	**34** R	VR	Ribeiro	Arnoia
14 R	VR	Ribeiro	Cenlle	**35** V	VS	Valdeorras	A Rúa
15 V	VS	Valdeorras	Rubiá	**36** R	VR	Ribeiro	Toén
16 V	VR	Valdeorras	Rubiá	**37** R	VR	Ribeiro	Toén
17 V	VR	Valdeorras	Rubiá	**38** R	VS	Ribeiro	Sober
18 RB	VS	Rías Baixas	Arbo	**39** V	VS	Valdeorras	Rubiá
19 RS	VS	Ribeira Sacra	Sober	**40** V	VS	Valdeorras	Vilamartín de Valdeorras
20 R	VR	Ribeiro	Sober	**41** R	VR	Ribeiro	Toén
21 V	VS	Valdeorras	O Barco de Valdeorras	**42** R	VR	Ribeiro	Toén

D.O.: Denomination of origin; RB: Rías Baixas; RS: Ribeira Sacra; R: Ribeiro; M: Monterrei; V: Valdeorras; VR: vineyard row; VS: vineyard strain; N: North; W: West.

At each sampling point, three subsamples (0–20 cm depth) were collected using an Edelman auger. The soil subsamples from each sampling point were mixed and homogenized to form a composite soil sample, which was stored in polyethylene bags. At the laboratory, the composite samples were air dried, sieved through a 2 mm mesh, and stored in polyethylene bottles until analysis.

2.2. Soil Characterization

Soil pH was determined both in deionized water and in a 0.1 M KCl solution at a soil/water ratio of 1:2.5 (w/v) [49]. Particle size analysis and soil texture were determined by the Day method [50]. The total carbon and nitrogen contents in the soil were determined using a Thermo Flash EA 1112 elemental analyzer. The sample was completely combusted in a reactor with a high-temperature oxidizing catalyst (900 °C). The elemental gases produced were analyzed by chromatography and a high-sensitivity thermal conductivity detector to determine carbon and nitrogen. The organic carbon content was determined by incinerating the sample in a muffle furnace at 450 °C for 4 h. The organic carbon content was calculated as the difference between the weight of the sample before and after ashing. The organic matter content was estimated by multiplying the total carbon content by the factor 1.724 proposed by Van Bemmelen for carbon contents below 5.8% and by 2 for higher contents. The dissolved organic carbon content was determined using a Shimadzu TOC-V CSH/CSN analyzer [51]. The effective cation exchange capacity (eCEC) was determined as the sum of Ca^{2+}, Mg^{2+}, Na^{+}, and K^{+} (exchangeable bases) and Al^{3+} in the exchange complex [52]. Cation contents were determined by atomic absorption spectroscopy (Thermo Solaar Series M, Thermo Fisher Scientific Inc., Waltham, Massachusetts, USA). To determine the Al^{3+} content in the exchange complex, a preliminary extraction with 1 M KCl was performed and the Al^{3+} content was determined by acid-base titration with 0.1 M NaOH. The results were expressed as $cmol_{(+)}\ kg^{-1}$ soil.

2.3. Total and Available Cu Content in Vineyard Soils

The total Cu content in vineyard soils was determined by acid extraction with aqua regia ($HCl{:}HNO_3$ ratio 3:1 v/v) using Teflon reactors in a microwave oven (190 °C and 9 bars, time 45 min). The Cu content was determined by atomic absorption spectroscopy (Thermo Solaar Series M, Thermo Fisher Scientific Inc.).

For the available Cu content in vineyard soils, 3 g of vineyard soils from each pot was weighed into a polyethylene tube and 30 mL of a solution of acetic, lactic, citric, malic, and formic acid was added for a total acid concentration of 10 mM and a molar ratio of 4:2:1:1:1. Soils with the acid mixture were shaken for 16 h, centrifuged, and filtered [53]. The extracts were analyzed by atomic absorption spectroscopy (Thermo Solaar Series M, Thermo Fisher Scientific Inc.) to determine the available Cu content.

For this purpose, the total concentration of the acids was 10 mM and the molar ratio was 4:2:1:1:1

The extraction efficiency (E_xE_f) of the acid mixture was determined to evaluate the percentage of available Cu relative to the total Cu content in the vineyard soils. The E_xE_f was calculated using the following equation:

$$E_xE_f = 100 \cdot \frac{Cu_a}{Cu_t} \tag{1}$$

where Cu_a and Cu_t are available Cu (mg kg^{-1}) and total Cu (mg kg^{-1}) content, respectively, for each vineyard soil.

2.4. Plant Material and Pot Experiments

Lolium perenne seeds were used for the pot experiments. Seedlings were grown in seedbeds, and 80 g of each of the vineyard soil samples listed in Table 1 were added to individual pots. For each soil, 5 pots were used. Then, 5 g of *L. perenne* seeds were added to each pot to ensure proper burial in the soil. After sowing, the pots were placed in a climate chamber with a light-dark cycle of 16 h and 8 h, respectively, and a constant temperature of 25 °C. For one month, the pots were irrigated with distilled water every two days to maintain the soil's moisture retention capacity. At the end of the month, the above-ground parts of each plant were harvested and the fresh biomass of *L. perenne* in each pot was recorded. The samples were then wrapped in aluminum foil and dried in an oven at 40 °C for 72 h. Finally, the dried samples were weighed to determine the dry biomass.

2.5. Determination of Cu in L. perenne Shoots

The dried shoot plant samples from each pot were digested with concentrated HNO_3 using the following procedure: 0.2 g of the plant sample was collected and placed in Pyrex glass tubes, together with 7.5 mL of concentrated HNO_3. The samples were allowed to contact the HNO_3 for 12 h at room temperature. The tubes were then sealed and placed in a digestion block at 105 °C for 2 h. After the digestion period, the tubes were allowed to cool for 1 h before repeating the process for another 2 h. The contents of the tubes were then filtered and transferred to polyethylene tubes filled to the mark with distilled water. The Cu content was determined by atomic absorption spectrometry (Thermo Solaar Series M, Thermo Fisher Scientific Inc.).

The bioaccumulation factor (BF) was determined to evaluate the relationship between the Cu content in the aerial part of *L. perenne* and the available Cu content in the vineyard soils. The BF was calculated using the following equation:

$$BF = \frac{\text{Cu content in shoots of } L.\ perenne}{\text{available Cu content in vineyard soil}} \cdot 100 \quad (2)$$

2.6. Statistical Analysis

All analyses of soils and plants were performed in triplicate. Statistical analysis of the data was performed using Past4.11 software (Oslo, Norway) for Windows. A bivariate Pearson correlation analysis was performed between the soil characteristics and the total and available Cu contents, as well as the parameters determined in the pot experiments with *L. perenne*. To interpret the correlation analysis, the criterion, where absolute values of R between 0.3–0.5 represent a medium effect and R $\geq$ 0.5 represents a large effect, was used [54]. Principal component analysis (PCA) was performed on the dataset to identify possible relationships between soil characteristics, total and available Cu content in soils, Cu content in the aerial part of *L. perenne*, and its biomass. For the PCA, soil samples were separated by sampling zone (vine strain or vine row) and grouped by designation of origin (D.O.). The correlation matrix was used and group differences (D.O.) were taken into account in the analysis.

3. Results and Discussion

3.1. General Characterization of the Vineyard Soils

The physico-chemical characteristics of the studied vineyard soils of the different Galician D.O.s are shown in Table 2. A high variability in soil composition was observed. Although most of the soils presented a sandy loam texture, significant variations in silt and clay content were found, resulting in the presence of silt loam, sandy clay loam, and loam textures in general. The presence of clay in the soil is particularly relevant due to its fundamental role in Cu retention [13]. Clays can retain cations, including Cu, within their crystalline structure or in cation exchange sites. Therefore, the prevalence of textures with a low clay content in vineyard soils may imply a lower capacity to bind Cu in the mineral fraction of the soil. This means that Cu could be more available in a free or easily exchangeable form, which could affect its availability to vines and potentially influence their growth [23]. In addition to its effect on Cu retention, soil texture also plays a critical role in its ability to retain water. Textures with higher proportions of clay and silt have a greater capacity to hold water than sandy textures. This is because finer particles, such as silt and clay, have a higher specific surface area and therefore a greater capacity to hold water. Consequently, soil texture has a significant impact on the amount of water available for proper vine growth and development, as it determines the water reserve available to the roots [55].

Table 2. General characteristics of vineyard soils (average values).

Soil	S.a.	pH_{H2O}	pH_{KCl}	C(%)	OM	N	C/N	DOC	DIC	Ca^{2+}	Mg^{2+}	K^{+}	Na^{+}	Al^{3+}	eCEC	Sand	Silt	Clay	Texture
					(%)		-	mg kg^{-1}				cmol$_{(+)}$ kg^{-1}					(%)		
1 M	VR	5.35	3.31	2.62	4.52	0.21	12.61	20.36	0.09	0.06	0.44	0.57	0.05	2.27	3.38	33.06	49.94	17.01	silt loam
2 RB	VS	7.37	5.42	4.48	7.72	0.27	16.77	26.20	0.79	13.48	1.40	0.79	0.09	0.00	15.75	46.54	35.87	17.59	loam
3 RS	VR	6.37	4.03	1.62	2.79	0.12	13.07	25.16	0.40	0.26	0.56	0.62	0.03	0.37	1.83	69.68	16.57	13.74	sandy loam
4 M	VR	6.46	4.18	1.07	1.84	0.08	12.63	16.12	0.50	0.18	0.70	0.44	0.05	0.00	1.38	72.99	14.27	12.75	sandy loam
5 M	VS	4.86	2.94	1.73	2.98	0.13	13.31	27.85	0.19	0.14	0.73	0.75	0.06	1.69	3.37	72.09	15.06	12.85	sandy loam
6 V	VR	8.02	7.32	2.84	4.90	0.37	7.70	28.58	11.47	18.00	0.76	0.38	0.04	0.00	19.18	44.82	36.10	19.08	loam
7 RS	VS	5.76	3.75	2.54	4.38	0.20	12.68	35.29	0.22	0.13	0.44	0.59	0.04	1.34	2.54	69.68	16.57	13.74	sandy loam
8 RB	VR	7.71	5.88	3.04	5.24	0.18	17.04	18.74	3.63	11.56	1.07	0.72	0.31	0.00	13.65	55.42	28.63	15.95	sandy loam
9 RB	VS	6.87	4.69	3.78	6.52	0.29	13.27	14.92	0.51	7.64	1.15	0.96	0.10	0.02	9.87	64.65	19.03	16.32	sandy loam
10 V	VR	5.45	2.96	1.30	2.24	0.11	11.91	17.70	0.01	0.10	0.40	1.05	0.03	0.89	2.48	32.52	52.00	15.49	silt loam
11 V	VS	7.73	6.09	3.94	6.79	0.14	28.23	23.70	2.06	6.16	1.25	0.57	0.03	0.00	8.01	65.45	21.25	13.30	sandy loam
12 R	VS	4.87	3.75	0.67	1.16	0.05	13.07	19.34	0.02	0.32	0.84	0.53	0.42	2.23	4.34	70.15	16.20	13.65	sandy loam
13 R	VR	4.83	3.81	0.62	1.07	0.05	12.53	18.14	0.00	0.29	0.66	0.52	0.44	1.93	3.84	70.15	16.20	13.65	sandy loam
14 R	VR	5.68	4.27	1.32	2.28	0.10	13.78	18.87	0.07	0.10	0.33	0.66	0.03	0.70	1.81	61.47	24.14	14.40	sandy loam
15 V	VS	5.77	3.99	1.39	2.40	0.15	9.50	25.07	0.05	0.42	0.42	0.61	0.03	1.12	2.60	49.56	24.31	26.13	sandy clay loam
16 V	VR	6.48	4.67	1.45	2.50	0.16	9.06	34.04	0.52	0.50	0.44	0.47	0.03	0.00	1.44	49.56	24.31	26.13	sandy clay loam
17 V	VR	6.48	4.26	1.63	2.81	0.16	10.19	30.21	0.24	4.68	1.42	0.45	0.06	0.00	6.61	39.11	43.88	17.01	loam
18 RB	VS	7.75	5.90	2.62	4.52	0.22	11.93	35.11	2.70	11.76	1.17	0.71	0.04	0.21	13.89	52.80	31.60	15.59	sandy loam
19 RS	VS	5.70	4.93	3.82	6.59	0.24	15.62	115.21	0.47	4.95	2.58	0.84	0.01	0.33	8.38	61.49	26.47	12.04	sandy loam
20 R	VR	5.55	5.33	1.96	3.38	0.17	13.10	54.36	1.93	4.77	1.66	0.94	0.01	0.34	7.38	62.16	23.10	14.74	sandy loam
21 V	VS	7.78	6.47	1.74	3.00	0.14	12.57	34.53	6.91	12.64	0.76	0.55	0.04	0.64	14.62	53.30	32.06	14.64	sandy loam
22 RS	VR	4.61	4.49	3.75	6.47	0.23	15.99	101.82	0.34	4.45	1.45	0.68	0.01	0.97	7.56	56.97	31.71	11.32	sandy loam
23 M	VR	6.77	4.81	1.50	2.59	0.12	12.84	20.46	2.81	0.47	0.52	0.67	0.04	0.63	2.32	54.81	30.78	14.40	sandy loam
24 RS	VR	6.19	4.63	2.97	5.12	0.21	14.15	35.18	1.33	7.20	0.55	0.54	0.07	0.00	8.35	68.25	17.84	13.91	sandy loam
25 RS	VR	6.19	3.81	1.95	3.36	0.14	14.19	28.93	0.15	0.28	0.51	0.67	0.03	0.47	1.95	61.92	24.12	13.96	sandy loam
26 V	VS	7.29	5.49	2.76	4.76	0.22	12.56	40.78	4.34	11.80	1.25	1.25	0.04	0.00	14.35	30.96	50.04	19.00	silt loam
27 RS	VS	7.60	5.98	1.87	3.22	0.14	13.54	31.28	4.08	6.80	1.87	1.80	0.13	0.05	10.65	65.83	19.52	14.66	sandy loam
28 RB	VR	6.47	4.08	0.90	1.55	0.10	9.50	20.18	0.03	0.12	0.25	0.66	0.02	0.15	1.20	35.17	47.55	17.28	loam
29 R	VR	5.29	4.70	1.35	2.33	0.11	12.74	43.68	0.32	3.12	0.54	0.79	0.01	0.33	4.79	61.09	24.18	14.73	sandy loam
30 RS	VS	5.94	5.59	5.22	9.00	0.33	15.61	127.07	6.58	13.96	2.53	0.96	0.01	0.33	17.79	56.97	31.71	11.32	sandy loam
31 RS	VS	5.64	4.88	3.06	5.28	0.22	14.01	75.12	0.53	4.29	1.12	0.79	0.01	0.33	6.54	59.69	27.55	12.76	sandy loam
32 R	VS	7.03	5.97	1.27	2.19	0.10	12.37	26.63	2.46	9.68	0.77	0.77	0.04	0.00	11.26	65.68	20.50	13.82	sandy loam
33 RS	VS	6.26	4.10	3.59	6.19	0.25	14.44	37.39	0.61	6.40	1.21	0.78	0.04	0.00	8.43	68.25	17.84	13.91	sandy loam
34 R	VR	6.29	4.28	1.05	1.81	0.08	12.65	6.72	0.29	0.20	0.32	0.43	0.05	0.32	1.31	64.53	20.51	14.96	sandy loam
35 V	VS	5.96	3.39	1.09	1.88	0.11	9.59	22.85	0.02	0.21	0.45	0.80	0.04	0.48	1.98	61.45	24.60	13.95	sandy loam
36 R	VR	6.10	3.57	1.10	1.90	0.09	12.51	20.50	0.92	0.22	0.69	0.71	0.05	0.92	2.59	70.15	16.20	13.65	sandy loam
37 R	VR	4.85	3.79	1.45	2.50	0.13	11.23	18.69	0.04	0.08	0.22	0.54	0.03	1.25	2.12	47.95	37.67	14.37	loam
38 R	VS	4.91	4.31	1.85	3.19	0.15	12.37	63.78	0.22	2.52	0.42	0.86	0.01	0.52	4.33	61.09	24.18	14.73	sandy loam
39 V	VS	6.11	5.85	3.61	6.22	0.37	9.80	62.94	1.34	18.88	0.65	2.58	0.12	0.00	22.23	42.22	35.88	21.89	clay loam
40 V	VS	6.01	3.89	2.06	3.55	0.09	22.81	18.72	0.05	0.19	0.88	0.62	0.02	0.20	1.92	57.14	29.42	13.44	sandy clay loam
41 R	VR	5.00	4.00	2.20	3.79	0.16	13.63	30.21	0.42	0.22	0.33	0.56	0.03	0.61	1.75	42.78	43.17	14.05	loam
42 R	VR	5.03	4.31	3.12	5.38	0.25	12.26	49.26	0.24	0.56	0.63	0.78	0.04	0.01	2.03	43.70	41.21	15.09	loam

RB: Rías Baixas; RS: Ribeira Sacra; R: Ribeiro; M: Monterrei; V: Valdeorras. S.a.: Sampling area; C (%): total carbon; OM: Organic matter; DOC: dissolved organic carbon; DIC: dissolved inorganic carbon; N: total nitrogen; eCEC: effective cation exchange capacity; VR: vineyard row; VS: vineyard strain.

The pH_{H2O} values ranged from acidic (4.61; soil 22-Ribeira Sacra) to alkaline (8.02; soil 6-Valdeorras). The pH is one of the chemical properties of the soil that has the most influence on the availability of Cu. Its regulation is influenced by organic matter content, clays, metal oxides, Al^{3+} concentration in the exchange complex, and parent rock composition. In addition, pH affects the solubility and mobility of nutrients in the soil, which determines the degree of nutrient toxicity or deficiency, as well as Cu toxicity [56]. In general, soils with low pH have higher Cu^{2+} solubility, which increases their mobility and bioavailability. This can lead to contamination problems in other adjacent parts of the ecosystem or toxicity in plants and/or other organisms [57,58].

The pH_{KCl} values were lower than pH_{H2O}, especially in the acidic soils. These values ranged from 2.94 (soil 5-Monterrei) to 7.32 (soil 6-Valdeorras). Acidic soils tend to have a lower pH_{KCl} than pH_{H2O}, which is related to a higher release of Al^{3+} cations that favor soil acidification. This decrease in pH can lead to an increase in the availability of Cu in the soil, which can have significant implications for plants and their growth since Cu is an essential nutrient but can also become toxic at high concentrations [59].

The presence of organic matter in vineyard soils plays a fundamental role in Cu retention, especially in the top soil layers [60]. When organic matter comes into contact with metals in the soil, chemical reactions occur that lead to the formation of organometallic compounds, which are capable of binding metals, including Cu [13,61]. The amount of organic matter present in the soil is a key indicator of its quality and fertility. A soil is considered to have a low organic matter content when this value is less than 2% [10]. In the context of this study, the soils analyzed showed organic matter values ranging from 1.07% (soil 13-Ribeiro) to 9% (soil 30-Ribeira Sacra). These results show the diversity of the organic matter levels found in the vineyard soils studied, which may influence the Cu retention capacity and, therefore, the availability/toxicity of this metal for the vines. It should be noted that an adequate amount of organic matter in the soil not only favors Cu fixation, but also improves soil structure, increases its water and nutrient retention capacity, and promotes microbiological activity [62]. Therefore, to optimize growing conditions and healthy vine development, it is important to maintain and increase the organic matter content of vineyard soils through appropriate management practices. Additionally, the presence of adequate soil organic matter can reduce the risk of Cu leaching and its potential negative environmental impacts [63]. Concerning the dissolved organic carbon (DOC), it plays a decisive role in overall soil organic matter and Cu retention properties [64]. The analyzed soils showed a wide variability, ranging from 6.72 $mg \cdot kg^{-1}$ (soil 34-Ribeiro) to 127.07 $mg \cdot kg^{-1}$ (soil 29-Ribeiro). In particular, Cu^{2+} can form a complex with DOM, which directly affects the availability of Cu for vines. In addition, the quality of DOM affects the availability of Cu in the soil more than its quantity [65]. On the other hand, dissolved inorganic carbon (DIC) concentrations ranged from 0.01 $mg \cdot kg^{-1}$ (soil 10-Valdeorras) to 11.47 $mg \cdot kg^{-1}$ (soil 6-Valdeorras). Furthermore, the DIC may affect the availability of nutrients in the soil and play a role in the processes of ion exchange and chemical balance in the root environment of vines.

The effective cation exchange capacity (eCEC) is a fundamental measure that reflects the ability of the surface of soil compounds, such as organic matter, oxides, and clays, to retain cations such as Ca^{2+}, Mg^{2+}, Na^{+}, and K^{+} on the surface negative charge. In general, soils are considered to have a strong limitation in cation retention when eCEC values are lower than 4 $cmol_{(+)}kg^{-1}$, and a slight limitation when values are between 4 and 7 $cmol_{(+)}kg^{-1}$. In this study, the eCEC values of the soils studied ranged from 1.20 (soil 28-Ribeira Sacra) to 22.23 (soil 39-Valdeorras), showing a wide variability in cation retention capacity.

In viticulture, the total nitrogen (N) content of the soil plays a fundamental role in the growth and development of the vine. A low N content is when it is <0.1%, while a high N content is >0.3%. The soils analyzed in this study varied between values below 0.1% (soils 4-Monterrei, 12-Ribeiro, 13-Ribeiro, 34-Ribeiro, 36-Ribeiro, and 40-Valdeorras) and above 0.3% in soil 6-Valdeorras. In soils with low N levels, vines may experience difficulties in their development and growth, which may affect their production. On the other hand, in soils with an N concentration higher than 0.3%, vines may show excessive vigor, which may result in lower grape quality [65].

3.2. Total and Available Cu Content and Extraction Efficiency (E_xE_f) in Vineyard Soils

The total and available Cu contents of the vineyard soils are presented in Table 3. The results show a wide variability among the selected soils, with total Cu concentrations ranging from 16.4 $mg \cdot kg^{-1}$ (soil 1-Monterrei) to 292.9 $mg \cdot kg^{-1}$ (soil 42-Ribeiro), and available Cu ranging from 0.4 $mg \cdot kg^{-1}$ (soil 15-Valdeorras) to 24.0 $mg \cdot kg^{-1}$ (soil 34-Ribeiro). These values indicate that there is considerable heterogeneity in the Cu content of the vineyard soils studied.

Table 3. Copper contents (total and available) in vineyard soils (mg·kg^{-1}) and extraction efficiency (E_xE_f).

Soil	Sampling Area	Total Cu	Available Cu	E_xE_f	Soil	Sampling Area	Total Cu	Available Cu	E_xE_f
		mg kg^{-1}	mg kg^{-1}	(%)			mg kg^{-1}	mg kg^{-1}	(%)
1 M	VR	16.4	4.2	25.8	22 RS	VR	129.4	4.3	3.3
2 RB	VS	33.1	3.7	11.0	23 M	VR	135.6	2.2	1.6
3 RS	VR	34.2	2.0	5.7	24 RS	VR	153.2	1.2	0.8
4 M	VR	42.3	21.3	50.4	25 RS	VR	164.3	19.4	11.8
5 M	VS	44.5	2.4	5.3	26 V	VS	166.9	1.4	0.8
6 V	VR	49.3	3.2	6.5	27 RS	VS	170.4	1.4	0.8
7 RS	VS	60.7	2.8	4.6	28 RB	VR	182.5	7.5	4.1
8 RB	VR	64.3	7.5	11.6	29 R	VR	187.6	8.3	4.4
9 RB	VS	65.7	2.4	3.6	30 RS	VS	192.3	8.1	4.2
10 V	VR	66.8	23.7	35.5	31 RS	VS	200.8	7.6	3.8
11 V	VS	69.5	11.1	15.9	32 R	VS	201.8	1.5	0.8
12 R	VS	73.6	10.0	13.6	33 RS	VS	204.0	3.4	1.6
13 R	VR	74.6	3.2	4.3	34 R	VR	204.0	24.0	11.8
14 R	VR	82.4	6.7	8.1	35 V	VS	216.0	13.7	6.4
15 V	VS	104.0	0.4	0.4	36 R	VR	219.4	15.2	7.0
16 V	VR	109.2	2.3	2.1	37 R	VR	229.8	4.2	1.8
17 V	VR	114.0	12.0	10.5	38 R	VS	230.8	11.5	5.0
18 RB	VS	117.5	2.3	1.9	39 V	VS	248.8	15.1	6.1
19 RS	VS	117.7	7.8	6.7	40 V	VS	268.4	18.7	7.0
20 R	VR	122.2	18.2	14.9	41 R	VR	285.5	7.6	2.7
21 V	VS	126.7	3.1	2.5	42 R	VR	292.9	8.0	2.7

RB: Rías Baixas; RS: Ribeira Sacra; R: Ribeiro; M: Monterrei; V: Valdeorras. VR: vineyard row; VS: vineyard strain.

Concerning the generic Cu reference limits for Galician soils [13], it should be noted that many of the vineyard soils analyzed in this study exceed the maximum content recommended for ecosystem protection, which is 50 mg kg^{-1} (soils from 7 to 42). In addition, many of these vineyard soils also exceed the phytotoxic limit of 100 mg kg^{-1} (soils from 15 to 42) [12].

The high Cu content in most of the soils studied is closely related to the accumulation of this metal as a result of the phytosanitary treatments carried out on the vines over the years with copper-based fungicides [66]. These treatments, although necessary to control fungal diseases and maintain the development and growth of the vines, can cause a high increase in the Cu content of the soil. It is therefore essential to take these reference levels into account when adopting pest and disease management measures with Cu-based fungicides, as exceeding them can have negative consequences for parts of the adjacent ecosystems as well as for the vineyards. Indeed, high levels of Cu in the soil can promote its leaching and affect the quality of groundwater and surface water [67]. In addition, an excess of Cu in the soil can be phytotoxic to other plants or crops in the vicinity of the vineyard [12].

The available Cu content of soils is much lower than the total Cu content. Available Cu levels < 1 mg kg^{-1} are considered deficient, while available levels > 25 mg kg^{-1} may be toxic to grapevines, especially in acidic soils [10] (extracted used: 0.005 M DTPA, 0.01 M $CaCl_2$, and 0.1 M TEA—pH 7.3—ratio 1 g soil /10 mL solution). None of the soils exceeded 25 mg kg^{-1}, while 15 soils were below 1 mg kg^{-1} of available Cu. In addition, the extraction efficiency (E_xE_f) of Cu in each vineyard soil was calculated to show the percentage of Cu released [53] (Table 3).

Pearson's bivariate correlation analysis for data of soils sampled in vineyard strains (VS) and vineyard rows (VR) indicates that there is no positive correlation between total and available Cu contents in the soils (Table 4). In fact, many of the soils with high Cu contents show low available Cu contents and therefore low E_xE_f (e.g., soils 26-Valdeorras, 27-Ribeiro Sacra, 32-Ribeiro and 37-Ribeiro), while soils with low total Cu contents show high available Cu contents and high E_xE_f (e.g., soils 4 and 10). It was found that there is a positive correlation between available Cu content and E_xE_f for soils sampled in vineyard strains (VS) (R = 0.63) and for soils sampled in vineyard rows (VR) (R = 0.62). Correlation analysis also indicates that there is no specific correlation between total or available Cu content and soil characteristics for vineyard soil samples. This suggests that it may be a combination of factors that influence Cu availability or total content in vineyard soils.

Table 4. Pearson's bivariate correlation analysis (R values) for data of soils sampled in vineyard strains (VS) and vineyard rows (VR) and data from assays with *L. perenne*.

SAMPLING AREA: VS																							
	pH_{H2O}	pH_{KCl}	C(%)	OM	N	C/N	DOC	DIC	Ca^{2+}	Mg^{2+}	K^{+}	Na^{+}	Al^{3+}	eCEC	Sand	Silt	Clay	Total Cu	Avail. Cu	E_xE_f	Bm	Cu a.p.	BF
Total Cu	−0.13	0.02	−0.06	−0.06	−0.03	−0.03	0.24	−0.06	0.04	−0.01	0.34	−0.27	−0.29	0.03	0.00	0.03	−0.08	-	-	-	-	-	-
Avail. Cu	−0.19	−0.27	−0.22	−0.22	−0.28	0.14	−0.08	−0.26	−0.25	−0.09	0.06	−0.09	−0.10	−0.26	0.02	0.04	−0.19	0.25	-	-	-	-	-
E_xE_f	−0.08	−0.26	−0.15	−0.15	−0.22	0.11	−0.23	−0.18	−0.23	−0.10	−0.13	0.04	0.10	−0.23	−0.06	0.12	−0.14	−0.41	0.63	-	-	-	-
Bm	0.26	0.35	0.50	0.50	0.56	−0.01	0.20	0.30	0.53	0.27	0.36	0.03	−0.25	0.54	0.03	−0.02	−0.02	0.13	−0.31	−0.30	-	-	-
Cu a.p.	−0.17	−0.30	−0.25	−0.25	−0.28	−0.11	−0.20	−0.23	−0.24	−0.19	−0.09	0.13	0.26	−0.23	−0.05	0.06	0.00	0.18	0.31	0.29	−0.17	-	-
BF	0.10	0.04	−0.07	−0.07	0.02	−0.23	−0.13	0.03	0.02	−0.11	−0.02	−0.02	0.08	0.01	−0.09	−0.06	0.53	−0.10	−0.55	−0.34	0.05	−0.08	-
SAMPLING AREA: VR																							
	pH_{H2O}	pH_{KCl}	C(%)	OM	N	C/N	DOC	DIC	Ca^{2+}	Mg^{2+}	K^{+}	Na^{+}	Al^{3+}	eCEC	Sand	Silt	Clay	Total Cu	Avail. Cu	E_xE_f	Bm	Cu a.p.	BF
Total Cu	−0.23	−0.03	−0.01	−0.01	0.02	−0.04	0.23	−0.10	−0.04	−0.08	0.28	−0.27	−0.29	−0.06	−0.11	0.16	−0.10	-	-	-	-	-	-
Avail. Cu	−0.18	−0.27	−0.22	−0.22	−0.27	0.14	−0.07	−0.25	−0.24	−0.09	0.06	−0.09	−0.10	−0.26	0.02	0.04	−0.19	0.22	-	-	-	-	-
E_xE_f	−0.04	−0.24	−0.16	−0.16	−0.22	0.11	−0.23	−0.16	−0.21	−0.08	−0.12	0.04	0.11	−0.20	−0.03	0.07	−0.13	−0.42	0.62	-	-	-	-
Bm	0.21	0.33	0.51	0.51	0.57	−0.02	0.21	0.27	0.48	0.25	0.36	0.02	−0.28	0.49	0.00	0.01	−0.02	0.16	−0.30	−0.30	-	-	-
Cu a.p.	−0.13	−0.28	−0.26	−0.26	−0.29	−0.11	−0.20	−0.22	−0.22	−0.17	−0.09	0.14	0.26	−0.20	−0.02	0.02	0.00	0.11	0.31	0.30	−0.18	-	-
BF	0.13	0.05	−0.08	−0.08	0.01	−0.22	−0.13	0.04	0.04	−0.09	−0.01	−0.01	0.08	0.03	−0.06	−0.09	0.53	−0.13	−0.55	−0.32	0.03	−0.07	-

C (%): total carbon; OM: Organic matter; DOC: dissolved organic carbon; DIC: dissolved inorganic carbon; N: total nitrogen; eCEC: effective cation exchange capacity; extraction efficiency ($\mathbf{E_xE_f}$); Biomass (Bm); Cu in aerial part of *L. perenne* (Cu a.p.); Bioaccumulation factor (BF). Color intensity: Red indicates a value of R = 1, blue indicates a value of R = −1, white R = 0.

A bivariate Pearson's correlation analysis was also carried out, separating the soils into those with total Cu contents below the limit considered phytotoxic (<100 mg kg^{-1}) and those with contents above 100 mg kg^{-1} (above the phytotoxic limit), regardless of the sampling zone (VS or VR) (Table 5). The rest of the results of the correlation analysis were similar to those obtained when the soils were separated by sampling zones. No specific correlation was observed between total and available Cu contents and a strong correlation between available Cu content and E_xE_f was observed, (R = 0.87 and R = 0.90), while there is also no specific correlation between total or available Cu content and soil characteristics.

Principal Component Analysis (PCA) was carried out to determine which variables could explain the different types of vineyard soils studied and their possible relationship with Cu contents. The results of the PCA are shown in Figures 1 and 2 and Tables 5 and 6. The PCA analysis of the data from the trials with soils sampled in the corridor zone (P) (Figure 3 and Table 6) allowed us to reduce the initial dimension of the dataset to two dimensions that explain 70.85% of the variance of the data (Component **1**: 42.25% and Component **2**: 28.71%) (Table 6). In the case of the soils sampled in the stock zone (C) (Figure 1 and Table 6), PCA also allowed us to reduce the initial dimension of the dataset to two dimensions that explain 76.51% of the variance of the data (Component **1**: 49.95% and Component **2**: 26.57%) (Table 6).

The available Cu content, as well as the E_xE_f and its relationship with the soil characteristics, in soils sampled in both VS and VR, can be explained by component **2**. A negative relationship between available Cu content and E_xE_f, organic matter content, DOC, and nitrogen content is observed. This means that in soils with low organic matter and nitrogen content, the availability of Cu in the soil can increase. This may be due to the fact that in soils poor in organic matter and nitrogen, there is less competition for adsorption sites and therefore less Cu retention, which facilitates its release and availability (Figures 1 and 2).

Table 5. Pearson's bivariate correlation analysis (R values) for data of soils with total Cu contents < 100 mg kg^{-1} (A) (below phytotoxic level) and > 100 mg kg^{-1} (B) (above phytotoxic level) and data from assays with *L. perenne*.

Soils with total Cu contents below phytotoxic level (<100 mg kg^{-1})																						
	pH_{H2O}	pH_{KCl}	C(%)	OM	N	C/N	DOC	DIC	Ca^{2+}	Mg^{2+}	K^+	Na^+	Al^{3+}	eCEC	Sand	Silt	Clay	Total Cu	Avail. Cu	E_xE_f	Bm	Cu a.p.	BF
Total Cu	−0.11	0.07	−0.26	−0.26	−0.35	0.19	−0.27	−0.05	−0.09	−0.01	0.14	0.41	−0.01	−0.09	0.33	−0.32	−0.31		0.19	−0.24	−0.09	−0.30	−0.44
Avail. Cu	−0.05	−0.18	−0.32	−0.33	−0.43	0.10	−0.47	−0.16	−0.25	−0.15	0.16	−0.10	−0.16	−0.29	−0.19	0.25	−0.24	0.19		0.87	−0.65	0.11	−0.78
E_xE_f	−0.05	−0.23	−0.25	−0.25	−0.30	0.00	−0.44	−0.17	−0.27	−0.18	−0.02	−0.21	−0.08	−0.31	−0.27	0.32	−0.13	−0.24	0.87		−0.62	0.37	−0.53
Bm	0.40	0.41	0.58	0.58	0.62	0.11	0.45	0.44	0.50	0.34	0.04	−0.07	−0.14	0.52	0.03	−0.09	0.31	−0.09	−0.65	−0.62		−0.52	0.36
Cu a.p.	−0.31	−0.22	−0.22	−0.22	−0.26	−0.11	−0.29	−0.29	−0.16	0.10	−0.35	0.45	0.40	−0.10	−0.03	0.04	−0.01	−0.30	0.11	0.37	−0.52		0.16
BF	−0.21	−0.20	0.19	0.19	0.30	−0.25	0.14	−0.16	−0.01	0.04	0.08	0.05	0.30	0.04	0.03	−0.07	0.21	−0.44	−0.78	−0.53	0.36	0.16	
Soils with total Cu contents above phytotoxic level (>100 mg kg^{-1})																							
	pH_{H2O}	pH_{KCl}	C(%)	OM	N	C/N	DOC	DIC	Ca^{2+}	Mg^{2+}	K^+	Na^+	Al^{3+}	eCEC	Sand	Silt	Clay	Total Cu	Avail. Cu	E_xE_f	Bm	Cu a.p.	BF
Total Cu	−0.40	−0.31	0.05	0.05	0.02	0.22	−0.13	−0.25	−0.14	−0.33	0.20	0.09	−0.06	−0.15	−0.04	0.15	−0.26		0.34	−0.02	0.24	0.17	−0.40
Avail. Cu	−0.27	−0.36	−0.15	−0.15	−0.18	0.20	−0.08	−0.34	−0.25	−0.09	0.03	−0.07	−0.01	−0.24	0.18	−0.12	−0.20	0.34		0.90	−0.19	0.08	−0.59
E_xE_f	−0.19	−0.24	−0.14	−0.14	−0.15	0.11	0.01	−0.27	−0.20	0.12	−0.05	−0.14	−0.03	−0.18	0.17	−0.12	−0.18	−0.02	0.90		−0.24	−0.04	−0.54
Bm	0.15	0.33	0.51	0.51	0.59	−0.08	0.21	0.22	0.50	0.23	0.41	0.31	−0.48	0.50	0.00	0.04	−0.08	0.24	−0.19	−0.24		−0.02	−0.01
Cu a.p.	0.00	−0.19	−0.19	−0.19	−0.21	−0.17	−0.13	0.07	−0.06	−0.18	−0.14	0.01	0.13	−0.08	0.25	−0.27	−0.01	0.17	0.08	−0.04	−0.02		0.04
BF	0.23	0.10	−0.14	−0.14	−0.05	−0.26	−0.21	0.10	0.04	−0.13	−0.05	0.15	0.17	0.03	−0.06	−0.15	0.56	−0.40	−0.59	−0.54	−0.01	0.04	

C (%): total carbon; OM: Organic matter; DOC: dissolved organic carbon; DIC: dissolved inorganic carbon; N: total nitrogen; eCEC: effective cation exchange capacity; extraction efficiency (**E_xE_f**); Biomass (Bm); Cu in aerial part of *L. perenne* (Cu a.p.); Bioaccumulation factor (BF). Color intensity: Red indicates a value of R = 1, blue indicates a value of R = −1, white R = 0.

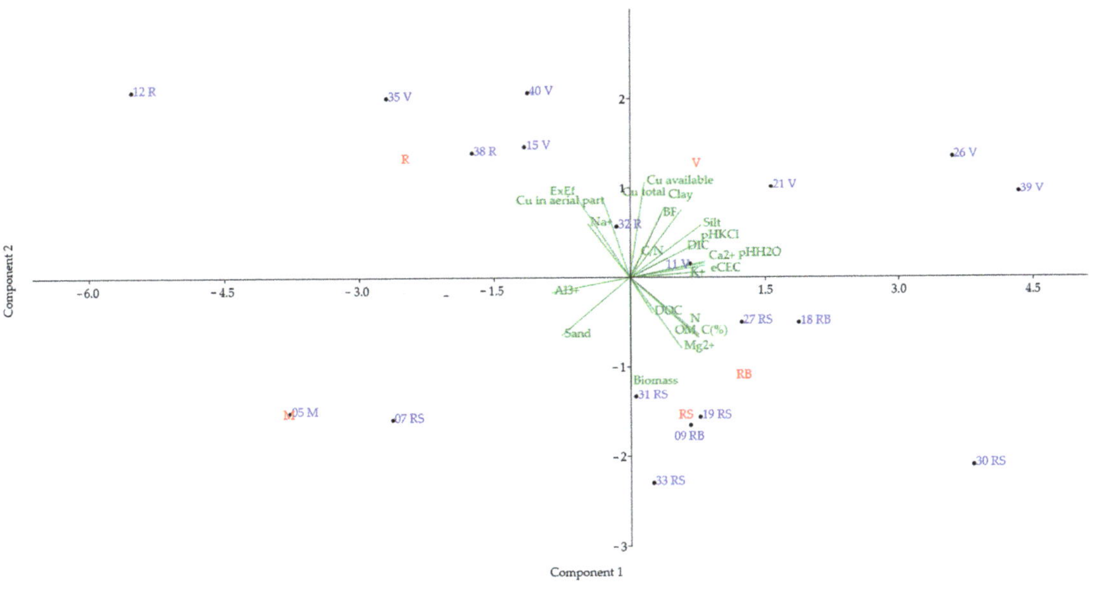

Figure 1. Scatter plot of principal component analysis (PCA) for soil data sampled in vineyard strains (VS) and data from assays with *L. perenne*.

3.3. Biomass of *L. perenne*, Copper Content, and the Bioaccumulation Factor

Table 7 shows the data related to the dry biomass obtained from the aerial part of *L. perenne*, as well as the Cu content in the aerial part of *L. perenne* obtained in the pot experiments. From the results of the dry biomass weights, it was found that the biomass content of *L. perenne* did not seem to be affected by the total Cu content in the soil. In fact, *L. perenne* biomass in soils with Cu concentrations above the phytotoxic limit (100 mg kg^{-1}) did not differ from soils with Cu values < 100 mg kg^{-1} (Figure 2). Furthermore, correlation analysis showed that *L. perenne* biomass for soils sampled in VR was related to organic matter content (R = 0.50), total N (R = 0.56), and eCEC (R = 0.54), with

Ca^{2+} (R = 0.53) being the cation of the exchange complex that contributed most to *L. perenne* biomass content. Similar results were found for soils sampled in VS, where correlation analysis indicated that *L. perenne* biomass was also related to organic matter (R = 0.51), total N (R = 0.57), eCEC (R = 0.49), and Ca^{2+} (R = 0.48). *Lolium perenne* is known for its ability to respond positively to soils with high N, organic matter, and cation exchange capacity, which affects its growth and development [68].

Figure 2. Image of the pot experiment with vineyard soils and *L. perenne* in a climatic chamber after one month of growth.

Table 6. Eigenvalues, percent variance explained, and loadings of each PCA component for data from soils sampled in vineyard rows (VR) and vineyard strains (VS) and data from assays with *L. perenne* (linked to Figures 1 and 2).

Soils Sampled in Vineyard Rows (VR)				
PC	**Eigenvalues**		**% Variance**	
1	9.694		42.148	
2	6.602		28.704	
3	4.591		19.961	
4	2.113		9.187	
	PC 1	**PC 2**	**PC 3**	**PC 4**
01 M	−0.996	−1.236	1.575	−2.04
03 RS	−1.088	0.709	0.088	−1.138
04 M	−1.88	−2.599	0.503	−1.008
06 V	6.897	0.597	3.817	−0.17
08 RB	3.803	2.62	−0.558	−2.253
10 V	−0.162	−2.797	−0.864	−0.111
13 R	−2.284	−0.959	−1.835	−0.988
14 R	−1.133	−0.439	−0.843	−0.569
16 V	0.444	−1.362	1.669	0.488
17 V	2.018	−0.574	0.966	0.639
20 R	0.929	1.467	0.339	1.837

Table 6. *Cont.*

	Soils Sampled in Vineyard Rows (VR)			
PC	**Eigenvalues**		**% Variance**	
22 RS	−0.869	4.436	2.025	0.986
23 M	−0.002	0.016	−0.086	−0.797
24 RS	−0.429	3.015	1.459	−1.004
25 RS	−0.965	0.02	−0.749	0.356
28 RB	1.04	−1.798	−1.276	0.29
29 R	−0.385	0.525	−1.048	1.197
34 R	−1.203	−2.035	−1.472	0.632
36 R	−2.298	−1.131	−1.351	1.14
37 R	−1.597	−1.095	−0.764	0.521
41 R	−0.681	0.199	−0.723	0.715
42 R	0.841	2.421	−0.872	1.276
	Soils sampled in vineyard strains (VS)			
PC	**Eigenvalues**		**% variance**	
1	11.487		49.945	
2	6.112		26.572	
3	2.988		12.992	
4	2.413		10.491	
	PC 1	**PC 2**	**PC 3**	**PC 4**
05 M	−3.78	−1.518	−0.729	−1.402
07 RS	−2.631	−1.589	−0.893	−0.893
09 RB	0.659	−1.654	−1.645	0.38
11 V	0.658	0.151	0.728	−1.347
12 R	−5.53	2.082	0.442	0.96
15 V	−1.164	1.466	−3.879	−1.874
18 RB	1.869	−0.505	−1.437	0.937
19 RS	0.767	−1.564	2.058	−0.358
21 V	1.563	1.015	−1.037	0.877
26 V	3.602	1.35	−0.988	0.577
27 RS	1.232	−0.501	−0.16	0.633
30 RS	3.834	−2.102	3.193	1.38
31 RS	0.055	−1.336	1.092	−0.187
32 R	−0.153	0.567	−0.544	1.561
33 RS	0.246	−2.295	−0.033	0.28
35 V	−2.694	2.01	0.576	−0.04
38 R	−1.743	1.399	1.421	−0.157
39 V	4.34	0.954	0.34	0.652
40 V	−1.13	2.069	1.494	−1.979

Regarding the ability of *L. perenne* to grow in vineyard soils with high Cu contents, it is important to consider that this species is considered tolerant to the presence of Cu in the soil [69]. The results of Cu in the aerial part and the bioaccumulation factor (the ratio between the Cu content in the aerial part of *L. perenne* and the available Cu content in the soil) show that the Cu contents accumulated in the aerial part of the plant are low, with percentages in the aerial part that, in most cases, are less than 1% of the available Cu content in the soil (Table 4). Only the soils 15-Valdeorras, 24-Ribeira Sacra, 26-Valdeorras, and 32-Ribeiro have a BF $\geq$ 1%. Correlation analysis showed that BF is negatively correlated with available Cu content, both in the soils sampled in VR ($R = 0.55$) and VS ($R = 0.55$). Thus, there is an inverse relationship between available Cu content in vineyard soils and Cu accumulation in the aerial part of *L. perenne*. In other words, Cu accumulation in the aerial part of *L. perenne* may decrease when there is higher Cu availability in the soil, suggesting that there may be defense mechanisms in *L. perenne* that prevent Cu accumulation in its aerial part [70].

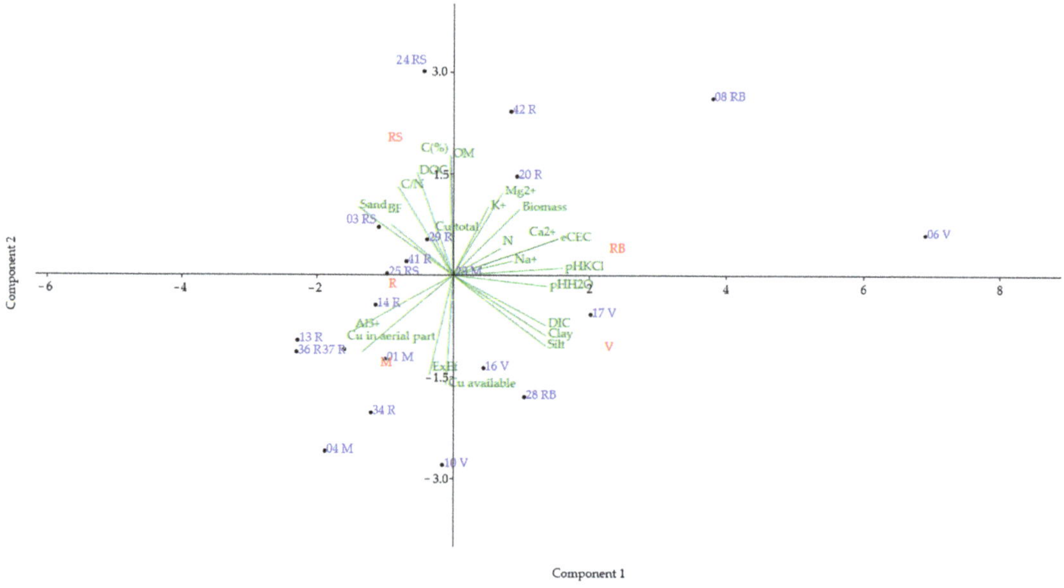

Figure 3. Scatter plot of principal component analysis (PCA) for soil data sampled in vineyard rows (VR) and data from assays with *L. perenne*.

Table 7. Dry biomass production (mg kg^{-1}) of *L. perenne*, Cu content in the aerial part, and bioaccumulation factor (BF).

Soil	Sampling Area	Biomass	Cu in Aerial Part	BF	Soil	Sampling Area	Biomass	Cu in Aerial Part	BF
		mg kg^{-1}	mg kg^{-1}	(%)			mg kg^{-1}	mg kg^{-1}	(%)
1 M	VR	0.769 ± 0.048	0.023 ± 0.002	0.55	**22 RS**	VR	0.690 ± 0.004	0.011 ± 0.001	0.26
2 RB	VS	0.731 ± 0.030	0.015 ± 0.001	0.41	**23 M**	VR	0.679 ± 0.036	0.013 ± 0.001	0.59
3 RS	VR	0.744 ± 0.010	0.009 ± 0.001	0.47	**24 RS**	VR	0.831 ± 0.127	0.015 ± 0.001	1.27
4 M	VR	0.621 ± 0.021	0.017 ± 0.004	0.08	**25 RS**	VR	0.724 ± 0.099	0.013 ± 0.001	0.07
5 M	VS	0.858 ± 0.078	0.011 ± 0.008	0.46	**26 V**	VS	0.837 ± 0.017	0.015 ± 0.001	1.05
6 V	VR	0.847 ± 0.048	0.009 ± 0.003	0.27	**27 RS**	VS	0.859 ± 0.103	0.012 ± 0.001	0.85
7 RS	VS	0.83 ± 0.103	0.008 ± 0.002	0.30	**28 RB**	VR	0.778 ± 0.099	0.014 ± 0.001	0.18
8 RB	VR	0.883 ± 0.077	0.013 ± 0.005	0.17	**29 R**	VR	0.699 ± 0.103	0.012 ± 0.001	0.15
9 RB	VS	0.847 ± 0.016	0.013 ± 0.003	0.55	**30 RS**	VS	0.99 ± 0.166	0.015 ± 0.004	0.18
10 V	VR	0.643 ± 0.042	0.014 ± 0.002	0.06	**31 RS**	VS	0.979 ± 0.042	0.013 ± 0.001	0.17
11 V	VS	0.788 ± 0.006	0.014 ± 0.001	0.12	**32 R**	VS	0.922 ± 0.011	0.015 ± 0.001	1.00
12 R	VS	0.695 ± 0.031	0.017 ± 0.001	0.17	**33 RS**	VS	1.061 ± 0.053	0.012 ± 0.001	0.37
13 R	VR	0.685 ± 0.053	0.016 ± 0.001	0.51	**34 R**	VR	0.765 ± 0.019	0.019 ± 0.001	0.08
14 R	VR	0.691 ± 0.012	0.013 ± 0.001	0.19	**35 V**	VS	0.643 ± 0.046	0.019 ± 0.001	0.14
15 V	VS	0.672 ± 0.052	0.012 ± 0.001	2.98	**36 R**	VR	0.708 ± 0.037	0.025 ± 0.008	0.16
16 V	VR	0.654 ± 0.019	0.016 ± 0.002	0.70	**37 R**	VR	0.727 ± 0.073	0.016 ± 0.002	0.37
17 V	VR	0.746 ± 0.027	0.013 ± 0.002	0.10	**38 R**	VS	0.645 ± 0.057	0.017 ± 0.001	0.15
18 RB	VS	0.855 ± 0.006	0.014 ± 0.001	0.60	**39 V**	VS	0.957 ± 0.047	0.013 ± 0.003	0.09
19 RS	VS	0.605 ± 0.227	0.012 ± 0.001	0.15	**40 V**	VS	0.569 ± 0.038	0.011 ± 0.001	0.06
20 R	VR	0.732 ± 0.009	0.014 ± 0.001	0.07	**41 R**	VR	0.689 ± 0.087	0.013 ± 0.001	0.17
21 V	VS	0.613 ± 0.049	0.012 ± 0.001	0.40	**42 R**	VR	0.961 ± 0.002	0.012 ± 0.001	0.14

RB: Rías Baixas; RS: Ribeira Sacra; R: Ribeiro; M: Monterrei; V: Valdeorras. VR: vineyard row; VS: vineyard strain.

The exclusion of Cu in the aerial part of *L. perenne* may be beneficial in reducing Cu transfer through the food chain. This makes this species a good choice as a ground cover without increasing the risk of Cu transfer to other organisms. Although these results suggest that *L. perenne* is not suitable for Cu phytoextraction in vineyard soils, it can play a valuable role as a cover crop due to its ability to grow in high Cu environments and improve edaphic soil conditions [69,71]. In addition, its use as a cover crop can have other benefits, such as improving soil structure by increasing porosity and water-holding capacity and preventing soil erosion.

The PCA analysis with the data of the assays with soils sampled in VR (Figure 3 and Table 6) allowed us to explain the effect of available Cu on the biomass of *L. perenne* and the Cu content in the aerial part of the plant. Component **1** for soils sampled in VR explained the negative relationship between *L. perenne* biomass and aerial Cu content, while component **2** explained the positive relationship between aerial Cu content and available Cu content. This means that a higher Cu content in the aerial part of *L. perenne* is associated with high available Cu contents in the soils sampled in VR. Furthermore, a higher Cu content in the aerial part has a negative effect on the biomass of *L. perenne*, a parameter that, as mentioned above, is not affected by the total Cu content in the soil (Figure 3 and Table 3). In line with these results, component **1** also explains a positive relationship between *L. perenne* biomass and soil characteristics such as eCEC and exchange complex cations such as Ca^{2+}, Mg^{2+}, and K^{+}, as well as with soil N content. In addition, Al^{3+} levels may be detrimental to the growth of *L. perenne*.

The PCA analysis with the data of the assays with soils sampled in VS (Figure 1 and Table 7) also served to explain the effect of available Cu on *L. perenne* biomass and Cu content in the aerial part of the plant. In particular, component **2** explains most of the results, where it has been observed that the biomass of *L. perenne* is negatively related to the Cu content in the aerial part of the plant and to the Cu available in the soil. In addition, the growth of *L. perenne* is mainly favored by the content of organic matter and N.

4. Conclusions

The vineyard soils of the different D.O.s of Galicia show great variability in their physicochemical characteristics and significant variations within each D.O., including differences in texture, pH, organic matter content, cation exchange capacity, and nitrogen content. The total and available Cu contents of the vineyard soils studied varied widely. Many of the soils analyzed exceeded the reference limits established for ecosystem protection and phytotoxicity, indicating that phytosanitary treatments with Cu-based fungicides used in vineyards have led to significant Cu accumulation in most of the soils. The total Cu content of the soil does not seem to affect the biomass of *L. perenne*. Even in soils with Cu concentrations above the phytotoxic limit, no differences in plant biomass were observed compared to soils with lower Cu levels. Plant biomass is related to soil characteristics such as organic matter content, N, and cation exchange capacity, indicating that these factors are more determinant for plant growth. Although *L. perenne* is considered to be tolerant to the presence of Cu in the soil, an inverse relationship was found between available Cu content in the soil and Cu accumulation in the aerial part of the plant. Therefore, although *L. perenne* is not suitable for phytoextraction of Cu from vineyard soils, it may play a valuable role as a phytoremediation species and as a ground cover in vineyard soils, helping to improve soil conditions and prevent mobility and transfer of Cu to other parts of the ecosystem or through the food chain. This study provides the basis for future scientific research aimed at achieving sustainable agriculture and developing techniques for the reclamation of vineyard soils affected by high Cu concentrations through the use of plant cover based on the use of *L. perenne*.

Author Contributions: Conceptualization, D.F.-C. and D.A.-L.; methodology, D.F.-C. and M.A.-E.; software, D.A.-L.; validation, R.V.-B., M.A.-E., D.F.-C. and D.A.-L.; formal analysis, R.V.-B.; investigation, R.V.-B., M.A.-E., D.F.-C. and D.A.-L.; resources, M.A.-E. and D.F.-C.; data curation, R.V.-B.; writing—original draft preparation, R.V.-B.; writing—review and editing, M.A.-E., D.F.-C. and D.A.-L.; visualization, M.A.-E. and D.A.-L.; supervision, D.F.-C. and D.A.-L.; project administration, D.F.-C.; funding acquisition, D.F.-C. All authors have read and agreed to the published version of the manuscript.

Funding: This work was funded by the COPPEREPLACE project, which has received funding from (grant agreement SOE4/P1/E1000) FEDER funds through the INTERREG- SUDOE program, and by Xunta de Galicia via the BV1 research group (ED431C 2017/62-GRC).

Data Availability Statement: All data are shown in the manuscript.

Acknowledgments: Daniel Arenas-Lago thanks the Ministerio de Ciencia e Innovación of Spain and the University of Vigo for the postdoc grant Juan de la Cierva Incorporación 2019 (IJC2019-042235-I).

Conflicts of Interest: The authors declare no conflict of interest. The funders had no role in the design of the study; in the collection, analyses, or interpretation of data; in the writing of the manuscript; or in the decision to publish the results.

References

1. Xunta de Galicia Autorizaciones Para Plantaciones de Viñedo (2016–2045). (In Spanish). Available online: https://mediorural.xunta.gal/es/temas/explotaciones/vinedo/autorizaciones-para-plantaciones-de-vina (accessed on 1 July 2023).
2. Denomination of Origen Ribeiro. Available online: https://www.ribeiro.wine (accessed on 14 August 2023).
3. Denomination of Origen Valdeorras. Available online: https://dovaldeorras.tv (accessed on 14 August 2023).
4. Denomination of Origen Rías Baixas. Available online: https://doriasbaixas.com (accessed on 14 August 2023).
5. Denomination of Origen Ribeira Sacra. Available online: https://ribeirasacra.org (accessed on 14 August 2023).
6. Denomination of Origen Monterrei. Available online: https://domonterrei.wine (accessed on 14 August 2023).
7. Rodríguez Guitián, M.A.; Ramil Rego, P. Clasificaciones Climáticas Aplicadas a Galicia: Revisión Desde Una Perspectiva Biogeográfica. *Recursos Rurais* **2007**, *3*, 31–53. [CrossRef]
8. Fraga, H.; Malheiro, A.C.; Moutinho-Pereira, J.; Cardoso, R.M.; Soares, P.M.M.; Cancela, J.J.; Pinto, J.G.; Santos, J.A. Integrated Analysis of Climate, Soil, Topography and Vegetative Growth in Iberian Viticultural Regions. *PLoS ONE* **2014**, *9*, e108078. [CrossRef] [PubMed]
9. Santos, I.V.; Bulle, C.; Levasseur, A.; Deschênes, L. Regionalized Terrestrial Ecotoxicity Assessment of Copper-Based Fungicides Applied in Viticulture. *Sustainability* **2018**, *10*, 2522. [CrossRef]
10. Vázquez-Blanco, R.; González-Feijoo, R.; Campillo-Cora, C.; Fernández-Calviño, D.; Arenas-Lago, D. Risk Assessment and Limiting Soil Factors for Vine Production—Cu and Zn Contents in Vineyard Soils in Galicia (Rías Baixas D.O.). *Agronomy* **2023**, *13*, 309. [CrossRef]
11. Macías, F. Proceso de Contaminación y Recuperación de Suelos. *Manten. Ing. Ind. Y De Edif.* **2000**, *135*, 15–25.
12. Kabata-Pendias, A. *Trace Elements in Soils and Plants*, 4th ed.; CRC Press: Boca Raton, FL, USA, 2010; ISBN 9781420093704.
13. Macías, F.; Calvo de Anta, R. *Niveles Genéricos de Referencia de Metales Pesados y Otros Elementos Traza en los Suelos de Galicia*; Xunta de Galicia: Santiago de Compostela, Spain, 2009; ISBN 978-84-453-4664-8.
14. Arenas-Lago, D.; Vega, F.A.; Silva, L.F.O.; Andrade, M.L. Copper Distribution in Surface and Subsurface Soil Horizons. *Environ. Sci. Pollut. Res.* **2014**, *21*, 10997–11008. [CrossRef]
15. Kumar, V.; Pandita, S.; Singh Sidhu, G.P.; Sharma, A.; Khanna, K.; Kaur, P.; Bali, A.S.; Setia, R. Copper Bioavailability, Uptake, Toxicity and Tolerance in Plants: A Comprehensive Review. *Chemosphere* **2021**, *262*, 127810. [CrossRef]
16. Adrees, M.; Ali, S.; Rizwan, M.; Ibrahim, M.; Abbas, F.; Farid, M.; Zia-ur-Rehman, M.; Irshad, M.K.; Bharwana, S.A. The Effect of Excess Copper on Growth and Physiology of Important Food Crops: A Review. *Environ. Sci. Pollut. Res.* **2015**, *22*, 8148–8162. [CrossRef]
17. Vareda, J.P.; Valente, A.J.M.; Durães, L. Assessment of Heavy Metal Pollution from Anthropogenic Activities and Remediation Strategies: A Review. *J. Environ. Manag.* **2019**, *246*, 101–118. [CrossRef]
18. Karimi, B.; Masson, V.; Guilland, C.; Leroy, E.; Pellegrinelli, S.; Giboulot, E.; Maron, P.A.; Ranjard, L. Ecotoxicity of Copper Input and Accumulation for Soil Biodiversity in Vineyards. *Environ. Chem. Lett.* **2021**, *19*, 2013–2030. [CrossRef]
19. Boso, S.; Alonso-Villaverde, V.; Gago, P.; Santiago, J.L.; Martínez, M.C. Susceptibility of 44 Grapevine (*Vitis vinifera* L.) Varieties to Downy Mildew in the Field. *Aust. J. Grape Wine Res.* **2011**, *17*, 394–400. [CrossRef]
20. Pham, N.T.H.; Babcsányi, I.; Balling, P.; Farsang, A. Accumulation Patterns and Health Risk Assessment of Potentially Toxic Elements in the Topsoil of Two Sloping Vineyards (Tokaj-Hegyalja, Hungary). *J. Soils Sediments* **2022**, *22*, 2671–2689. [CrossRef]
21. Pietrzak, U.; McPhail, D.C. Copper Accumulation, Distribution and Fractionation in Vineyard Soils of Victoria, Australia. *Geoderma* **2004**, *122*, 151–166. [CrossRef]

22. Eon, P.; Robert, T.; Goutouly, J.P.; Aurelle, V.; Cornu, J.Y. Cover Crop Response to Increased Concentrations of Copper in Vineyard Soils: Implications for Copper Phytoextraction. *Chemosphere* **2023**, *329*, 138604. [CrossRef]
23. Campillo-Cora, C.; Fernández-Calviño, D.; Pérez-Rodríguez, P.; Fernández-Sanjurjo, M.J.; Núñez-Delgado, A.; Álvarez-Rodríguez, E.; Arias-Estévez, M.; Nóvoa-Muñoz, J.C. Copper and Zinc in Rhizospheric Soil of Wild Plants Growing in Long-Term Acid Vineyard Soils. Insights on Availability and Metal Remediation. *Sci. Total Environ.* **2019**, *672*, 389–399. [CrossRef] [PubMed]
24. Monteiro, A.; Lopes, C.M. Influence of Cover Crop on Water Use and Performance of Vineyard in Mediterranean Portugal. *Agric. Ecosyst. Environ.* **2007**, *121*, 336–342. [CrossRef]
25. Alengebawy, A.; Abdelkhalek, S.T.; Qureshi, S.R.; Wang, M.Q. Heavy Metals and Pesticides Toxicity in Agricultural Soil and Plants: Ecological Risks and Human Health Implications. *Toxics* **2021**, *9*, 42. [CrossRef]
26. Fagnano, M.; Agrelli, D.; Pascale, A.; Adamo, P.; Fiorentino, N.; Rocco, C.; Pepe, O.; Ventorino, V. Copper Accumulation in Agricultural Soils: Risks for the Food Chain and Soil Microbial Populations. *Sci. Total Environ.* **2020**, *734*, 139434. [CrossRef]
27. Komárek, M.; Száková, J.; Rohošková, M.; Javorská, H.; Chrastný, V.; Balík, J. Copper Contamination of Vineyard Soils from Small Wine Producers: A Case Study from the Czech Republic. *Geoderma* **2008**, *147*, 16–22. [CrossRef]
28. Madejón, P.; Ramírez-Benítez, J.E.; Corrales, I.; Barceló, J.; Poschenrieder, C. Copper-Induced Oxidative Damage and Enhanced Antioxidant Defenses in the Root Apex of Maize Cultivars Differing in Cu Tolerance. *Environ. Exp. Bot.* **2009**, *67*, 415–420. [CrossRef]
29. Aarya, K.R.; Mathew, A.M. Effect of Copper Based Fungicides on the Properties of Red Soils. *Int. J. Sci. Res. Eng. Dev.* **2020**, *3*, 856–860.
30. Ortega, P.; Sánchez, E.; Gil, E.; Matamoros, V. Use of Cover Crops in Vineyards to Prevent Groundwater Pollution by Copper and Organic Fungicides. Soil Column Studies. *Chemosphere* **2022**, *303*, 134975. [CrossRef]
31. Giffard, B.; Winter, S.; Guidoni, S.; Nicolai, A.; Castaldini, M.; Cluzeau, D.; Coll, P.; Cortet, J.; Le Cadre, E.; d'Errico, G.; et al. Vineyard Management and Its Impacts on Soil Biodiversity, Functions, and Ecosystem Services. *Front. Ecol. Evol.* **2022**, *10*, 850272. [CrossRef]
32. Dumitriu Gabur, G.D.; Gabur, I.; Cucolea, E.I.; Costache, T.; Rambu, D.; Cotea, V.V.; Teodosiu, C. Investigating Six Common Pesticides Residues and Dietary Risk Assessment of Romanian Wine Varieties. *Foods* **2022**, *11*, 2225. [CrossRef] [PubMed]
33. Komárek, M.; Čadková, E.; Chrastný, V.; Bordas, F.; Bollinger, J.C. Contamination of Vineyard Soils with Fungicides: A Review of Environmental and Toxicological Aspects. *Environ. Int.* **2010**, *36*, 138–151. [CrossRef]
34. Fouillet, E.; Delière, L.; Flori, A.; Rapidel, B.; Merot, A. Diversity of Pesticide Use Trajectories during Agroecological Transitions in Vineyards: The Case of the French DEPHY Network. *Agric. Syst.* **2023**, *210*, 103725. [CrossRef]
35. Ruggieri, F.; Coulon-Leroy, C.; Mazé, A. How Can Collective Action Support the Agroecological Transition in Geographical Indication Vineyards? Insights from the Loire Valley Wine Area. *Sustainability* **2023**, *15*, 9371. [CrossRef]
36. Coll, P.; Le Cadre, E.; Blanchart, E.; Hinsinger, P.; Villenave, C. Organic Viticulture and Soil Quality: A Long-Term Study in Southern France. *Appl. Soil Ecol.* **2011**, *50*, 37–44. [CrossRef]
37. Cataldo, E.; Fucile, M.; Mattii, G.B. A Review: Soil Management, Sustainable Strategies and Approaches to Improve the Quality of Modern Viticulture. *Agronomy* **2021**, *11*, 2359. [CrossRef]
38. Abad, J.; de Mendoza, I.H.; Marín, D.; Orcaray, L.; Santesteban, L.G. Cover Crops in Viticulture. A Systematic Review (2): Implications on Vineyard Agronomic Performance. *OENO One* **2021**, *55*, 1–27. [CrossRef]
39. Abad, J.; Hermoso De Mendoza, I.; Marín, D.; Orcaray, L.; Santesteban, L.G. Cover Crops in Viticulture. A Systematic Review (1): Implications on Soil Characteristics and Biodiversity in Vineyard. *OENO One* **2021**, *55*, 295–312. [CrossRef]
40. Pardini, A.; Faiello, C.; Longhi, F.; Mancuso, S.; Snowball, R. Cover Crop Species and Their Management in Vineyards and Olive Groves. *Adv. Hortic. Sci.* **2002**, *16*, 225–234.
41. Scavo, A.; Fontanazza, S.; Restuccia, A.; Pesce, G.R.; Abbate, C.; Mauromicale, G. The Role of Cover Crops in Improving Soil Fertility and Plant Nutritional Status in Temperate Climates. A Review. *Agron. Sustain. Dev.* **2022**, *42*, 93. [CrossRef]
42. Abad, F.J.; Marín, D.; Imbert, B.; Virto, I.; Garbisu, C.; Santesteban, L.G. Under-Vine Cover Crops: Impact on Physical and Biological Soil Proprieties in an Irrigated Mediterranean Vineyard. *Sci. Hortic.* **2023**, *311*, 111797. [CrossRef]
43. Santibáñez, C.; Verdugo, C.; Ginocchio, R. Phytostabilization of Copper Mine Tailings with Biosolids: Implications for Metal Uptake and Productivity of Lolium Perenne. *Sci. Total Environ.* **2008**, *395*, 1–10. [CrossRef]
44. Dineen, M.; McCarthy, B.; Ross, D.; Ortega, A.; Dillon, P.; Van Amburgh, M.E. Characterization of the Nutritive Value of Perennial Ryegrass (*Lolium perenne* L.) Dominated Pastures Using Updated Chemical Methods with Application for the Cornell Net Carbohydrate and Protein System. *Anim. Feed Sci. Technol.* **2021**, *272*, 114752. [CrossRef]
45. Healy, M.G.; Ryan, P.C.; Fenton, O.; Peyton, D.P.; Wall, D.P.; Morrison, L. Bioaccumulation of Metals in Ryegrass (*Lolium perenne* L.) Following the Application of Lime Stabilised, Thermally Dried and Anaerobically Digested Sewage Sludge. *Ecotoxicol. Environ. Saf.* **2016**, *130*, 303–309. [CrossRef]
46. Thilakarathna, M.S.; Serran, S.; Lauzon, J.; Janovicek, K.; Deen, B.; Th Ilakarathna, M.S.; Serran, S.; Janovicek, K.; Deen, B.; Lauzon, J. Management of Manure Nitrogen Using Cover Crops. *Agron. J.* **2015**, *107*, 1595–1607. [CrossRef]
47. Giese, G.; Velasco-Cruz, C.; Roberts, L.; Heitman, J.; Wolf, T.K. Complete Vineyard Floor Cover Crops Favorably Limit Grapevine Vegetative Growth. *Sci. Hortic.* **2014**, *170*, 256–266. [CrossRef]

48. Smit, H.J.; Tas, B.M.; Taweel, H.Z.; Tamminga, S.; Elgersma, A. Effects of Perennial Ryegrass (*Lolium perenne* L.) Cultivars on Herbage Production, Nutritional Quality and Herbage Intake of Grazing Dairy Cows. *Grass Forage Sci.* **2005**, *60*, 297–309. [CrossRef]
49. Guitián, F.; Carballas, T. *Técnicas de Análisis de Suelos*; Pico Sacro: Santiago de Compostela, Spain, 1976.
50. Day, R. Particle Fractionation and Particle-Size Analysis. In *Methods of Soil Analysis, Part 1*; Black, C., Evans, D.D., White, J.L., Eds.; American Society of Agronomy: Madison, WI, USA, 1965; pp. 545–567.
51. Sánchez-Monedero, M.A.; Roig, A.; Martínez-Pardo, C.; Cegarra, J.; Paredes, C. A Microanalysis Method for Determining Total Organic Carbon in Extracts of Humic Substances. Relationships between Total Organic Carbon and Oxidable Carbon. *Bioresour. Technol.* **1996**, *57*, 291–295. [CrossRef]
52. Hendershot, W.H.; Duquette, M. A Simple Barium Chloride Method for Determining Cation Exchange Capacity and Exchangeable Cations. *Soil Sci. Soc. Am. J.* **1986**, *50*, 605–608. [CrossRef]
53. Feng, M.H.; Shan, X.Q.; Zhang, S.Z.; Wen, B. Comparison of a Rhizosphere-Based Method with Other One-Step Extraction Methods for Assessing the Bioavailability of Soil Metals to Wheat. *Chemosphere* **2005**, *59*, 939–949. [CrossRef] [PubMed]
54. Cohen, J. *Statistical Power Analysis for the Behavioral Sciences*; Routledge: Abingdon, UK, 2013.
55. Trought, M.C.T.; Dixon, R.; Mills, T.; Greven, M.; Agnew, R.; Mauk, J.L.; Praat, J.P. The Impact of Differences in Soil Texture within a Vineyard on Vine Vigour, Vine Earliness and Juice Composition. *J. Int. Sci. Vigne Vin* **2008**, *42*, 62–72. [CrossRef]
56. Trentin, E.; Cesco, S.; Pii, Y.; Valentinuzzi, F.; Celletti, S.; Feil, S.B.; Zuluaga, M.Y.A.; Ferreira, P.A.A.; Ricachenevsky, F.K.; Stefanello, L.O.; et al. Plant Species and PH Dependent Responses to Copper Toxicity. *Environ. Exp. Bot.* **2022**, *196*, 104791. [CrossRef]
57. Vega, F.A.; Covelo, E.F.; Andrade, M.L. A Versatile Parameter for Comparing the Capacities of Soils for Sorption and Retention of Heavy Metals Dumped Individually or Together: Results for Cadmium, Copper and Lead in Twenty Soil Horizons. *J. Colloid Interface Sci.* **2008**, *327*, 275–286. [CrossRef]
58. Fernández-Calviño, D.; López-Periago, E.; Novoa-Muñoz, J.C.; Arias-Estévez, M. Short-Scale Distribution of Copper Fractions in a Vineyard Acid Soil. *Land Degrad. Dev.* **2008**, *19*, 190–197. [CrossRef]
59. Fernández-Calviño, D.; Nóvoa-Muñoz, J.C.; López-Periago, E.; Arias-Estéves, M. Changes in Copper Content and Distribution in Young, Old and Abandoned Vineyard Acid Soils Due to Land Use Changes. *Land Degrad. Dev.* **2008**, *19*, 165–177. [CrossRef]
60. Parat, C.; Chaussod, R.; Lévéque, J.; Dousset, S.; Andreux, F. The Relationship between Copper Accumulated in Vineyard Calcareous Soils and Soil Organic Matter and Iron. *Eur. J. Soil Sci.* **2002**, *53*, 663–670. [CrossRef]
61. Duplay, J.; Semhi, K.; Errais, E.; Imfeld, G.; Babcsanyi, I.; Perrone, T. Copper, Zinc, Lead and Cadmium Bioavailability and Retention in Vineyard Soils (Rouffach, France): The Impact of Cultural Practices. *Geoderma* **2014**, *230–231*, 318–328. [CrossRef]
62. Marín-Martínez, A.; Sanz-Cobeña, A.; Bustamante, M.A.; Agulló, E.; Paredes, C. Effect of Organic Amendment Addition on Soil Properties, Greenhouse Gas Emissions and Grape Yield in Semi-Arid Vineyard Agroecosystems. *Agronomy* **2021**, *11*, 1477. [CrossRef]
63. Jez, E.; Pellegrini, E.; Contin, M. Copper Bioavailability and Leaching in Conventional and Organic Viticulture under Environmental Stress. *Appl. Sci.* **2023**, *13*, 2595. [CrossRef]
64. Filipović, L.; Defterdarović, J.; Chen, R.; Krevh, V.; Gerke, H.H.; Baumgartl, T.; Kovač, Z.; Ondrašek, G.; Ružičić, S.; He, H.; et al. Leached Copper Correlation with Dissolved Organic Carbon in Sloped Vineyard Soil. *Water* **2023**, *15*, 800. [CrossRef]
65. Ouédraogo, F.; Cornu, J.Y.; Janot, N.; Nguyen, C.; Sourzac, M.; Parlanti, E.; Denaix, L. Do DOM Optical Parameters Improve the Prediction of Copper Availability in Vineyard Soils? *Environ. Sci. Pollut. Res.* **2022**, *29*, 29268–29284. [CrossRef] [PubMed]
66. Villanueva-Rey, P.; Vázquez-Rowe, I.; Quinteiro, P.; Rafael, S.; Gonçalves, C.; Moreira, M.T.; Feijoo, G.; Arroja, L.; Dias, A.C. Regionalizing Eco-Toxicity Characterization Factors for Copper Soil Emissions Considering Edaphic Information for Northern Spain and Portuguese Vineyards. *Sci. Total Environ.* **2019**, *686*, 986–994. [CrossRef]
67. Fernández-Calviño, D.; Pateiro-Moure, M.; López-Periago, E.; Arias-Estévez, M.; Nóvoa-Muñoz, J.C. Copper Distribution and Acid-Base Mobilization in Vineyard Soils and Sediments from Galicia (NW Spain). *Eur. J. Soil Sci.* **2008**, *59*, 315–326. [CrossRef]
68. Baragaño, D.; Gallego, J.L.R.; Baleriola, G.; Forján, R. Effects of Different In Situ Remediation Strategies for an As-Polluted Soil on Human Health Risk, Soil Properties, and Vegetation. *Agronomy* **2020**, *10*, 759. [CrossRef]
69. Fernández-Calviño, D.; Cutillas-Barreiro, L.; Núñez-Delgado, A.; Fernández-Sanjurjo, M.J.; Álvarez-Rodriguez, E.; Nóvoa-Muñoz, J.C.; Arias-Estévez, M. Cu Immobilization and *Lolium perenne* Development in an Acid Vineyard Soil Amended with Crushed Mussel Shell. *Land Degrad. Dev.* **2017**, *28*, 762–772. [CrossRef]
70. Kgaladi, M.; Masotla, L.; Melato, F.A.; Mokgalaka-Fleischmann, S. Extraction Potential of *Lolium perenne* L. (Perennial Rye Grass) for Metals in Landfill Soil: Its Tolerance and Defense Strategies. *Minerals* **2023**, *13*, 873. [CrossRef]
71. De Conti, L.; Cesco, S.; Mimmo, T.; Pii, Y.; Valentinuzzi, F.; Melo, G.W.B.; Ceretta, C.A.; Trentin, E.; Marques, A.C.R.; Brunetto, G. Iron Fertilization to Enhance Tolerance Mechanisms to Copper Toxicity of Ryegrass Plants Used as Cover Crop in Vineyards. *Chemosphere* **2020**, *243*, 125298. [CrossRef]

horticulturae

MDPI

Article

Thermal Requirements, Phenology, and Maturation of Juice Grape Cultivars Subjected to Different Pruning Types

Camilo André Pereira Contreras Sánchez *, Daniel Callili , Débora Cavalcante dos Santos Carneiro, Safira Pataro Sampaio da Silva, Ana Carolina Barduzzi Scudeletti, Sarita Leonel and Marco Antonio Tecchio

Departament of Horticulture, Campus Botucatu, Faculty of Agricultural Sciences, São Paulo State University (UNESP), São Paulo 01049-010, Brazil; daniel_callili@hotmail.com (D.C.); debora.cavalcante-santos@unesp.br (D.C.d.S.C.); safira.pataro@unesp.br (S.P.S.d.S.); ana.scudeletti@unesp.br (A.C.B.S.); sarita.leonel@unesp.br (S.L.); marco.a.tecchio@unesp.br (M.A.T.)

* Correspondence: camilo.apc.sanchez@unesp.br

Abstract: The purpose of this study was to assess the impacts of pruning methods (short and mixed pruning) on the duration of phenological phases and thermal requirements of juice grape cultivars ('Bordô', 'BRS Cora', and 'BRS Violeta'), as well as to confirm the development of chemistry throughout berry ripening. The duration of the following phenological phases was measured in days after pruning over two production cycles: budburst, full-bloom, setting, veraison, and ripening. Degree days were used to compute the thermal requirements. Soluble solids, pH, titratable acidity, and maturation index were all measured as the berries ripened. There was no impact on the length of the phenological cycle or the thermal necessities of the vines due to the various types of pruning. In terms of cultivars, 'BRS Violeta' was found to be earlier than 'Bordô' and 'BRS Cora'. In terms of chemical evolution, the berries achieved 16 °Brix at 21 and 28 days following the veraison, and the greatest values obtained were 17.0 ('BRS Violeta') and 18.4 °Brix ('BRS Cora'). Furthermore, 'BRS Cora' produced more acidic berries. However, regardless of cultivar, the grapes were of high quality, making them a viable option for Brazilian subtropical viticulture.

Keywords: subtropical viticulture; BRS Violeta; BRS Cora; Bordô; mixed pruning; short pruning

1. Introduction

Brazil's consumption of whole grape juice has climbed by 10% in the previous four years, with per capita consumption reaching 1.36 L in 2020 [1]. To meet this expanding need, one potential option may be to expand grape planting for juice production in non-traditional producing regions, such as the State of São Paulo, which is largely located in subtropical climates. Thus, in order to adopt sustainable cultivation measures in viticulture, studies on cultural management and the introduction of more acclimatized cultivars that meet the needs of the market are essential for viticulture in São Paulo.

'Isabel', Concord', and 'Bordô' are three of the most important juice grape cultivars [2]. However, these traditional cultivars do not adapt well to the subtropical climate of Brazil, resulting in low output and quality [3]. 'Bordô', a rustic cultivar, offers strong resistance to the crop's major fungal diseases, as well as a high concentration of soluble solids and a low acidity [4].

Other cultivars for juice production created in the nation include Embrapa's hybrids 'BRS Cora' and 'BRS Violeta' [5,6]. These cultivars are currently rather popular due to their agronomic and organoleptic properties. 'BRS Cora' has a nice flavor, a high soluble solid content, and is recommended for the enhancement of juices with poor color. It also has high production potential and a medium cycle. 'BRS Violeta', on the other hand, has an early cycle, great production, high sugar content, and low acidity [3]. However, it is crucial to note that numerous factors, such as the types of pruning employed, can impact the features of the vines, and, as a result, the quality of the grape [7,8].

Short, long, and mixed pruning are the three types of pruning employed in viticulture. Short pruning leaves 1 to 3 buds on the producing branches. Long pruning entails leaving 4 to 12 buds on the pruned branches. The combination of short and long pruning on the same plant, altering short and long branches on the same spur, is known as mixed pruning [9]. However, it is stressed that the effect of pruning management may vary depending on cultivar bud fertility and training system [8], necessitating investigations of cultivars in various edaphoclimatic conditions.

Studying the phenology of grapevines is crucial for introducing new cultivars in non-traditional areas as it helps to identify the length of plant growth phases based on climate and seasonal variations [10,11]. Furthermore, knowing the duration of the phenological phases helps to increase the sustainable management practices of the vineyard, as well as in scheduling grape processing firms to receive and process raw material in order to minimize product quality loss [12].

Another useful technique for producers involves the monitoring of the evolution of berry maturity, which allows them to determine the optimum period for harvesting, i.e., when the berries achieve desirable levels of soluble solids and acidity, using chemical components. According to Lima and Choudhury [13], adequate maturity of the vine fruits is vital for the quality of the juices because physical, physiological, and biochemical changes occur during the ripening of the grape.

As a result, the current study sought to assess the phenological behavior, temperature requirements, and maturity progression of the 'Bordô', 'BRS Cora', and 'BRS Violeta' cultivars subjected to short and mixed pruning under subtropical conditions.

2. Materials and Methods

2.1. Treatments, Experimental Design and Experimental Area

The experimental design was a randomized block design in a 3 × 2 factorial scheme (6 treatments), consisting of three juice grape cultivars (Bordô, BRS Cora, and BRS Violeta) and two types of pruning (short pruning and mixed pruning). The experiment was conducted with 96 vines, divided into four blocks, with four plants per plot.

During the summer seasons of 2017 and 2018, the trial took place in an experimental vineyard at the Experimental Farm of the School of Agriculture (FCA) of UNESP, in São Manuel, São Paulo, Brazil (22°44′50″ S, 48°34′00″ W; altitude 765 m). The climate in this region is classified as Cfa by the Köppen classification, indicating a subtropical climate with hot summers. According to the Brazilian soil classification system [14], the soil is a Dystroferric Red Latosol with a sandy texture, with base saturation characteristics greater than 50% in most of the first 100 cm of the B horizon, and is also classified as Hapludox Typical dystrophic. According to the Soil Taxonomy of the USDA (United States Department of Agriculture) [15], the Typical Dystrophic Hapludox presents base saturation characteristics close to 35% up to 150 cm from the B horizon.

Throughout the experiment, a weather sensor 100 m away recorded daily precipitation (mm) as well as maximum, minimum, and average temperatures (°C). During the months of production in the experiment, the lowest average temperature was 17.3 °C in 2017 and 17.6 °C in 2018, while the highest average temperature was 29.6 °C in 2017 and 29.9 °C in 2018. In 2017, the cumulative rainfall was 589 mm, while in 2018, it was 738 mm, with a propensity to concentrate throughout the spring and summer months (Figure 1).

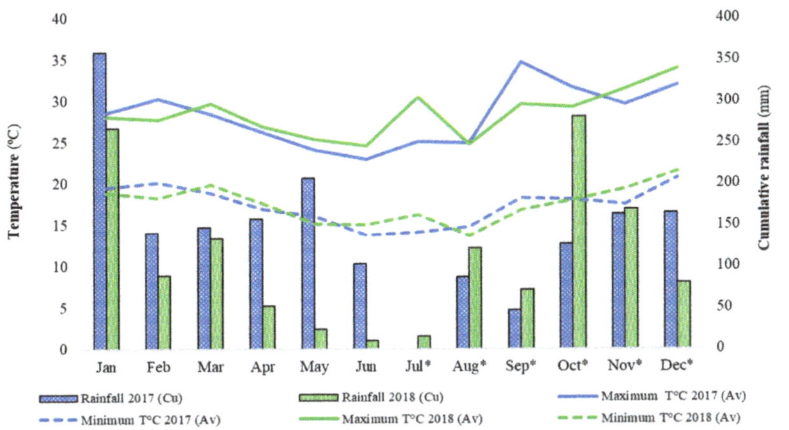

Figure 1. The temperature and cumulative rainfall data from the experimental site in São Manuel, located in the state of São Paulo, Brazil, during the productive period of 2017 and 2018. The bars indicate the total amount of rainfall, while the lines represent the minimum and maximum temperatures. * Productive period.

2.2. *Vine Growing Conditions and Production Pruning*

The vines were trained in a low espalier system with wires at 1.0, 1.3, and 1.6 m in height. The row spacing was 2.0 m and the plant spacing was 0.8 m (density of 6250 plants per hectare). The cultivars 'Bordô', 'BRS Cora', and 'BRS Violeta' were grafted onto the 'IAC 572 Jales' rootstock. For irrigation, drippers installed 50 cm from the ground were used, with a higher incidence of use during the budding and fruiting period, as described by Conceição et al. (2017) [16]. An 18% polyethylene screen was utilized for bird protection. Furthermore, all cultural, fertilization, and phytosanitary control was carried out in accordance with conventional regional cultivation methods.

The first and second cycles of production pruning were completed on 22 July 2017 and 20 July 2018, respectively. All production pruning maintained one to two buds per spur in short pruning and three to five buds in mixed pruning. Additionally, 2.5% hydrogen cyanamide was provided after pruning to stimulate and standardize bud sprouting.

2.3. *Phenology, Thermal Requirement, and Ripening of Grape*

The phenological phases were assessed using the criteria suggested by Coombe [17]. Visual observations were made three times a week to assess the length of each phenological stage in days after pruning (DAP). Pruning to budburst, full-bloom, setting, veraison, and full maturation (harvest) are the subperiods.

The harvest dates were set when the berries attained at least 16 °Brix or a minimum ripeness index (SS/AT) of 15. In the first and second cycles, samplings were taken up to 28 and 35 days after the berries began to mature, respectively, because, after this period, the berries began to exhibit signs of rot due to the significant rainfall in December.

For the thermal requirement, the total of the degree day (DD) was estimated from pruning to harvest, that is, the entire cycle, using the Winkler [18] equation:

$$DD = \Sigma\ [\text{average temperature} - 10\ ^\circ\text{C})] \times \text{days after pruning} \quad (1)$$

The evolution of soluble solids (SS), pH, titratable acidity (TA), and the maturity index (SS/AT) were used to measure berry ripeness. At the veraison, ten bunches from each plot were randomly picked and assessed until the grapes were fully matured. Every 7 days, that is, at 0, 7, 14, 21, 28, 35 days after the berries began to mature, they were gathered and analyzed.

Direct refractometry of grape must was used to estimate SS via the use of a digital refractometer (Reichert®, model r2i300, Depew, NY, USA), and the findings were represented in °Brix. The pH of the grape must was established by directly reading it (Tecnal® model Tec-10 potentiometer, Piracicaba, Brazil). Titration with 0.1 N NaOH to the equivalency point of pH = 8.2 yielded TA, which was stated as a percentage of tartaric acid. The maturity index was computed as the ratio of SS to TA.

2.4. Statistical Analyses

Two production cycles were statistically analyzed. To examine the impacts of cultivars and pruning methods, as well as their interactions, all data were subjected to analysis of variance. The Tukey test at 5% probability was used to compare averages for phenology and thermal requirement, and regression analysis was used to examine the chemical evolution of cultivars during ripening using the statistical program SISVAR®, version 5.7 (Lavras, MG, Brazil).

3. Results and Discussion

3.1. Phenological Stages and Thermal Requirement

There was no significant interaction ($p > 0.05$) observed between the types of pruning and cultivars (Table 1). The absence of a significant interaction ($p > 0.05$) between the types of pruning and cultivars highlights the need for a thorough examination of the underlying factors influencing the observed results. However, a significant difference was found among cultivars regarding the duration of phenological phases and thermal requirement (Table 1).

Table 1. Interaction between phenological phases and degree days (DD) of juice grape cultivars subjected to different types of pruning in two production periods.

Cultivar/ Pruning	Season	**Budburst**		**Full-Bloom**		**Setting**	
		Short	**Mixed**	**Short**	**Mixed**	**Short**	**Mixed**
Bordo	I	23 ± 1.5	22 ± 1.7 b	52 ± 3.0	51 ± 2.0 b	58 ± 3.1	57 ± 2.4 b
BRS Cora	I	25 ± 1.8	27 ± 0.6 a	54 ± 0.6	54 ± 0.9 a	58 ± 0.6	61 ± 2.5 a
BRS Violeta	I	23 ± 1.9	24 ± 1.8 b	53 ± 1.4	53 ± 1.4 a	58 ± 1.4	59 ± 0.7 b
p-value	I	0.29		0.22		0.35	
Bordo	II	23 ± 1.5 b	23 ± 1.5 b	57 ± 3.7	57 ± 3.7	61 ± 1.6 b	61 ± 1.6 b
BRS Cora	II	26 ± 1.0 a	26 ± 2.0 a	58 ± 3.1	58 ± 3.0	67 ± 1.5 a	67 ± 1.4 a
BRS Violeta	II	23 ± 0.8 b	23 ± 0.6 b	57 ± 2.3	56 ± 3.2	64 ± 2.6 b	64 ± 2.6 b
p-value	II	0.90		0.93		0.99	
Cultivar/ Pruning	Season	**Veraison**		**Harvest**		**Full Demands**	
		Short	**Mixed**	**Short**	**Mixed**	**Short**	**Mixed**
Bordo	I	110 ± 2.1 a	110 ± 1.8 a	130 ± 4.4 a	131 ± 2.3 a	1574 ± 57.6 a	1588 ± 28.6 a
BRS Cora	I	109 ± 4.7 a	109 ± 2.3 a	133 ± 3.5 a	130 ± 2.9 a	1601 ± 45.2 a	1567 ± 39.1 a
BRS Violeta	I	97 ± 2.4 b	97 ± 1.3 b	118 ± 3.1 b	118 ± 2.0 b	1405 ± 44.2 b	1408 ± 29.4 b
p-value	I	0.99		0.33		0.34	
Bordo	II	99 ± 1.4 b	99 ± 2.3 b	128 ± 1.7 b	130 ± 2.9 a	1470 ± 22.3 ab	1492 ± 38.3 a
BRS Cora	II	105 ± 2.1 a	104 ± 2.3 a	133 ± 2.2 a	131 ± 2.1 a	1517 ± 61.1 a	1519 ± 28.9 a
BRS Violeta	II	99 ± 1.5 b	99 ± 1.6 b	125 ± 2.3 b	125 ± 1.2 b	1440 ± 31.8 b	1437 ± 14.8 b
p-value	II	0.7		0.20		0.65	

Data are expressed as mean ± standard deviation ($n = 6$). Values followed by different letters on the same column indicate significant differences (Tukey test, $p > 0.05$).

Specifically, 'BRS Cora' exhibited prolonged durations of phenological phases and accumulated degree days irrespective of the type of pruning applied. The extended durations of phenological phases and increased degree-day accumulation exhibited by 'BRS Cora' suggest that this cultivar may possess unique physiological characteristics or growth patterns that differ from the other cultivars under study. This observation highlights the significance of analyzing these specific components independently (Table 2).

Table 2. Phenological stages and degree days (DD) of juice grape cultivars subjected to different types of pruning in two production seasons.

Phenological Stages (DAP)	Season	Scion				Pruning		
		Bordô	BRS Cora	BRS Violeta	p-Value	Short	Mixed	p-Value
Budburst	I	22 ± 1.6 b	28 ± 4.2 a	23 ± 1.8 b	<0.01	23 ± 1.8 b	25 ± 4.9 a	<0.01
	II	23 ± 1.4 c	26 ± 1.5 a	23 ± 0.7 b	<0.01	24 ± 1.9	24 ± 2.0	0.62
Full-bloom	I	51 ± 2.6 b	54 ± 0.7 a	53 ± 1.3 ab	<0.01	53 ± 1.9	53 ± 2.1	0.46
	II	56 ± 3.5 ab	58 ± 2.9 a	56 ± 2.7 b	<0.05	57 ± 3.0	57 ± 3.2	0.85
Setting	I	58 ± 2.8 b	60 ± 2.3 a	59 ± 1.1 ab	<0.01	59 ± 1.9	59 ± 3.0	0.52
	II	61 ± 1.6 c	67 ± 1.4 a	64 ± 2.5 b	<0.01	64 ± 3.0	64 ± 3.0	0.99
Veraison	I	110 ± 1.9 a	110 ± 5.8 a	97 ± 1.9 b	<0.01	106 ± 7.1	106 ± 7.4	0.97
	II	99 ± 1.9 b	104 ± 2.1 a	99 ± 1.5 b	<0.01	101 ± 3.2	101 ± 3.0	0.93
Harvest	I	131 ± 3.4 a	131 ± 3.4 a	118 ± 2.5 b	<0.01	127 ± 7.4	126 ± 6.3	0.67
	II	129 ± 2.4 b	132 ± 2.2 a	125 ± 1.7 c	<0.01	129 ± 4.0	129 ± 3.4	0.81
Full demands (DD)	I	1581 ± 44.0 a	1584 ± 44.1 a	1406 ± 35.8 b	<0.01	1527 ± 100.2	1521 ± 86.7	0.67
	II	1480 ± 32 b	1518 ± 45 a	1438 ± 24 c	<0.01	1476 ± 51.2	1481± 44.5	0.64

Data are expressed as mean ± standard deviation (n = 6). Values followed by different letters on the same line indicate significant differences (Tukey test, $p > 0.05$).

Regarding the time between pruning and sprouting, a significant difference ($p < 0.05$) was observed only in the first harvest, where vines subjected to short pruning sprouted two days earlier than those subjected to mixed pruning (23.2 versus 25.5 DAP) (Table 2).

In general, in the first season, the average duration of the phases from pruning to budburst, full-bloom, setting, veraison, and full maturation was 24, 53, 59, 106, and 127 days, respectively. The second season took 24, 57, 64, 100, and 129 days from pruning to budburst, full-bloom, setting, veraison, and harvest.

Although short and mixed pruning had no effect on the duration of the three assessed cultivars' phenological phases, Sozim [19] observed that mixed pruning provided precocity in comparison to long pruning. These findings imply that not all grapevine cultivars respond to any method of pruning since bud fertility, which may be described as the ability to distinguish between vegetative and productive buds, is a factor. However, it is important to note that research conducted to evaluate the impact of pruning methods on grapevine phenology is limited.

In all cycles, there was a substantial difference in all phenological phases tested for the cultivars. In general, it was found that the 'BRS Cora' was later in the early phenological stages. In terms of the duration from pruning to harvest, the duration associated with the cultivar 'BRS Violeta' was much shorter than that of 'Bordô' and 'BRS Cora' in the first production cycle, with 118 and 131 days, respectively. 'BRS Violeta' (125 days) demonstrated 3 days of precocity in comparison to 'Bordô' and 7 days in comparison to 'BRS Cora' in the second season (Table 2).

Mariani [20] noticed that the cycle of 'BRS Violeta' took place over 102 days in research conducted in comparable climatic conditions to the ones in the current study, i.e., Cfa climate, therefore recording an earlier timeframe in comparison to the results obtained in the present study. Camargo [21], on the other hand, state that the cycle from pruning to harvesting of 'BRS Violeta' takes 120 days in tropical climatic conditions. It should be noted that the longer season duration of the cv. 'BRS Violeta' reported in this study (118 and 125 days) was attributable to the lower average temperature attained during the assessed production cycles.

Camargo and Maia [5] state that the production cycle for 'BRS Cora' lasts around 130 to 140 days in tropical climates. In the Cfb climate, which is temperate with mild summers, the average duration is 157 days. As a result, the climatic differences that occur in each year and location have a direct impact on the duration of the grapevine cycle, with high temperatures shortening the cycle and low temperatures prolonging it [22,23]. In general, the total thermal requirement of the vines in the first season was 1523 GD and 1478 GD in the second cycle (Table 2). 'BRS Cora' and 'Bordô' required more degree-day accumulation in the first season than 'BRS Violeta', with 1584, 1581, and 1406 GD, respectively. In the second cycle, 'BRS Cora' had the greatest thermal need (1517 GD), while 'BRS Violeta' had the lowest thermal requirement (1438 GD) (Table 2).

According to Ahmed [24], differences in climatic circumstances alter the accumulation of degree days regarding grapes between places and seasons. It is worth noting that this notion is useful for anticipating plant development in a variety of conditions, as well as being a useful indication for studying vine behavior in each production region [25].

3.2. Quality Parameter

During grape development, there was no significant interaction between the manner of pruning and the chemical factors ($p > 0.05$) (Figures 2–4). Consequently, the mean of quality parameter for each cultivar was calculated based on the number of days after veraison (Figure 5).

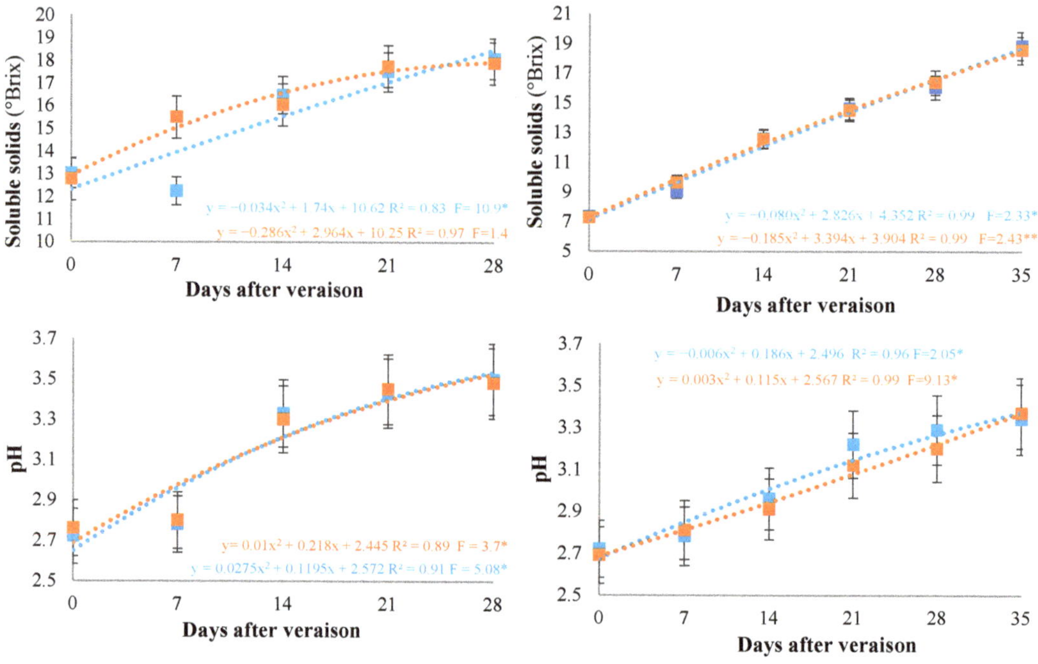

Figure 2. *Cont.*

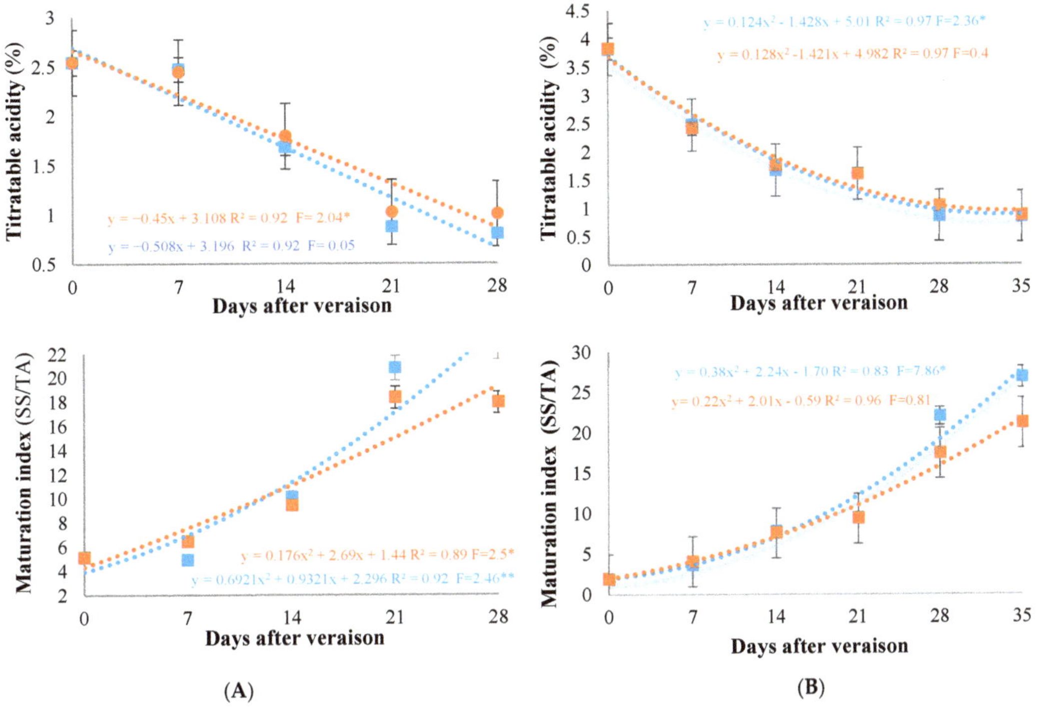

Figure 2. The evolution of titratable acidity, pH, soluble solids, and maturation index during ripening of 'Bordo' grapes grown in 2017 (**A**) and 2018 (**B**). Short and mixed pruning are represented by blue and orange colors, respectively. * $p > 0.01$ ** $p > 0.05$. Error bars with 95% CI.

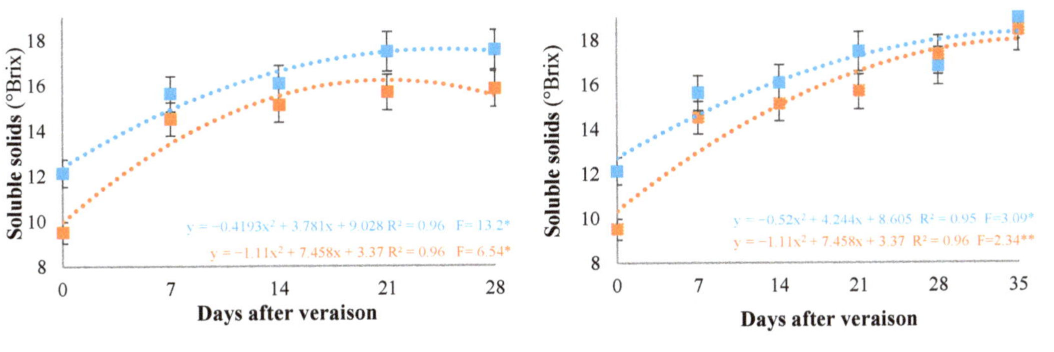

Figure 3. *Cont.*

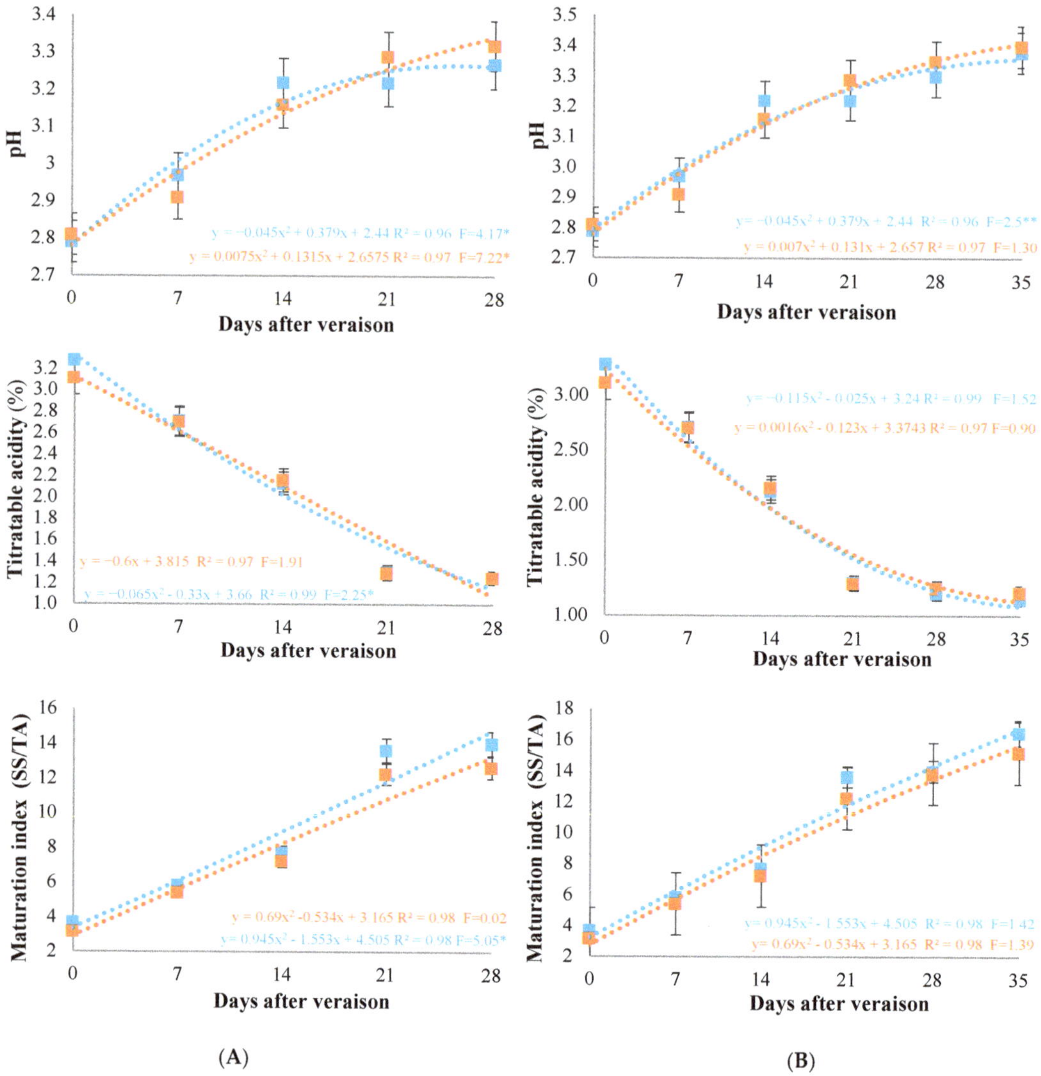

Figure 3. The evolution of titratable acidity, pH, soluble solids, and maturation index during ripening of 'BRS Cora' grapes grown in 2017 (**A**) and 2018 (**B**). Short and mixed pruning are represented by blue and orange colors, respectively. * $p > 0.01$ ** $p > 0.05$. Error bars with 95% CI.

According to Brazilian law, the minimum value needed for harvesting grapes for processing is 14 °Brix [26]. However, in the present study, the minimal value determined for collecting was 16 °Brix. In this case, the cultivars 'Bordô', 'BRS Cora', and 'BRS Violeta' reached 16 °Brix 21 days after the veraison, and the maximum point obtained by these cultivars was at 28 days after the veraison, with values of 17.8, 17.9, and 17.4 °Brix, respectively (Figure 5A).

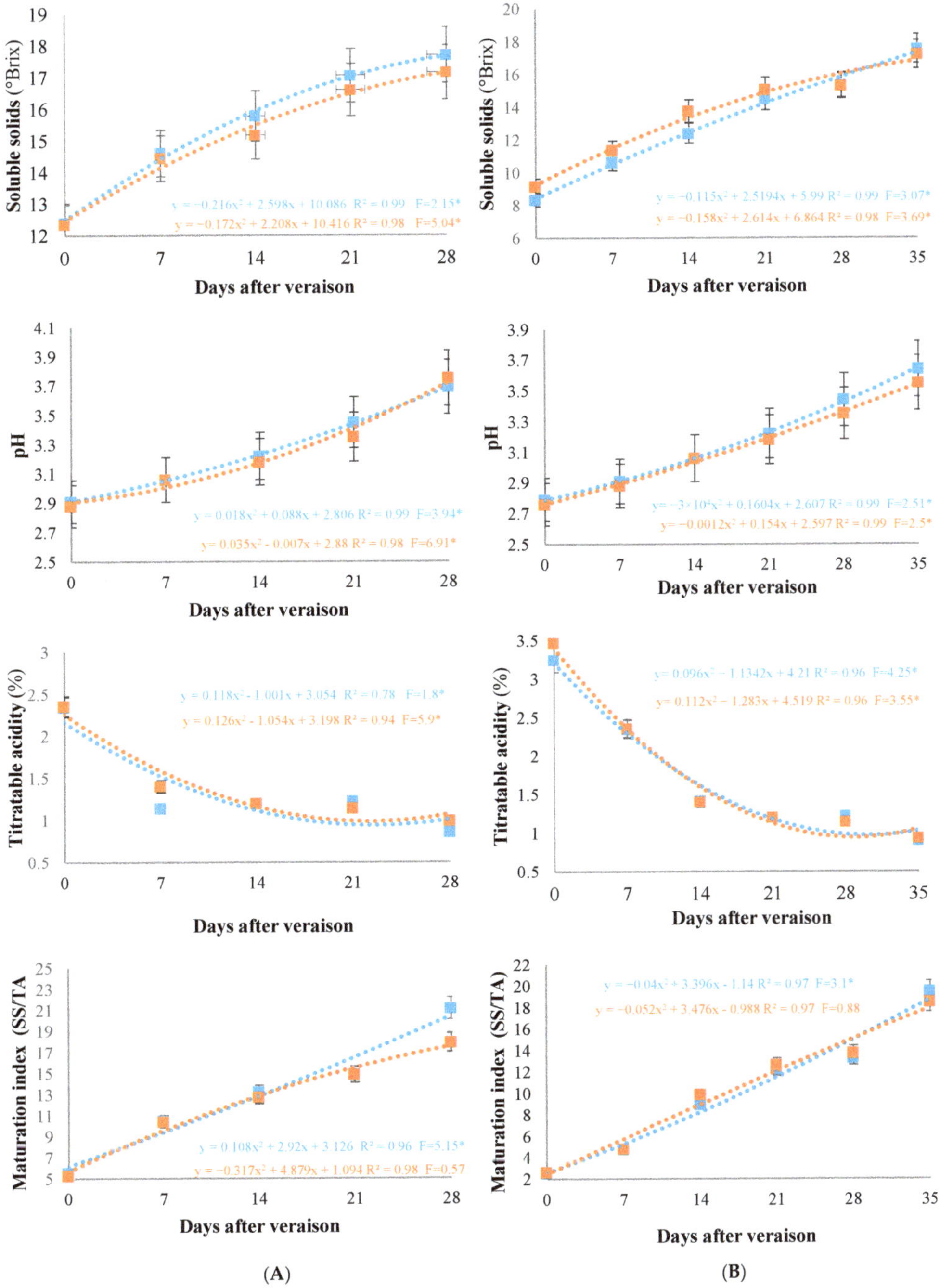

Figure 4. The evolution of titratable acidity, pH, soluble solids, and maturation index during ripening of 'BRS Violeta' grapes grown in 2017 (**A**) and 2018 (**B**). Short and mixed pruning are represented by blue and orange colors, respectively. * $p > 0.01$. Error bars with 95% CI.

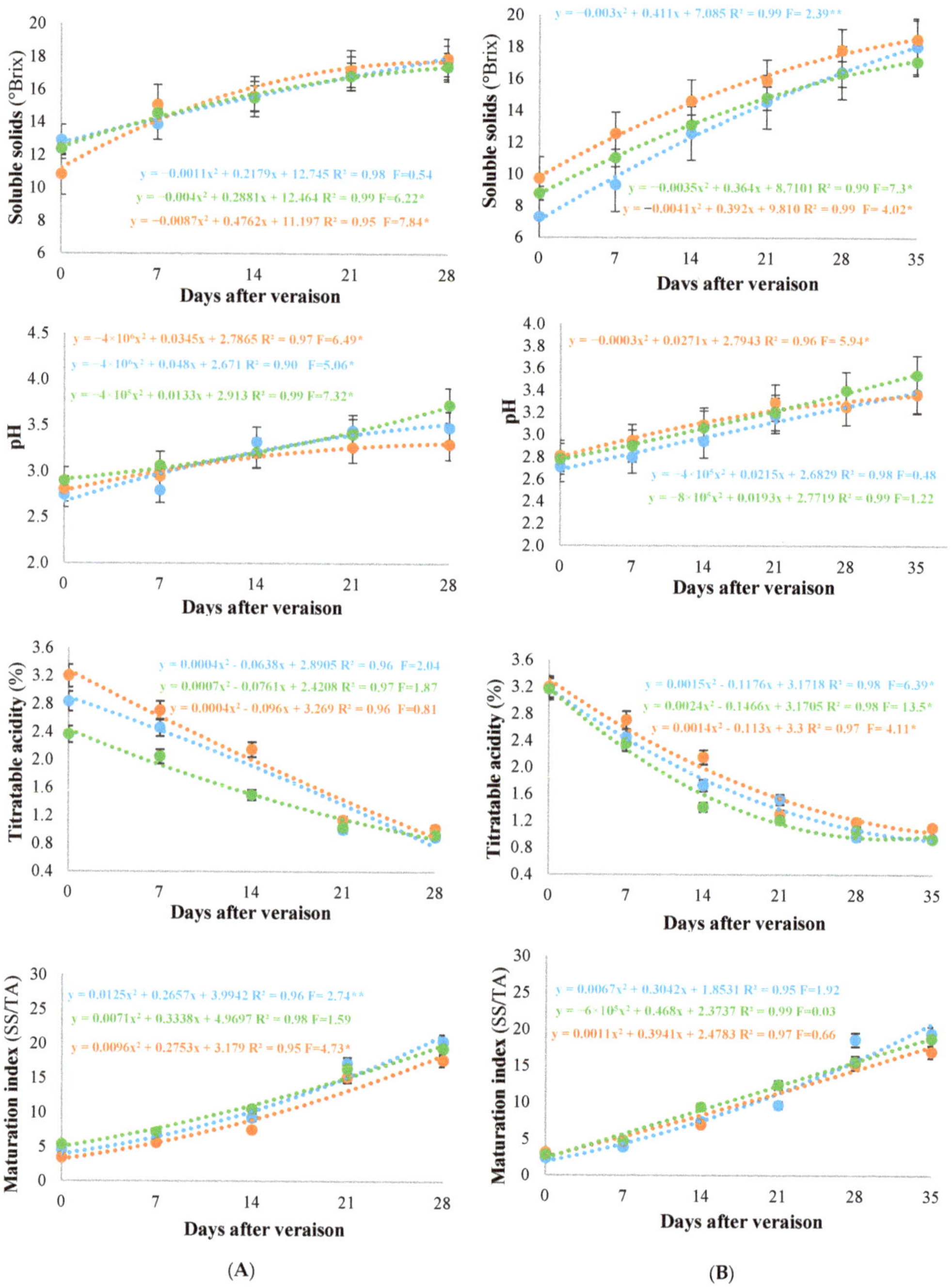

Figure 5. The progress of titratable acidity, pH, soluble solids, and maturation index during the ripening of juice grapes cultivated in 2017 (**A**) and 2018 (**B**) 'Bordô', 'BRS Cora', and 'BRS Violeta' are represented by the colors blue, orange, and green, respectively. * $p > 0.01$ ** $p > 0.05$. Error bars with 95% CI.

However, in the second season, 16 °Brix was achieved 28 days after veraison. In this scenario, the highest point of the 'Bordô', BRS Cora', and 'BRS Violeta' cultivars was 18.0, 18.4, and 17.0 °Brix 35 days from the commencement of fruit ripening, respectively (Figure 5B). The soluble solids readings reported for 'BRS Cora' are consistent with the descriptions given by the authors of [5], who state that the soluble solid content of this cultivar should range between 18 and 20 °Brix. However, the 'BRS Violeta' provided lower readings in comparison to the information supplied by the authors of [6], whose descriptions imply that the soluble solid content should be between 19 and 21°Brix.

Ref. [2] reported that 'Bordô' and 'BRS Violeta' had lower amounts of soluble solids than in the current investigation in a study with grape cultivars for processing. When compared to the result achieved in this investigation, the authors of [27] found lower levels for 'BRS Cora'. The percentage of sugar in grapes is inextricably tied to the quality of the grape must used to make juices and wines [2]. As a result of the high number of soluble solids identified in the current study, it is proposed that these cultivars shown good adaptability to the local edaphoclimatic conditions.

In general, the pH of the cultivars ranged from 3.30 (BRS Cora) to 3.72 (BRS Violeta) 28 days after the start of ripening in the first season (Figure 5B). In turn, 35 days after veraison, the pH values found in the second season were 3.36 for 'BRS Cora', 3.37 for 'Bordô', and 3.54 for 'BRS Violeta' (Figure 5B). The rise in pH in the berries was caused by the salinization of organic acids and an increase in potassium cation [28].

Although pH is not required by Brazilian law, it is an important characteristic to examine since it is directly connected to anthocyanin stability and the color intensity of grape juice or red wine [29,30]. Furthermore, the determination of pH in grapes for juice is an essential component since, when the value is low, it helps balance the sweet and acidic flavors [31].

At 28 days after veraison, the titratable acidity ranged from 0.90 to 1.02% in the first season (Figure 5A). In the second season, 35 days following the veraison, the titratable acidity levels varied from 0.93 to 1.10% (Figure 5B). In this scenario, 'BRS Cora' had more acidic berries, that is, higher titratable acidity and lower pH, than 'Bordô' and 'BRS Violeta' in both seasons.

The values observed in this study are higher on average than those discovered by the authors of [2] in 'Bordô' (0.90 versus 0.53%) and 'BRS Violeta' (0.93 versus 0.73%). However, for 'BRS Cora', the observed results are lower compared to those reported by the authors of [12,27], who conducted their studies in a tropical climate. Several physiological processes contribute to the decrease in titratable acidity (TA) during grape ripening. However, temperature variation, light intensity, and rainfall can also impact this reduction, leading to variations in vine metabolism. These factors can either promote or impede the genetic potential of the grapes [32].

Because the maturation index is generated from the connection between sugar and acidity levels (SS/AT), the low acidity of 'Bordô' and 'BRS Violeta' resulted in higher maturation index values, with roughly 20.3 and 19.3 in the first season and 19.4 and 18.7 in the second season, (Figures 5A and 5B) respectively. However, 'BRS Cora' is distinguished by its ability to retain high amounts of soluble solids and acidity until full maturity, as documented by various scholars [12,27,32].

Given that the maturation index identifies the optimal time for harvesting, that is, when there is a better balance of sugars and acids, and that Brazilian legislation requires that the range for grapes destined for juice processing be between 15 and 45, it is recommended that the harvest of these cultivars takes place between 21 and 28 days after the veraison, which may vary depending on the climatic conditions of each harvest.

Considering the fact that grape juice quality is linked to the chemical quality of the berries, the current study confirmed that 'Bordô,' 'BRS Cora', and 'BRS Violeta' had acceptable chemical characteristics, making them a suitable choice for Brazilian subtropical viticulture.

Considering the results of the study and the specific context in which the research was carried out, it is advisable to interpret the results with caution. Although the study provides information on the thermal requirements, phenology, and maturation of the vines, it is essential to recognize the potential influence of external factors, such as climatic conditions and application rates of hydrogen cyanamide, on the observed results.

4. Conclusions

The duration of the phenological phases and the thermal requirements of the juice grape cultivars were unaffected by the different pruning types. This is important for sustainable viticulture, as it allows for the adoption of pruning practices that are more appropriate to the local context without compromising the development of culture.

There is a need to expand studies on the different pruning methods in grape cultivars intended for processing purposes, as well as in emerging grape-growing regions, such as those with subtropical climatic conditions, in order to comprehend their influences on phenology, thermal requirements, and maturation.

In comparison to 'BRS Cora' and 'Bordô', 'BRS Violeta' was the most precocious cultivar; this is useful for producers who need to anticipate the harvest and commercialization of grapes, which can be important in terms of competitiveness.

The cultivar with the highest soluble solid content, highest acidity, and lowest pH was BRS Cora'. This is very important for the juice industry, as fruits with these characteristics are needed for the production of high-quality juices.

The soluble solid concentration of the 'Bordô', 'BRS Cora', and 'BRS Violeta' cultivars was over 16 °Brix, indicating acceptable chemical quality.

Harvesting the grapes between 21 and 28 days after veraison is recommended in order to imbue them with a better balance of sugars and acids.

Author Contributions: Conceptualization, M.A.T.; methodology, C.A.P.C.S.; software, D.C.; validation, M.A.T. and S.L.; formal analysis, C.A.P.C.S., D.C., D.C.d.S.C., S.P.S.d.S. and A.C.B.S.; investigation, C.A.P.C.S., D.C., D.C.d.S.C., S.P.S.d.S. and A.C.B.S.; resources, M.A.T.; data curation, C.A.P.C.S. and D.C.; writing—original draft preparation, C.A.P.C.S. and D.C.; writing—review and editing, C.A.P.C.S., D.C., S.L. and M.A.T.; visualization, M.A.T. and S.L.; supervision, M.A.T.; project administration, M.A.T.; funding acquisition, M.A.T. All authors have read and agreed to the published version of the manuscript.

Funding: This research was funded by the State of São Paulo Research Foundation (FAPESP) (process nos. 2015/16440-5; 2020/12152-3); the Research Productivity Grant (process no. 307377/2021-0); and the Coordination for the Improvement of Higher Education Personnel (CAPES) through scholarship.

Data Availability Statement: Not applicable.

Acknowledgments: The authors would like to thank the State of São Paulo Research Foundation (FAPESP) for their financial support (process nos. 2015/16440-5; 2020/12152-3); CNPq for the Research Productivity Grant (process no. 307377/2021-0); and the Coordination for the Improvement of Higher Education Personnel (CAPES) for the scholarship. The authors would also like to thank the fruit growing research group (FRUT!) for their assistance, as well as São Paulo State University (UNESP), Faculty of Agricultural Sciences, Campus Botucatu for their help in carrying out the study.

Conflicts of Interest: The authors declare no conflict of interest.

References

1. Mello, M.R.; Machado, C.E. *Comunicado Técnico n° 226Vitivinicultura Brasileira: Panorama 2021*; Embrapa Uva e Vinho: Bento Gonçalves, Brazil, 2022; p. 17.
2. Tecchio, M.A.; Silva, M.J.R.; Callili, D.; Hernandes, J.L.; Moura, M.F. Yield of white and red grapes, in terms of quality, from hybrids and *Vitis labrusca* grafted on different rootstocks. *Sci. Hortic.* **2020**, *259*, 108846. [CrossRef]
3. Camargo, U.A.; Maia, J.D.G.; Ritschel, P. Cultivares de videira para processamento. In *Implantação do Vinhedo, Cultivares e Manejo da Planta*, 1st ed.; Silveira, S.V., Hoffmann, A., Garrido, L.R., Eds.; EMBRAPA: Brasília, Brazil, 2015; Volume 3, pp. 26–46.
4. Rizzon, L.A.; Meneguzzo, J. *Suco de Uva*; Embrapa Informação Tecnológica: Brasília, Brazil, 2007; p. 45.

5. Camargo, U.A.; Maia, J.D.G. *Comunicado Técnico n° 53: BRS Cora—Nova Cultivar de Uva para Suco, Adaptada a Climas Tropicais*; Embrapa Uva e Vinho: Bento Gonçalves, Brazil, 2004; pp. 1–7.
6. Camargo, U.A.; Maia, J.D.G.; Nachtigal, J.C. *Comunicado Técnico n° 63:BRS Violeta—Nova Cultivar de Uva para Suco e Vinho de Mesa*; Embrapa Uva e Vinho: Bento Gonçalves, Brazil, 2005; pp. 1–8.
7. Malgarim, M.B.; Afinovicz, A.P.; Euleuterio, M.D.; Pierin, F.F.; Dias, J.S.; Sozim, M. Diferentes tipos de poda na produçãoda videira cv. Bordô. *Semin. Cienc. Agrar.* **2009**, *30*, 1203–1206. [CrossRef]
8. Yilmaz, T.; Fennell, A. Spur and Short cane pruning influence bud viability, yield and fruit quality. *Proc. S. D. Acad. Sci.* **2021**, *100*, 81–91.
9. Aliquó, G.; Catania, A.; Aguado, G. La poda de la vid. In *Secretaria de Agricultura, Ganadería, Pesca y Alimentación*; Instituto Nacional de Tecnología Agropecuaria (INTA): Mendoza, Argentina, 2010; pp. 9–14.
10. Hernandes, J.L.; Pedro Júnior, M.J.; Santos, A.O.; Tecchio, M.A. Fenologia e produção de cultivares americanas e híbridas de uvas para vinho, em Jundiaí-SP. *Rev. Bras. Frutic.* **2010**, *32*, 135–142. [CrossRef]
11. Barros, L.B.; Margoti, G.; Fowler, J.G.; De Mio, L.L.M.; Biasi, L.A. Thermal requirement and phenology of different cultivars of Vitis labrusca on different rootstocks. *Semin. Cienc. Agrar.* **2015**, *36*, 2433–2442. [CrossRef]
12. Silva, M.J.R.; Paiva, A.P.M.; Pimentel Junior, A.; Sánchez, C.A.P.C.; Callili, D.; Moura, M.F.; Leonel, S.; Tecchio, M.A. Yield performance of new juice grape varieties grafted onto different rootstocks under tropical conditions. *Sci. Hortic.* **2018**, *241*, 194–200. [CrossRef]
13. Lima, M.A.C.; Choudhury, M.M. Características dos cachos de uva. In *Uva de Mesa: Pós-Colheita*, 2nd ed.; Lima, M.A.C., Ed.; Embrapa Semiárido-Informação Tecnológica: Petrolina, Brazil, 2007; pp. 21–30.
14. Santos, H.G.; Jacomine, P.K.T.; Anjos, L.H.C.; Oliveira, V.A.; Lumbreras, J.F.; Coelho, M.R.; Almeida, J.A.; Cunha, T.J.F.; Oliveira, J.B. *Sistema Brasileiro de Classificação de Solos*, 3rd ed.; Embrapa: Brasília, Brazil, 2013; 353p, Available online: https://www.embrapa.br/solos/sibcs (accessed on 10 May 2023).
15. Soil Survey Staff. *Soil Taxonomy: A Basic System of Soil Classification for Making and Interpreting Soil Surveys*, 2nd ed.; USDA, NRCS: Washington, DC, USA, 1999. Available online: https://www.nrcs.usda.gov/wps/portal/nrcs/main/soils/survey/class/taxonomy/ (accessed on 27 November 2022).
16. Conceição, M.A.F.; Tecchio, M.A.; Souza, R.T.; Silva, M.J.R.; Moura, M.F. Estimativa dos coeficientes de cultivo (kc) de videiras para suco. *Agrometeoros* **2017**, *25*, 203–210. [CrossRef]
17. Coombe, B.G. Growth Stages of the Grapevine: Adoption of a system for identifying grapevine growth stages. *Aust. J. Grape Wine Res.* **1995**, *1*, 104–110. [CrossRef]
18. Winkler, A.J. *Viticultura*; Compañia Editorial Continental: Coyoacán, Mexico, 1965; 792p.
19. Sozim, M.; Ayub, R.A.; Malgarim, M.B. Efeito do tipo de poda na produção e qualidade da videira cv, Vênus. *Sci. Agrar.* **2007**, *8*, 169–172. [CrossRef]
20. Mariani, J.A.; Nava, G.A. Productivity and chemical quality of grapevines cultivated in agroecological system under subtropical climate. *Rev. Bras. Frutic.* **2018**, *40*, e924. [CrossRef]
21. Camargo, U.A.; Maia, J.D.G.; Ritschel, P. *Embrapa Uva e Vinho: Novas Cultivares Brasileiras de Uva*; Embrapa Uva e Vinho: Bento Gonçalves, Brazil, 2010; p. 64.
22. Brixner, G.F.; Schöffel, E.R.; Lago, I.; Radünz, A.L.; Krüger, A.P. Risco de geada e duração dos subperíodos fenológicos da 'Cabernet Sauvignon' na Região da Campanha. *Rev. Bras. Eng. Agríc. Ambient.* **2014**, *18*, 210–216. [CrossRef]
23. Chiarotti, F.; Morgoti, G.; Fowler, J.G.; Cuquel, F.; Biasi, L.A. Caracterização fenológica, exigência térmica e maturação da uva 'Bordô' em Bocaiuva do Sul, PR. *Agrária* **2014**, *9*, 338–342. [CrossRef]
24. Ahmed, S.; Roberto, S.R.; Shahab, M.; Colombo, R.C.; Silvestre, J.P.; Koyama, R.; Souza, R.T. Proposal of double-cropping system for "BRS Isis" seedless grape grown in subtropical area. *Sci. Hortic.* **2019**, *251*, 118–126. [CrossRef]
25. Callili, D.; Silva, M.J.R.; Sánchez, C.A.P.C.; Watanabe, C.Y.; Macedo, B.M.P.; Domingues Neto, F.J.; Teixeira, L.A.J.; Tecchio, M.A. Rootstock and potassium fertilization, in terms of phenology, thermal demand and chemical evolution, of berries on Niagara Rosada grapevine under subtropical conditions. *Bragantia* **2022**, *81*, e2022. [CrossRef]
26. Brasil. Complementação dos Padrões de Identidade e Qualidade do Vinho e Derivados da Uva e do Vinho. In *Instrução Normativa N° 14 de 8 de Fevereiro de 2018*; Diário Oficial da União; Ministério da Agricultura, Pecuária e Abastecimento: Brasília, Brazil, 2018; pp. 1–27. Available online: https://www.gov.br/agricultura/pt-br/assuntos/noticias/mapa-atualiza-padroes-de-vinho-uva-e-derivados/INMAPA142018PIQVinhoseDerivados.pdf (accessed on 15 January 2023).
27. Tecchio, M.A.; Silva, M.J.R.; Cunha, S.R.; Callili, D.; Sánchez, C.A.P.C.; Souza, J.R.; Moura, M.F. Productive performance and physicochemical quality of grapes for processing grown on different rootstocks. *Pesqui. Agropecu. Bras.* **2022**, *57*, e02071. [CrossRef]
28. Manfroi, L.; Miele, L.; Rizzon, L.A.; Barradas, C.I.N.; Souza, P.V.D. Evolução da maturação da uva 'Cabernet Franc' conduzida no sistema lira aberta. *Ciênc. Agrotec.* **2004**, *28*, 306–313. [CrossRef]
29. Wrolstad, R.E.; Durst, R.W.; Lee, J. Tracking color and pigment changes in anthocyanin products. *Trends Food. Sci. Technol.* **2005**, *16*, 423–428. [CrossRef]
30. Yamamoto, L.Y.; Assis, A.M.; Roberto, S.R.; Bovolenta, Y.R.; Nixford, S.L.; García-Romero, E.; Gómez-Alonso, S.; Hermanoín-Gutiérrez, I. Application of abscisic acid (S-ABA) to cv. Isabel grapes (*Vitis vinifera* × *Vitis labrusca*) for color improvement: Effects on color, phenolic composition and antioxidant capacity of their grape juice. *Food Res.* **2015**, *77*, 572–583. [CrossRef]

31. Gurak, P.D.; Cabral, L.M.C.; Leão, M.H.M.R.; Matta, V.M.; Freitas, S.P. Quality evaluation of grape juice concentrated by reverse osmosis. *J. Food Eng.* **2010**, *96*, 421–426. [CrossRef]
32. Ribeiro, T.P.; Lima, M.A.C.; Alves, R.E. Maturação e qualidade de uvas para suco em condições tropicais, nos primeiros ciclos de produção. *Pesqui. Agropecu. Bras.* **2012**, *47*, 1057–1065. [CrossRef]

horticulturae

Review

The Health of Vineyard Soils: Towards a Sustainable Viticulture

Fernando Visconti [1,*], Roberto López [2] and Miguel Ángel Olego [3,*]

1 Department of Ecology and Global Change, Desertification Research Centre-CIDE (CSIC, UVEG, GVA), Carretera CV-315, km 10.7, 46113 Moncada, Spain
2 Department of Applied Chemistry and Physics, Faculty of Biology and Environmental Sciences, Universidad de León, Campus de Vegazana, 24071 León, Spain; rlopg@unileon.es
3 Research Institute of Vine and Wine, Universidad de León, Avenida de Portugal, 41, 24071 León, Spain
* Correspondence: fernando.visconti@csic.es (F.V.); molem@unileon.es (M.Á.O.)

Abstract: Soil health encompasses the effects the uppermost part of the land have on human well-being in a broad sense, because soil is where most food ultimately comes from, and because it more inconspicuously fulfils other ecological functions, as important as feeding, for our planet's welfare, which is ours. Viticulture exploits the soil's resources from which wine, its most valuable produce, boasts to obtain some of its unique quality traits, which are wrapped within the terroir concept. However, using conventional methods, viticulture also has harsh impacts on the soil, thus jeopardizing its sustainability. How long will the terroir expression remain unchanged as vineyard soil degradation goes on? While this question is difficult to answer because of the complex nature of terroirs, it is undeniable that conventional soil management practices in viticulture leave, in general, ample room for improvement, in their impact on vineyards as much as on the environment. In response, viticulture must adopt practices that enable the long-lasting preservation of its grounds for both on-farm and off-farm benefits. In this regard, the increase in the soil's organic matter alongside the enhancement of the soil's biological community are key because they benefit many other soil properties of a physical, chemical, and biological nature, thus determining the soil's healthy functioning, where the vines may thrive for a long time, whereas its surroundings remain minimally disturbed. In the present review, the importance of soil health as it relates to vineyards is discussed, the soil degradation factors and processes that threaten winegrowing areas are presented, successful soil-health enhancement practices are shown, and future research trends are identified for the benefit of researchers and stakeholders in this special agricultural industry.

Keywords: environmental health; environmental quality; soil biology; soil degradation; soil fertility; soil organic carbon; soil organic matter; soil quality; soil water; winegrowing

1. Introduction

Grapevine growing is an important agricultural activity, fueling an international trade in fresh grapes and wine worth, respectively, G€9.4 and G€37.8 [1]. However, agricultural revenues come at an expense in terms of depletion of natural resources [2], with viticulture standing out in this regard because of its need for the intensive use of machinery, tillage, and chemicals under conventional management conditions [3]. These conventional practices decrease porosity, aeration, aggregation, and the organic matter of soils and increase their toxicity to soil organisms, thus directly and indirectly damaging the soil's biological community, and eventually decreasing the availability of water, oxygen, nutrients, and mutualists to vine roots.

Since soils are living systems, the soil degradation vineyards experience because of conventional practices may be described as a health loss. Degraded soils are nonhealthy soils and, consequently, are unable to satisfactorily meet their functions both on-farm and off-farm. First, through conventional running practices, viticulture jeopardizes the health of the base, from which its most valued produce, i.e., wine, boasts to obtain a good deal of its distinctive traits, i.e., terroir. Moreover, through conventional practices, viticulture contributes to the jeopardizing of environmental health, which is a growing concern for humanity.

Vineyard management practices that preserve the soil health, and even help it recover, have been studied and continue to be investigated. In this article, we present some of these studies to address the concept of soil health as it relates to the more classical concepts of soil quality and soil fertility (Section 2). Then, we show how biological fertility is at the core of the concept of soil health, since soil organisms are what make soils alive (Section 3). In this regard, soils are healthy as long as they harbor a diverse, balanced, and therefore functional biological community. Vineyard soils can benefit from such healthy associations of organisms due to their ability to work and rework the organic matter continuum that starts from the vegetative production of vines, as well as from the primary production due to accompanying plants. This reworking of the organic matter continuum releases inorganic compounds, which are plant-absorbed and transformed again into vine tissues. Moreover, in this cycling process, the soil's biological community influences the vines through biochemical signaling and develops the soil organic matter (SOM).

As SOM is tackled (Section 4), it will be shown that it is on one side of a coin that has the soil's biological activity on the other for vine nutrient supply. However, they also complement each other for soil structure development in such a way that SOM keeps upright what soil organisms build. Then, as the soil's physical fertility in vineyards is presented (Section 5), the relevance of keeping a sound soil structure will be put under the spotlight regarding soil compaction avoidance, erosion control, and the soil's water-holding capacity for vines. Unfortunately, what mostly threatens soil structure in vineyards, and therefore primarily menaces soil and water conservation therein, is tillage, which will therefore be given particular attention (Section 6).

The threat of tillage to soil's physical fertility is particularly severe in sloping vinelands under a Mediterranean climate, which leaves the soil highly vulnerable to water erosion. As for the soil's biological fertility, the toxicity of phytosanitary products is a direct menace (Section 7), which also indirectly threatens its physical and chemical counterparts. Phytosanitary products in ample use in vineyards, e.g., copper (Cu) compounds, are xenobiotic, and thus they are toxic to soil organisms. On the contrary, fertilizers and other amendments are usually not so toxic. However, synthetic fertilizers and amendments may change the diversity and balance of the soil's biological community, thus disrupting its functionality and activity, with impairing effects on soil health (Section 8). Given all these damaging results of conventional vineyard management on soil health, one may ask how long terroir will remain unchanged as soil degrades, and conversely, one may ask how terroir will evolve as soil heals (Section 9).

Another pertinent question is how all the viticulture stakeholders can check the health status of vineyard soils to know if they are degrading or healing, and hence how to take action. There are several options for this task, from the most traditional to the most modern, which will be addressed (Section 10). Finally, some conclusions will be drawn (Section 11).

2. Soil Health, Soil Quality, and Soil Fertility

The concept of health has been traditionally applied to living beings, and particularly humans, for describing a state of complete physical, mental, and social well-being, which goes beyond the absence of disease or infirmity [4]. As beings that require air for breathing, water for drinking, and food for eating from our environment, and furthermore, that relate to each other, our health depends on the state of our surrounding atmosphere, hydrosphere, biosphere, and societal networks. Therefore, the state of these spheres may be understood in terms of health as long as they affect ours. At this point, the concept of environmental health naturally arises, and since the environment comprises four interdependent compartments, air, water, land, and biota (Figure 1), one may speak in terms of their respective health. However, it is not only appropriate to speak in terms of soil health due to its effects on human health, it is also so because soils are living systems on their own [5].

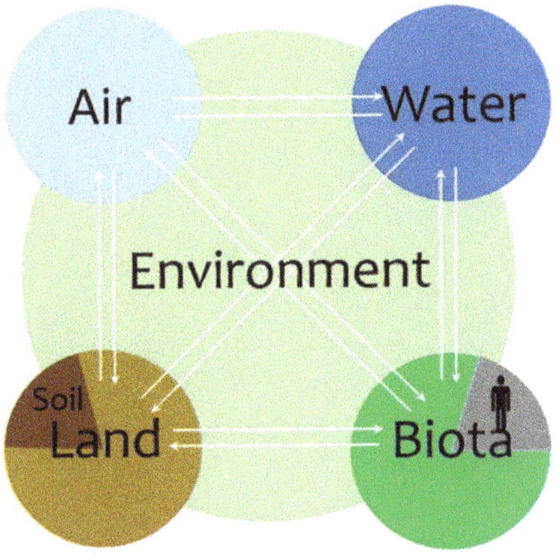

Figure 1. Energy and matter exchange relationships between the four environmental compartments and the places soil and humanity occupy within, respectively, land and biota. Circles and sectors sizes are not at scale.

A healthy soil is one with a continued capacity to function within its ecological boundaries to sustain productivity, maintain environmental quality, and promote plant and animal health [6,7], including human health [8]. Therefore, soils are healthy if they provide enough plant support for crops to yield adequate amounts of safe and nourishing food, and for woods to produce timber and other raw materials. However, soils are also healthy if they meet their other ecological functions: the protection of humans and the environment and, moreover, the constitution of a gene reservoir [9]. In this regard, the protection of humans and the environment is provided by the soil's capacities for (i) water cycle regulation through precipitation collection and its transmission underground, (ii) buffering of chemicals through filtering and transformation reactions, and (iii) weather and climate regulation through matter and energy exchanges with the other environmental compartments. Moreover, as a gene reservoir, soil also contributes to human health since it may house more biodiversity than the above-ground biota and, therefore, it is a gene source for facing present and future microbial diseases [10,11].

Quality concerning soil could be understood in terms of fitness for the purpose of producing adequate amounts of safe and nourishing food, as well as materials for the industry, i.e., in terms of fertility. This was the original meaning; however, it has evolved, and soil quality is now understood, along with air and water quality, as contributing to the multicomponent interrelated and holistic concept of environmental quality [12]. Therefore, soil quality cannot be understood only in terms of soil fertility, but in terms of balancing the biomass production function with the ecological functions. Consequently, soil quality has been defined as the fitness of a specific kind of soil, to function within its capacity and within natural or managed ecosystem boundaries, to sustain plant and animal productivity, maintain or enhance water and air quality, and to support human health and habitation [13].

Soils that satisfactorily fulfil all their ecological functions score highly in soil-quality and soil-health standards. Therefore, both concepts almost overlap. On the contrary, the soil fertility [14] is clearly more restricted, since it is a plant support-centered concept, because humanity directly and indirectly depends on terrestrial plants for over 98% of calorie intake [15] as well as for many raw materials (Figure 2). Nevertheless, it is crucial to acknowledge that the nutritional quality of agricultural crops is intricately linked to the overall quality and health of soils [15]. Human health and quality of life cannot progress any further within an environment that does not meet minimum health and quality standards. Among these, soil is key, because it continuously exchanges inert matter and living matter, as well as different forms of energy, with the other three environmental compartments [16–18]. The great similitude between soil-health and soil-quality concepts is more obvious within the framework of the "One World—One Health" paradigm in which the tight linkages between human, animal, and ecosystem health are recognized [14]. The burst of the COVID-19 pandemic taught us this lesson the hard way [19].

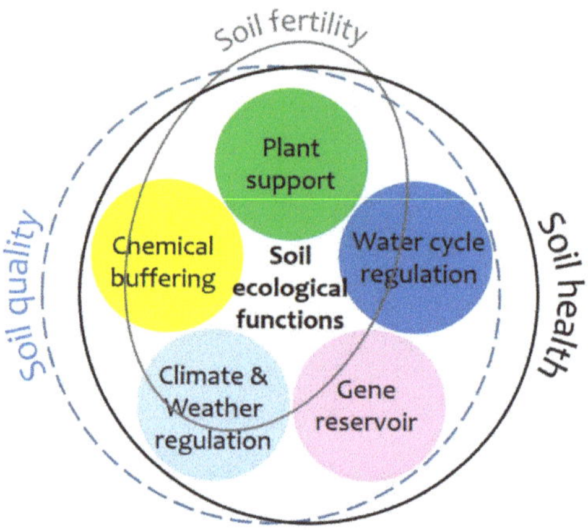

Figure 2. The concepts of soil health, soil quality, and soil fertility as defined by the soil ecological functions they totally or partially embrace.

3. The Soil Biological Fertility of Vineyards

The traditional concept of soil fertility has focused primarily on the chemical and physical fertility, while the soil biological fertility has been often sidestepped. Fortunately, this is changing as agricultural and forest systems are seen from a holistic point of view. In this regard, Abbot and Murphy [20] provided a comprehensive definition of soil biological fertility that underscores the direct connection between the soil biological community and the plant nutrient supply, thus recognizing the linkage of the soil biology with the soil chemistry and physics. In doing so, they showed how the physical, chemical, and biological soil fertility approaches may integrate the concept of soil health. Therefore, soil's biological fertility is understood as the capacity of the organisms residing in the soil to (i) directly contribute to the nutritional requirements of crops for productivity and harvest quality, and (ii) indirectly, but concurrently, to contribute to the maintenance of all the biological processes that have a positive impact on the soil's chemical- and physical conditions with on-farm and off-farm enhancement effects. Unfortunately, our understanding of the soil's biological fertility remains relatively limited in two critical aspects: (i) how to optimize the advantages derived from soil's biological processes, and (ii) whether it is economically or environmentally sustainable to take advantage of the benefits of the various biological processes unfolding in soils [20].

Although the soil organisms constitute only a small fraction of the total soil weight, and even a small fraction of the SOM itself, they play numerous pivotal roles in soil systems [21]. The immense diversity of life forms we can find in soils is responsible for several critical processes that support soil functions. These processes include: on the one hand, (i) the decomposition of organic debris, (ii) the transformation of chemical compounds between different forms, (iii) the facilitation of plant-nutrient acquisition through symbiotic associations, and (iv) the production of soil-stabilizing agents that promote the aggregation of mineral particles, while, on the other hand, they also include (v) the degradation of synthetic organic chemicals like pesticides and herbicides, and (vi) the production of antibiotics that can aid in suppressing soil-borne diseases [20]. Consequently, soils with a functional biological community provide both a nurturing and a healing environment for crops and other agroecosystem plant-beneficial organisms, thus making the soil healthy. As our understanding of the soil biology improves, we will gain insight into how to take advantage of all these biological processes to enhance yields and incomes for producers, as well as environmental health for everyone [22].

Unfortunately, the intensification of agriculture contributes to soil biodiversity losses, with vineyard terrains being one of the most harmed because they are very intensively managed and erosion-prone [23,24]. In this context, Karimi et al. [25], in their meta-analysis of the ecotoxicological impact of viticultural practices on soil biodiversity, demonstrated that soil microorganisms are more abundant in organic viticulture compared to conventional viticulture. This is due to the detrimental effects associated with conventional vineyard management, which involves (i) frequent tillage and use of weed-killers to keep soils bare both below and between vines, (ii) the application of mineral fertilizers, and (iii) the use of several pesticides, particularly against fungal pathogens. The adverse impacts on overall soil biodiversity lead to degradation of its biological fertility, with harmful consequences for the ecological services soils provide [26]. On the contrary, conservation tillage, organic fertilization, and cover cropping, among other practices, may increase the diversity of the soil's biological community and hence boost their associated functions for developing more sustainable vineyards [3]. Over time, viticulturists are increasingly adopting these conservation agriculture practices (CAPs). However, concerns persist for winegrowers, as these practices, particularly cover cropping, if improperly carried out, may increase water and nutrient stresses on vines markedly enough to harm, at least, grape production objectives [27].

Fostering a suitable soil habitat is paramount to enhance its biological fertility [20]. Through the application of different CAPs, viticulturists can modulate microbial activity inasmuch as they directly or indirectly provide the adequate physical and chemical conditions for the microbial populations to thrive (Table 1) [20]. In this respect, a sufficient supply of organic debris to the soil is the foundation, though not the only method, for enhancing its biological fertility. At least, to have a non-compacted soil is equally important since the soil-pore network plays a critical role in determining the spatial and temporal distribution of soil biota and nutrients, facilitating flow paths for solutes and gases [20]. In this regard, further research is needed to unravel the impact of soil management practices on the soil's microbial diversity, and on changes in the evenness of the soil's microbial community.

Table 1. Impact of soil agricultural management practices on the microbial biomass (↑ increase in population; ↓ decrease in population; ↑↓ controversial results) [20,28–31].

Soil Practice	Effect on Soil Microbial Population
Retention of crop residues	↑
Application of organic amendments to soil	↑↓
Minimum tillage practices or absence of ploughing	↑↓
Irrigation of crops in rain-limited ecosystems	↑
Application of fungicides to soil	↓
Application of herbicides to soil	↑↓
Liming of soil to raise pH on acidic soils	↑
Unbalanced fertilization	↓

There is a growing body of evidence indicating that a comprehensive understanding of soil requires the integration of relevant physical and biochemical approaches [32]. Therefore, incorporating a pedological perspective into soil biology becomes essential for assessing the significance of soil's biological processes and identifying fundamental principles that can be universally applied across various soil types and environments [20]. In the context of viticultural soils, the impact of soil biodiversity on soil functions, encompassing changes in both chemical and physical soil fertility, as well as ecosystem services, holds significant importance. Additionally, it is crucial to acknowledge that grapevines are typically cultivated as scions grafted onto rootstocks, which are selected based on the prevailing and expected soil and climate conditions [33]. Consequently, the molecular and microbial interactions between the scion and the soil via the rootstock must be considered in order to enhance the sustainability and resilience of viticulture [34].

3.1. The Impact of Soil Organisms on Chemical Fertility

Maintaining a suitable soil habitat by fostering the supply to the soil organisms of adequate quantities of organic materials of enough quality, is crucial for soil microbial communities to efficiently release nutrients in root-absorbable forms and, therefore, available to plants [20]. In this regard, rhizodepositions from vines, but also from other accompanying plants in the agroecosystem, both spontaneous and seeded as cover crops, should be considered as organic materials of the highest quality that can effectively contribute to increased soil carbon (C), and maybe also nitrogen (N) [35].

In soil agroecosystems, most nutrient transformations involve a multitude of biochemical reactions that are part of the biota metabolisms [36]. Pertaining to this, key microbial processes associated with soil nutrient cycling encompass (i) nutrient release through mineralization of organic compounds and solubilization of nutrient-containing minerals, e.g., Bettenfeld et al. [37] established that some plant growth-promoting bacteria species like *Pseudomonas*, *Enterobacter*, *Arthrobacter*, and *Bacillus* improve grapevine nutrition, at least in part, by increasing the vine-available phosphorus (P) through the transformation of organic P into soluble phosphate ions, (ii) nutrient enrichment through symbiotic and non-symbiotic N_2 fixation, etc., (iii) nutrient transformations through nitrification of NH_4^+ to NO_3^-, reduction of Fe^{3+} to Fe^{2+} and MnO_2 to Mn^{2+}, oxidation of S and S^{2-} to SO_4^{2-}, etc., and (iv) nutrient losses through denitrification of NO_3^- to N_2O and N_2 [20]. Consequently, the activity of soil microorganisms plays a pivotal role in regulating the amount and availability to grapevines of nutrients such as phosphate and nitrate and hence it controls chemical fertility [38].

As indicated, the soil biota affects the concentrations of nutrients in available forms; however, the other way round, the soil nutrient richness leads to intricate metabolic interactions within microbial communities, resulting in changes in the identity and relative abundances of soil taxa [39]. In this regard, it seems that P is the nutrient that mostly affects soil microbiome traits, probably because N is not applied in vineyards at high enough rates to clearly show up [40].

Since vineyard fertilization is complex and highly dependent on plant material, soil characteristics, and the production objectives [41], enhancement of grapevine nutrition by microorganisms is an interesting line of viticulture research.

3.2. The Impact of Soil Organisms on Physical Fertility

The profound impact soil fauna activities have on the soil's physical properties include (i) excavation for food or living spaces, (ii) ingestion of soil materials, (iii) production of fecal pellets, (iv) utilization of excreta, mucus, or salivary secretions, and (v) collection of plant, animal, and fungal debris mixing with the soil mineral particles, with or without prior digestion [20]. The combined activities of soil fauna with those of plant roots, besides fungi and bacteria, result in the formation of biological aggregates, to which earthworms particularly contribute [42].

The soil aggregates thus formed serve as microhabitats that support the proliferation of soil microorganisms. These microorganisms further contribute to soil aggregate formation by producing specific exudates and by releasing lysates when they die [43]. While soil fauna strongly affects the distribution and organization of soil particles, the microbiota crucially contributes to their soil-structuring works through several direct and indirect processes. These include: (i) moving and aligning primary particles along cell or hyphal surfaces, (ii) adhering particles together through sticking compounds, e.g., exudates, involved in colony cohesion, (iii) enmeshing and binding aggregates through fungal hyphae, actinomycete filaments, and associated mycelia, and (iv) coating pore walls with hydrophobic compounds [36]. Bacteria contribute to aggregate formation by producing negatively charged polysaccharides, and polyuronic and amino acids that stick to soil particles with the aid of cations. Fungi, including saprophytic, ectomycorrhizal, and arbuscular mycorrhizal, also play essential roles in forming soil aggregates through physical mechanisms such as the formation of hyphae and mycelia that entangle soil particles, and through physicochemical mechanisms that glue soil particles together by means of exudates, which include polysaccharides like glomalin, mucilages, hydrophobins, and extracellular compounds from hyphae [43].

3.3. *Key Interactions in the Rhizosphere on Soil Biological Fertility*

The rhizosphere is the soil region closer to the plant roots and is, therefore, greatly influenced by the rhizodepositions, which encompass exudates, as well as detached root tissues and cells and the lysates they release when break up [44,45]. Accordingly, the rhizosphere is the primary interface for interaction between a plant and the soil environment [46], possessing distinct biological, physical, and chemical characteristics compared to the bulk soil (Table 2). The rizhosphere high contents of labile organic materials boost the microbial populations, which are jointly referred to as the rhizomicrobiome [45]. The rhizomicrobiome is particularly important in regulating plant growth and health, and makes the rhizosphere to become the most dynamic part of the soil, where multitude of biogeochemical processes take place. A comprehensive understanding of these processes is vital for sustaining the microorganisms that inhabit soils and preserve soil health [47].

Table 2. Distinctive characteristics between rhizosphere and bulk soil (↑ this soil characteristic is increased; ↓ this soil characteristic is decreased) [20,48–53].

Characteristic	Rhizosphere	Bulk Soil
Concentrations of easily degradable substances (sugars, organic acids, and amino acids)	↑	↓
Diversity of microorganisms	↓	↑
Abundance of specific microbial taxa	↑	↓
Phytotoxicity of heavy metals	↓	↑
Enzyme activities	↑	↓
Nutrient availability	↑	↓
Soil pH	↑ (alkalinisation)	↓
Water retention because of changes in soil structure	↑	↓

The nature of the soil, environmental conditions, and plant genotype, particularly the characteristics of its root rhizodepositions, determine the microbial diversity of the rhizosphere [45]. It is quite likely that plants and their associated rhizomicrobiome have coevolved in natural ecosystems, leading to the mutual adaptation of each other for survival [54]. Furthermore, the availability of plant nutrients in the rhizosphere is influenced by a combination of soil properties, plant characteristics, and the interactions between plant roots and the rhizomicrobiome [20]. As a result, the rhizosphere and the characteristics of the rhizomicrobiome feature a delicate balance between stress and fitness, and could serve as valuable biological indicators of plant health status [34] and, therefore, soil health.

The interaction between the vine roots and the soil microbiota is essential for the control of soil phytopathogens and for the acquisition of the nutrients required for growth, and the rhizosphere is at the core of these interactions [34]. Indeed, the richness and diversity of bacterial and fungal communities in the root rhizosphere are significantly influenced by the grape variety [55], and also by the rootstock genotype as an important driver of the bacterial community composition [56]. Therefore, a new research approach named rhizosphere engineering is increasingly coming into focus to enhance grapevine growth and health, in which microbial communities are modified by adding specific fertilizers, nutrients or by bio-inoculation with certain bacteria and/or fungi, suggesting a potential for reaching a more sustainable development of pesticide-reduced viticulture in the future [57]. Another potential area of future research would be to investigate the grapevine rhizosphere as a microbial source of inoculum of grape berry microbiota since a hypothetical endophytic way of colonization has been proposed for the fermentative yeast *Saccharomyces cerevisiae* to be transported from the soil via roots and xylem [34].

3.4. *The Impact of Mycorrhizae on Soil Health*

The rhizomicrobiome of plants colonized by mycorrhizae exhibits significant differences compared to that of non-colonized plants, and thus deserves some depth of comment. Mycorrhizal plants have been shown to have higher proportions of organic acids, including amino acids, in their roots in contrast to non-mycorrhizal plants [20]. Mycorrhizal fungi are considered effective candidates for biofertilization programs due to several advantageous features, making them a valuable resource for soil health in the context of sustainable agriculture. These advantages include their ubiquity in nature, ease of use and production as carrier-based inoculants with a long shelf life, capacity to increase the surface area of plant roots, phosphate solubilization capability, improvement of micronutrient uptake, assistance in plant stress tolerance, and mitigation of biotic stress [58].

Arbuscular mycorrhizal fungi (AMF) benefits start from their ability to extend the volume of soil explored by grapevines, allowing an adequate uptake of nutrients such as N and P [41]. The fungal mycelium increases the exploitable soil volume, allowing access to soil pores that would be otherwise non-explorable by root hairs. Regarding the grapevine uptake of other nutrients, Schreiner [59] showed an enhanced Cu and sulfur (S) uptake by AMF-inoculated grapevine cuttings from two different soils, and suggested that the selection of appropriate inocula for specific functions would be the best strategy in biofertilizer procedures. Furthermore, in grapevines grafted onto rootstocks that are susceptible to iron chlorosis, AMF inoculation may be a way to alleviate its symptoms on calcareous soils, but further studies are needed [41]. Interestingly, the modulating effect of mycorrhization on vine nutrient absorption rate and selectivity seem to unfold through its effect on the bacterial microbiota in the root endosphere, where the populations of several beneficial taxa increase. Additionally, grapevines gain enhanced protection against pathogens when colonized by AMF, although the safeguarding effects of AMF may not apply uniformly to all pathogens [60] and should be further investigated.

In addition to vine nutrition and protection, as well as soil aggregation, AMF have other benefits for vines. In this regard, Gazioglu [61] also showed how AMF can mitigate the physiological and morphological changes induced by salt stress in grapevine cuttings, while Cardinale et al. [62] showed how mycorrhization can mitigate the impact of drought stress on plant survival and growth. The drought-ameliorating effect of AMF on grapevines seems to develop through various mechanisms, including higher osmolyte build-up and improved photosynthetic activity [41]. Therefore, AMF could form the basis of an effective management strategy to counteract drought effects on grapevines, and thus further investigations in this regard are of the utmost importance for enhancing many vineyards, particularly under Mediterranean and related semi-arid conditions, which are predicted to become dryer under the climate change scenario.

Counteracting the beneficial effects of AMF on crop plants is tillage, because this practice seems to effectively break up the mycelia networks, as the results of Lumini et al. [63] suggest. Indeed, these authors found lower AMF biodiversity in a tilled vineyard soil regarding another covered with spontaneous plants. In spite of the above, the question arises whether shallow tillage can affect AMF communities which grow relatively deeply in the soil [41]. Since vines' roots can remarkably deepen, more research is needed in this regard.

To sum up, given the crucial role AMFs are likely to play in alleviating abiotic and biotic stress in grapevines, more studies are needed to identify the diversity of AMF species already present in vineyards across different edaphoclimatic conditions [60] and how their beneficial effects may be enhanced.

4. Soil Organic Matter as a Key Property of Soil Health in Vineyards

Soil biological activity and SOM are strongly interrelated and heavily depend on one another. The SOM comprises the part of the soil condensed material that features carbon in formal oxidation states under four ($z_C < +4$). The SOM has gone through one to several cycles of biosynthesis that started from the plant or bacterial assimilation of simple inorganic compounds such as CO_2, H_2O, N_2, HPO_4^{2-}, and SO_4^{2-}, and is eventually headed to release the same inorganic compounds to the environment [64,65]. The SOM often constitutes a little share of the soil condensed material featuring a worldwide average of 6.5% on a dry-weight base in the top 10 cm [66]. However, an ample range of SOM between 0.2 and 60% may be estimated for the topsoil, depending on soil and climate conditions [67]. Despite the general small portion the SOM represents regarding the soil's minerals, the changes in SOM remarkably affect most soil properties, from physical to chemical and biological [68,69]. Additionally, despite its low share of soil weight, SOM down to 1 m depth also stores 1800 Gt of C, i.e., more than the atmosphere (570 Gt) and biota (740 Gt) combined [70,71], which might increase by 20–50 Gt if best CAPs were globally adopted [72]. Therefore, the SOM is key for soils to be healthy through at least (i) the maintenance and enhancement of the soil fertility and hence sustainability of cropping and farming systems; and (ii) the regulation of climate [73].

4.1. Soil Organic Matter and Soil-Health Attributes

From a physical perspective, SOM significantly increases the aggregation of the soil's mineral particles [74–77]. Indeed, as the SOM increases, soil particles define a more organized and interconnected network of pores of variable sizes featuring, in general, lower bulk density [78]. This soil pore space is built and rebuilt time and again by plant roots [79,80] and soil fauna and fungi [81–83]. However, this dynamic soil pore space is kept from collapsing by the SOM ability to increase the magnitude of the non-directional Van der Waals attractive forces among the soil particles [84,85]. Through such an organized systems of pores, gases can continually diffuse in and out of the soil [86–88], water can infiltrate [89,90], and, moreover, be held in part against gravity and evaporation, and thus be kept available for plants and soil organisms [91].

From a chemical perspective, SOM is a reservoir of many essential plant nutrients. Indeed, as the SOM increases, the soil contents of the macronutrients N, P, and S, generally rise [92–94]. However, since N, P and S are covalent-bonded in the SOM in many different compounds [95–97], their availability to plants is constrained, and their release in the soil environment as plant-absorbable species is mediated by the soil organisms' metabolism [94] as previously addressed (Section 3). This condition contrasts with that of the other macronutrients, i.e., potassium (K), magnesium (Mg), and calcium (Ca), whose plant-available contents increase more readily with SOM. This is because SOM is negatively charged at even the lowest soil pH values, whereas K, Mg, and Ca are always found as cations in the soil environment, i.e., K^+, Mg^{2+} and Ca^{2+}, and thus SOM is able to hold them readily exchangeable by other cations, including root-exudated protons [98,99].

Regarding the micronutrients, as the SOM increases, their availability may increase or decrease depending on the elements' redox properties as well as on soil aeration, pH, and the fraction of SOM that is water-soluble [100]. Specifically, iron (Fe) is a nutrient whose available soil concentration seems to increase in general with SOM, because SOM decreases the soil redox potential (Eh), thus fostering the reduction of the (oxyhidr)oxide-immobilized Fe(III) to the water-soluble Fe(II) species [101–103]. Additionally, SOM also provides low-weight water-soluble organic ligands that further stabilize Fe(II) in solution and keep them plant-available [102,104]. For manganese (Mn), the same processes have been described [102] and, in general, it seems that Mn availability increases as SOM rises, but the effect seems weaker than with Fe [100]. Finally, zinc (Zn), Cu, boron (B), and molybdenum (Mo) present, in general, poorer redox chemistry in the soil environment [102]. Therefore, the availability of Zn, Cu, B, and Mo is not so controlled by the low Eh the SOM may foster, but rather by SOM adsorption abilities. Consequently, since SOM also forms surface complexes with Zn, Cu, B, and Mo, in general SOM decreases the availability of these elements to plants [102,105–107], which would thus be released at the pace the SOM is mineralized.

From a chemical perspective, in addition to behaving as a nutrient reservoir, SOM also behaves as a pollutant buffer by adsorbing xenobiotic organic compounds (XOC) [108–110]. In addition to adsorption, SOM may also foster XOC degradation through the enhancement of the activity of the soil organisms, hence behaving as a factor in environmental remediation [111]. In the case of heavy metals, the effects of SOM may or may not be beneficial, since SOM can immobilize metals as surface organic complexes, but it can also render them more available in the soil-pore water following reduction to more soluble chemical species and complexation with low-weight organic ligands [112,113]. Nevertheless, in general, upon passing through soil, the volatilization and, importantly, the leaching and plant uptake of many harmful chemicals is slowed down or even avoided if degradation is feasible and enough time passes by.

From a biological perspective, SOM is strongly and positively related to the functionality of the soil biological community. Indeed, as the SOM content increases, the abundance, activity, diversity, food web structure, and stability of the soil biological community are all enhanced [114–116]. A functional biological community is what distinguishes a healthy soil [6]. A functional biological community builds and rebuilds the soil's pore space [81]. In doing this, the soil organisms feed on the plant debris, mostly dead roots, which are the primary source of SOM [117], resynthesize its compounds time and again, and eventually release and mix its remains with the mineral particles. The mixture of SOM with the mineral particles presents more cohesion than the separated organic or mineral particles. Therefore, such an arrangement of organic and inorganic materials is able to hold up the soil pore space against several forces that may act upon it: disruption by wetting, impact, overburden, shear, etc., thus showing in the increase in the stability of soil aggregates with SOM [118–120]. Moreover, as the organisms feed on SOM, some nutrients, which were bonded to it, are released for the benefit of other soil organisms and plants. In addition, some pollutants, remarkably XOCs, may be degraded by the organisms' metabolism, or tightly adsorbed into the organic–mineral assemblages [111], and hence their volatilization, leaching, and plant uptake are diminished.

As presented, SOM is key to assure the ability of soils to sustain (i) plant growth and development, (ii) the biogeochemical cycles of water, as well as C, oxygen, N, etc., and (iii) the habitat structure of the soil biological community. The durable running of these ecological functions makes soils able to indefinitely keep the productivity and health-fostering properties of biological systems due to the quality preservation of agricultural produce, as well as of water and air, thus ultimately promoting the animal and human welfare [121,122].

4.2. *Dynamics of Winegrowing Systems and Levels of Soil Organic Matter*

Since the Neolithic age, the transformation of lands to agriculture has been accompanied by remarkable losses of SOM, whose C has been eventually oxidized to CO_2 and released to the atmosphere [72]. In the case of viticultural systems, the SOM depletion has been especially severe [123,124] because vineyards have been overall traditionally developed on sloping lands subjected to sustained conventional tillage, thus becoming highly erodible [24,125–128]. Indeed, in areas under the Mediterranean climate, i.e., where a good deal of the World's wine growing takes place, soil water erosion in vineyards attains high rates [129] due to the Mediterranean-characteristic intense rainfalls [130,131], evident in remarkable topsoil thinning [123] and even soil truncation [132]. Since SOM is usually highest in the topsoil and exponentially decreases with depth [133–135], topsoil thinning and soil truncation mean a severe loss of SOM. Therefore, it is not surprising that SOM levels in vineyards have decreased much more than under other land uses, including similar agricultural ones [123], and globally are estimated to range between 0.1% and 6.3%

with a mean of 1.7% [136]. Such SOM levels are considered low in general for balanced textured soils [137], and therefore they jeopardize soil health in viticulture [3,138].

Water erosion causes SOM losses throughout the World [139], and in turn low SOM levels may increase water erosion rates because of the loss of topsoil aggregate stability [140]. Therefore, SOM loss and water erosion feed back to each other. As a way to counteract the harms to soil health in vineyards driven by SOM loss, several CAPs have been proposed [3,141]. The CAPs act by reducing the soil erosion and, consequently, by reducing or even reversing SOM loss, or by directly increasing the SOM content. The CAPs used in viticulture can be classified into the following four types: (i) soil mulching with organic materials like pruning residues and cereal straws, (ii) soil incorporation of foreign organic matters like composts, vermicomposts, biochars, and manures, (iii) soil inoculation of beneficial organisms like AMF, and (iv) the use of cover crops in vineyard alleys.

Straw mulches can foster SOM increase if timely applied, as well as adequately rated and replaced as indicated by Gómez de Barreda et al. [142], who found in a flat Eastern Spain vineyard that straw mulching increased SOM in the 0–12 cm layer up to 2.3%, i.e., a +32% increase as compared to bare soil. Interestingly, the average SOM increase during the three years of application was a linear function of the straw rate up to, at least, the maximum they tested, which was 50 Mg ha^{-1}. Similarly, Qin et al. [143] found that, in a flat Eastern China vineyard, straw mulching applied at rates of between 7 and 21 Mg ha^{-1} significantly increased SOM in the 0–20 cm layer from 1.6 to 1.8%, i.e., a +10% increase as compared to bare soil. However, the linear increment of SOM with the straw rate was observed only for rates of 14 Mg ha^{-1} onwards during the first year of application, thus apparently remaining steady, and independent of the straw rate during the second one.

The application of organic amendments is another widely adopted method to enhance SOM in vineyards [144,145]. In this regard, Morlat and Chaussod [144] found in a flat Central France vineyard that SOM significantly increased up to almost saturation in the 0–30 cm layer from 1.5 to 2.2%, i.e., a +50% increase, in 22 years since the start of cattle manure and mushroom compost additions, at rates of, respectively, 20 Mg ha^{-1} yr^{-1} and 16 Mg ha^{-1} yr^{-1} (in terms of fresh weight). Besides, Gaiotti et al. [145] found in another flat NE Italy vineyard that SOM significantly increased in the 0–50 cm layer from 2.3 to 2.6% and from 2.3 to 2.9%, i.e., +14% and +26% increases, in 5 years since the starting of additions of, respectively, pruning wastes and cattle manure, at rates of 4 Mg ha^{-1} yr^{-1}.

The inoculation of beneficial organisms in vineyards offers a promising method for SOM enhancement. Particularly, mycorrhizal fungi improve the relationships between vines and soil [146] with enhancing consequences on photosynthetic activity and C assimilation [147]. The allocation of vine photosynthates to the mycorrhizae, as well as their dampening of the rhizosphere-priming effect on SOM decomposition, might additionally contribute to higher SOM levels in vineyards [148]. However, there is limited knowledge of net changes in SOM levels as a consequence of plant mycorrhization in actual cropping conditions. Therefore, to what extent mycorrhizae may contribute to increase SOM in vineyards, if any at all, should be further investigated [149].

The use of cover crops has the potential to improve soil health by enhancing SOM and its dynamics through soil protection and the boosting of the functionality of the soil's biological community [150]. Ruíz-Colmenero et al. [151] found, in a Central Spain vineyard, decreases of mean soil loss of 78% and 87% regarding traditional tillage due to two grass treatments, the former mown and the latter permanent. The soil conservation gains under both treatments were accompanied by a significant build-up of SOM from 1.2 to 1.6%, i.e., a +28% increase, in the 0–10 cm layer, after four years of cover crop establishment. Furthermore, Guzmán et al. [149] also found in a survey of 16 vineyards in a hilly area in Southern Spain that cover crops, either spontaneous or sown, significantly increased SOM in the 0–10 cm layer from 1.4 to 1.8%, i.e., a +29% increase, as compared to bare soil, after three years of cover crop establishment. Away from the sloping Mediterranean vineyards, Fleishman et al. [152] found in Pennsylvania a significant increase in SOM from 3.8 to 5.2%, i.e., a +37% increase, in the 0–20 cm layer, after four years of cover crop establishment [153,154].

The beneficial effect of cover crops for SOM enhancement regardless of vineyard slope is that cover crops are a source of fresh organic matter in the form or rhizodepositions [155,156]. However, the use of cover crops under arid to dry subhumid climates presents some drawbacks, since they decrease the availability of soil water for vines [157], thus generally increasing the water stress upon them. Conversely, mulches diminish the water stress on vines [158]. The effects of cover crops on vine water status under such dry climate conditions have consequences on vine vigor and also on grape yields under rain-fed water regimes [157]. However, the yield losses may be made up for by the increase in the grape must quality for winemaking [159]. The use of one management practice or the other to increase SOM levels, whereas preserving grape yields and quality, depends on the soil and climate conditions. The effects of all these CAPs on SOM, as well as on vine water status, and ultimately grape yield and quality, with the specific aim of increasing this latter, i.e., quality, should be further investigated.

5. Soil Physical Fertility in Vineyards

Soil organic matter affects soil's physical properties, and a good physical condition of soil enhances SOM. A viticultural soil is considered to have good physical quality when it possesses the strength to not collapse against several forces that may act upon it, thus maintaining a sound structure that facilitates unhindered root growth, as well as the burrowing and settlement of soil fauna. Additionally, such a soil exhibits good water-transmission capacities and water-holding characteristics that allow for the proper distribution and storage of incoming rainfall and irrigation, its solutes, and gases. These qualities are essential both for optimal crop performance and for minimizing environmental degradation [160].

Maintaining soil structure is crucial for soil health as it governs the soil-water cycle and provides the medium for plant roots to anchor, respire and absorb water and nutrients [161]. Moreover, soil structure plays a pivotal role in regulating the physicochemical heterogeneity of the soil, thus influencing the distribution of microbial communities. The heterogeneity of soil aggregates promotes diverse biological niches with varying physicochemical properties, i.e., variation in aeration and hence Eh, pH, etc., thus microbial activities across different aggregate sizes differ [162]. In this context, biological activity plays a crucial role in resisting the physical degradation of soil since it importantly contributes to the formation and preservation of soil aggregates and pores, and thus soil structure [161].

Given that soil structure, as an abiotic factor, significantly influences microbial communities by creating distinct niches that support specific microbial adaptations, the interaction between soil structure and microbes emerges as a decisive factor shaping soil microbial diversity and, consequently, soil health [39]. It is evident, therefore, that soils suitable for grape production should possess well-developed structural integrity, stability against water and external mechanical stresses, and resilience, enabling them to recover their structure following disturbances [163].

5.1. Soil Compaction, Water Infiltration, and Aeration

Soil compaction, a significant concern for intensively used agricultural soils, is a physical process resulting in increased soil bulk density, and decreased soil porosity [164]. The ecological and economical damages associated to soil compaction are issues of global importance [165], which are particularly insidious in modern mechanized agricultural exploitations, particularly where soils are low in SOM, and high in clay and/or silt [166]. Unfortunately, soil compaction issues are anticipated to escalate due to the worsening weather patterns expected under climate change, including higher average temperatures and extended dry periods [167]. The effects due to higher temperatures are driven by the expected SOM losses as the climate warms up [168]. The effects due to longer droughts arise because the drier the soil, the more severe the compaction consequences on water infiltration, root development, etc. [169].

Soil compaction not only decreases porosity, it also changes pore length, average diameter, and tortuosity, thus adversely impacting ground penetrability by roots and soil fauna, as well as water infiltration, holding capacity, percolation, and drainage [166]. As a consequence, soil ecological, hydrological, and agronomic functions are harmed because of the following effects: (i) reduced overall plant productivity due to lack of water and nutrient availability to roots or either, because of water logging and scarce aeration; (ii) increased risk and severity of flooding and erosion due to decreased infiltration and increased surface runoff; (iii) elevated emissions of greenhouse gases like N_2O and CH_4 resulting from reduced soil aeration, but also more CO_2 resulting from increased fuel consumption for tillage; (iv) decreased fertilizer efficiency due to the constrained mobility of their soluble species along with water; and (v) diminished rhizodepositions and hence less C input from plant roots due to heightened soil mechanical resistance to root penetration and impaired aeration [167].

The impact of compaction on the soil microbiome structure and activity varies considerably across different textures, compaction levels, and water contents. However, in general, the alterations in the soil-pore system characteristics induced by compaction create less favorable conditions for beneficial microorganisms [170]. Increasing soil compaction, and hence rising bulk density, constrains gas and water fluxes [170,171], eventually diminishing air-filled porosity, i.e., soil aeration. Anaerobic soil conditions lead to changes in the microbial community, causing most aerobic organisms to become quiescent or perish [172]. This favors microbial populations capable of tolerating anaerobic conditions. These involve both facultative and obligate anaerobic microorganisms including Fe- and sulfate-reducing microorganisms, which develop at the expense of fungal populations [162,172]. Moreover, under such low-aeration conditions soil C and N dynamics change and SOM concentration and availability of soil N also changes [173] thus impacting soil chemical fertility and plant nutrition.

5.2. Soil Erosion

Soil erosion by water is a critical threat to soil resources, particularly in Mediterranean regions, due to their specific climatic and geomorphologic conditions. Soil erosion rates in vineyards are higher than in other land uses due to both natural and anthropogenic influences [24]. Natural factors include rainfall intensity, slope steepness, rock fragment cover, and soil texture. Specifically, slope steepness, which characterizes many vineyards under the Mediterranean climate, is the factor that emerges, along with the late-summer and early-autumn intense rainfalls, as a key determinant of water-induced soil loss [126]. Indeed, soil erosion dramatically materializes during the intense rainfalls that feature the Mediterranean wet season's inception, when plant cover is naturally at its minimum following the characteristic summer drought. Soil erosion is furthermore exacerbated in vineyard soils due to the use of conventional soil-management practices, including harsh tillage operations [174–176], frequent machinery traffic, and herbicide applications to get rid of inter-row and below-vine vegetation. Moreover, often young vineyards are particularly vulnerable, and therefore they experience unacceptably large soil erosion rates [177].

The conventional vineyard soil management practices significantly elevate erosion rates, potentially leading to severe and possibly irreversible situations [123,132]. For instance, a study in the conventionally managed steep Prosecco wine-producing area in Italy estimated erosion rates as high as 44 Mg ha^{-1} yr^{-1}, which exceeds 31 times the maximum soil erosion threshold established for Europe [178]. Therefore, conventional vine management practices, particularly in areas with steep slopes, shallow soils, and many years of ploughing, such as the Prosecco area, negatively impact soil health. Herbicide application, which is used as an alternative to tillage for weed control in vineyards, possibly reduces erosion regarding tillage, but maintains it at non-negligible rates [179]. High soil-erosion rates in vineyards lead to increased nutrient losses [180] and less soil water availability and, therefore, to less vine vigor [181], as well as lower grape yield and berry quality, featuring higher total soluble solids and pH [182]. To counteract this soil fertility decline, viticulturists resort to greater external input use, ultimately compromising viticultural sustainability [181]. Climate change, with its higher-intensity rainfall events and interspersed droughts, threatens to increase soil erosion rates [183]. Therefore, encouraging viticultural practices that enhance soil and water conservation becomes increasingly important in this scenario.

Improved soil management practices aimed at increasing soil cover and SOM levels can bolster viticultural soil resistance to erosion, thereby improving their health [161]. To achieve sustainability in vineyards prone to soil erosion, it is imperative to adopt soil and water conservation practices. Particularly, terracing and the use of inter-row cover cropping can help. Terracing is a traditional practice that stands out as one of the most effective for erosion control, effectively protecting farms from soil loss. For instance, a study conducted by Pijl et al. [183] in 50 vineyards in Northern Italy found that three terracing methods (contour terracing, broad-base terracing, and oblique terracing) presented higher sediment flux mitigation ability than non-terraced practices. This was especially notable considering that terraced vineyards often occupy steeper slopes. Among the terracing methods, regular contour terracing consistently outperformed the others in terms of reducing uphill erosion, sediment flux, and downhill deposition. In spite of these beneficial effects, ineffective terrace establishment and management can lead to declines in soil fertility, as well as slope instability issues that may eventually result in terrace abandonment, amplified soil degradation, and even hydrogeological risks [184].

In addition to terracing, which is an engineering practice that tries to hold the soil in place by physical containment, cover crops can also be used to decrease soil erosion [184,185] and to improve water-holding capacity and overall soil condition, hence reducing reliance on industrial fertilizers [151,176,186,187], although not as effectively as terracing [188]. Pappalardo et al. [178] estimated that by using a 100% ground-covering inter-row cover crop, the erosion in the Prosecco wine-growing area could be curtailed by 67%. More specifically, a six-year experiment conducted by Stanchi et al. [179] in an Italian sloping vineyard revealed that erosion rates were almost negligible under permanent grass cover, whereas tractor passage significantly accelerated erosion showing up in an increased downhill accumulation of sediments high in SOM, N, and clay. These findings strongly point towards the conservation of soil fertility and health which results from the use of cover crops in vineyards.

In addition to behaving as a physical shield against raindrop impacts, the positive effects of vegetation covers that contribute to soil erosion control include increased aggregate stability and water infiltration rates because of the facilitation of soil aggregate formation [186], and enhancement of soil porosity due to gains in SOM and biological activity [126]. However, the use of cover crops in vineyards under dry climates may limit grape yields [187]. Notwithstanding, inter-row cover cropping in vineyards may also help improve grape quality, as Pou et al. [189] found. By precisely selecting the cover crop species and the time of sowing and reaping with regards to the local climate and vine phenology, the drawbacks of this soil management practice can be minimized and their benefits maximized in vineyards under arid to dry-subhumid climates [157].

Another soil management practice for soil conservation against water erosion in vineyards is the use of mulches. In this regard, Prosdocimi et al. [176] found in Eastern Spain vineyards an 80% decrease in median soil loss due to the protection effect of barley straw mulch rated at 750 kg ha^{-1}. Furthermore, Mirás-Avalos et al. [190] found in a Western Spain vineyard a 57% decrease in average soil loss as a consequence of the protection effect of a mixture of straw and gorse residues. However, for soil erosion control, the issue with mulches, and straw mulching in particular, is the ease with which they may be dragged downhill by wind, e.g., by the mountain breeze, or rainfall.

6. Tillage and Soil Health in Vineyards

In agriculture, tillage has traditionally served as the primary means for weed control, playing a pivotal role in getting crops rid from the competition for water and nutrients of these "plants in the wrong place". Therefore, tillage has been considered as an important means for achieving high crop production and economic profitability. In addition to weed control, tillage is usually viewed by agriculturists as an essential means for attaining several cultural targets, such as (i) grinding and soil mixing of leaves, pruning wood and other organic materials, (ii) promotion of water infiltration, (iii) boosting of SOM mineralization and nutrient release in plant available forms, (iv) thorough mixing of nutrients in the soil, (v) disruption of the habitats and the life cycle of soil-borne pests, and (vi) breakup of soil compacted layers that constrain root penetration and water movement throughout the soil depth [191].

Therefore, tillage has been considered as promoting several soil functions beneficial for cropping, i.e., the soil fertility in the short term. However, tillage comes with built-in disadvantages, as the soil disturbance it causes affects soil properties essential for the agroecosystem functioning in the mid- to long-run [192]. Regarding the soil physical fertility, tillage disrupts aggregates and ruins soil structure, paradoxically fostering the appearance of compacted soil layers [193], whose breakup demands more tillage operations, which are deeper and of higher intensity to make up for the low physical fertility status that results [194]. Tillage also exposes the soil to the rainfall drop impacts as well as to the runoff, hence increasing the soil vulnerability to water erosion. Moreover, although right after tillage the water infiltration capacity in vineyards may increase [158], tillage eventually eases water evaporation from soils [195], thus likely contributing to vine water-stress.

Tillage also influences soil chemical fertility but often not exactly in the way it was intended. Notably, it has been observed that topsoil total N, P, and K concentrations decrease, particularly under the harsher tillage operations in comparison to no tillage, as a meta-analysis by Nunes et al. has shown [196]. This effect on the primary nutrients is explained because tillage expedites SOM oxidation, thus boosting nutrient release in plant-available forms, which are, nevertheless, also more easily lost by leaching and gas emission [197] whether the adequate conditions are met, which it seems the most likely [196].

Regarding the soil's biological fertility, the more intensive and frequent the tillage operations, the higher the destruction of the fungal hyphae and the soil fauna habitats. As a consequence, tillage eventually favors the development of bacterial communities over eukaryotic soil organisms, thus leading to imbalance in the soil biome [20] with harmful consequences on the functionality of the soil's biological community and, consequently, on soil health. Conversely, minimum cultivation practices, like reduced tillage, decrease the destruction of soil fungi and fauna. Besides, minimum cultivation practices also result in less surface area contact between organic residues and soil microbial communities, which further boosts fungi development [20] and hence the balance among soil biological communities. As a consequence of all this, reducing tillage intensity significantly increases soil organic C (SOC) content and enhances biological activity in the topsoil and at greater soil depths, as shown by Nunes et al. [198] for US agricultural soils. Therefore, seeking more sustainable alternatives to tillage, such as cover cropping or mulching, not only decreases soil erosion but also promotes soil biodiversity. Moreover, in vineyard soils, these CAPs increase SOM content, enhance soil porosity, improve structural stability, and increase water holding capacity [23].

6.1. How Agricultural Traffic and Tillage Affect Soil Compaction

Tillage alters the physical characteristics of soil, significantly affecting its aeration, nutrient levels, and biological activity, but also water-holding capacity and hence temperature regulation. The effects on soil water content and flow are closely linked to the resulting compaction [199] caused by the forces exerted by the tractor wheels and tillage equipment, especially when they operate under a particular range of intermediate soil water contents, which render the soil highly susceptible to deformation [200,201]. Almost all types of tires result in increased soil compaction within the wheel tracks, although there may be differences depending on tire pressure [202]. Fortunately, as distance increases from the wheel tracks, soil compaction rapidly decreases, particularly in the subsoil [203].

Specifically for vineyards, studies have shown that subsoil compaction is predominantly caused by the machinery wheel load and not by the tillage operations themselves [204]. In addition to the soil condition during the pass, i.e., mainly the soil water content, factors such as the tire pressure and the number of passes over the same track are the primary drivers of local soil compaction [205]. Interestingly, high tire-inflation pressure is associated with greater soil compaction because it reduces the area contact between tire and soil [202,203,206]. Moreover, excessive numbers of machinery passes reinforce compaction and can spoil the positive effects of practices aiming at soil-structure enhancements, such as the use of organic amendments [203]. In intensive viticulture areas, where soil compaction due to tillage operations and machinery traffic is higher, there may be long-term effects on vineyard productivity [207]. Simply, soil compaction hinders root distribution both vertically and horizontally, thus interfering with water- and nutrient vine uptake, eventually hampering the achievement of yield and grape quality objectives, which may only be improved through regenerating soil management practices [208].

6.2. Cultivation Techniques to Avoid Soil Compaction

To avoid and mitigate soil compaction, the most suitable practices include minimal (or zero) tillage and machinery traffic. However, if machinery is used in vineyards, recognizing symptoms of soil compaction and developing methods to minimize its occurrence should be a top priority. The primary factors to bear in mind for preventing issues of soil compaction include kind, number, and places of soil tillage operations, as well as soil moisture and machinery weight and management in the field [199]. Therefore, in viticultural soils, compaction can be minimized through various strategies, including: (i) less passes over the vineyard; (ii) optimized weight equipment; (iii) adequate tire pressure; and (iv) use of cover crops and organic mulches.

To reduce soil compaction issues, simultaneous execution of multiple farm operations to minimize passes is recommended [203]. Additionally, controlled farming traffic, where machinery operations are confined to specific traffic lanes, using satellite guidance systems if needed, is an effective way to mitigate soil compaction [209]. Concerning tire inflation, avoiding excessively high pressures is important. The effective use of these techniques can stabilize soils against pore space collapse, and even be used alongside some reduced tillage operations to break up recalcitrant compacted soil layers with low probability of recompaction in the future, hence improving soil health and viticulture sustainability [210]. In this regard, the concomitant use of aggregating agents like organic amendments and gypsum may help to avoid recompaction [203].

The use of grassed vine inter-rows can be recommended as a corrective strategy to reduce soil compaction [207,211], as it lowers bulk density and root penetration resistance, increases soil porosity, and improves soil structure and infiltration rates [197]. However, it is important to note that the ameliorating effects of cover crops may take time to show up [212]. Therefore, cover crops with deep, strong roots capable of penetrating compacted soils should be preferred to combat soil compaction [203]. Additionally, the choice of the appropriate cover crops should be based on the specific characteristics of the vineyard, as well as others, as discussed in the next subsection.

6.3. Effects of Cover Crops and Mulching on Soil Health

Cover crops and living mulches, defined as any crop or crop mixture planted or allowed to spontaneously blossom into or after a main crop and then, respectively, killed or maintained alive, offer numerous advantages in grape production [213]. Despite their competition for water and nutrients, especially in the upper soil layers, where most of the annual vine roots are active [214], cover crops should be adopted, as soil and water conservation are becoming even more important in the face of climate change [215]. Regarding both synthetic and organic mulching, these have proven highly effective in conserving water around grapevine roots, a critical factor for viticulture in arid to dry sub-humid regions, thereby enhancing water use efficiency amid changing climate patterns [216,217].

The utilization of cover crops is closely linked to reduced tillage, demonstrating their joint effectiveness in increasing root biomass [218,219], and hence labile SOC, microbial biomass, and, eventually, SOC stocks [220]. Additionally, cover crops play an important role in improving soil N availability, thereby contributing to more efficient N uptake and utilization [221], reducing soil erodibility in terms of both lower surface-runoff rates and lower sediment load in runoff [176], controlling weed populations, maintaining grapevine yields, enhancing soil quality, and ultimately promoting the sustainability of grapevine growing [222].

The population and diversity of soil biota depends on the plant species that are chosen for the cover crop [20,217]. Indeed, increasing plant species richness may foster biodiversity across different trophic levels [223]. Remarkably, evidence from world-wide representative studies highlights that soil microbial diversity rises as a result of plant diversity in agroecosystems [224] with significant increasing effects on soil microbial biomass, bacterial biomass, fungal biomass, the fungi-to-bacteria ratio, and microbial respiration with eventual beneficial effects on soil nutrient cycling, particularly of C and N [225]. As for vineyards, Pingel et al. [40], in their research on cover crops in the German winegrowing region of Rhinehesse, observed a positive response in bacterial diversity to plant richness, with a positive but non-significant trend in fungal diversity. This increase in bacterial diversity may be attributed to the heterogeneous resources provided by cover crop plants, such as root exudates, leaf litter, and dead roots, as well as to the creation of diverse soil habitats by their root systems [224]. Importantly, these changes in the soil microbiome influenced by cover cropping practices may have implications for the grape microbiome and, consequently, wine quality [138].

Selecting the appropriate type of cover crops poses a challenge in vineyard soil management. Several characteristics should be considered when choosing cover crops [215]: (i) establishment capacity and resistance to repeated trampling; (ii) homogeneity and duration; (iii) ability for effective weed control; (iv) perennial habit to reduce the need for reseeding, that is to reduce costs; (v) reduced aerial development to minimize maintenance and vineyard interventions; (vi) seasonal growth pattern: featuring a lag in summer growth followed by autumn recovery; (vii) low evapotranspiration rates; (viii) deepening root growth patterns; and (ix) soil aggregate stability potential. Moreover, cover crop management, including the frequency and timing of maintenance operations and the choice of plant species, must be tailored to local climate and weather conditions, striking a balance between wine production, biodiversity, and ecosystem service provision [187]. Related to this, temporary plant cover, offering services like soil protection and water soil profile replenishment during winter, represents an attractive alternative to permanent ground cover, mitigating water and N competition with grapevines in dry years [220]. In this regard, the introduction of livestock in vineyards for grazing might help manage some cover crops, specifically considering that grazing animals can speed up the recycling of grass nutrients to benefit vines [226].

7. Impacts of Viticultural Phytosanitary Products on Soil Health

The profitable development of viticulture requires the use of crop protection techniques against several phytopathogens and parasites, which without control may severely harm grapevines and ultimately ruin the grape harvest [227]. Among these vine-damaging biotic agents, it is worth highlighting the fungi, which cause some well-known grapevine cryptogamic diseases such as downy mildew, powdery mildew, and grey mold. Additionally, there are many arthropods that particularly harm vines by herbivory or by being vectors of bacterial diseases, among which the leafhopper *Scaphoideus titanus* Ball, vector of the flavescence dorée disease, the grape berry moth *Lobesia botrana*, and the mealybug stand out [228–230]. In addition to these more vine-specific and widely spread fungi and arthropods, there are the weeds. Although weeds do not impair vines as much as the previous organisms, they are traditionally seen as competitors for water and nutrients, as hindering agricultural operations and, more grievously, as shelters for plagues of mollusks, bugs, viruses and other opportunists. Indeed, under uncontrolled circumstances, weeds can become highly detrimental for vine production, and thus viticulturists fight them with the aid of tillage operations and chemicals under conventional management conditions.

A 'pesticide' is something that prevents, destroys, or controls a harmful organism ('pest') or disease, or protects plants or plant products during production, storage, and transport [231]. However, often this something is of chemical nature and, specifically, of xenobiotic chemical nature. The use of chemicals for crop protection dates back to at least the third millennium B.C. [232]. However, since the onset of the Green Revolution, the use of agrochemicals for pest-fighting highly intensified and spread to all crops [233]. These kinds of agrochemicals are known as pesticides to be distinguished from fertilizers, although more recently, the nouns 'phytosanitary' and 'plant-protection products' have gained some ground. Whichever the name, the chemically based crop protection treatments showed some early failings in the form of the emergence of pest resistances, resurgence of primary pests, and upsurges of secondary pests [234,235]. Not least, the concerns about the effects of the xenobiotic chemicals that are used for crop protection on both the environment and human health have grown in recent decades [236,237]. Consequently, the techniques for crop protection have shifted from purely chemically based approaches to the combined use of these, along with biological and various cultural practices, under the umbrella of the integrated pest management concept [234]. However, in viticulture, despite the efforts to find alternatives to chemicals, the protection of vines still relies a lot on their use [238], even under organic production schemes [239,240].

Pesticides are classified into, mainly, fungicides, insecticides, bactericides, and herbicides, depending on their organism target [241]. The use of pesticides is particularly intensive in viticulture [242], with fungicides dominating the share of phytosanitary treatments [238]. Particularly in European agriculture, viticulture is by far the main user of fungicides with a market share of 86%, despite vines only spanning 3.3% of the cropland [243]. Among fungicides, of particular importance in viticulture are the Cu-based products. In addition to these, insecticides, mainly organo-phosphates, are also widely used in viticulture for the control of the leafhopper, the berry moth, and the mealybug.

7.1. Copper-Based Phytosanitary Products

Copper-based phytosanitary products in viticulture come in different compounds and formulations that range from the classical Bordeaux ($CuSO_4$ and CaO in water) and Burgundy ($CuSO_4$ and Na_2CO_3 in water) mixtures to the more modern Cu-oxychloride ($3Cu(OH)_2{\cdot}CuCl_2$), $CuSO_4{\cdot}3Cu(OH)_2$, Cu_2O, $Cu(OH)_2$, etc. The effectivity against the vine-typical fungal diseases, though particularly against downy mildew, of the Cu-based fungicides is difficult to match by whichever more environment-friendly strategy [239,240,244–246]. Therefore, Cu has been applied for vine protection since 1885 [247], and in rates that were as high as 50 kg Cu ha^{-1} yr^{-1} before legislations for their control were passed, e.g., the European Commission [231]. Because of the long-lasting and intensive Cu use in vineyards, its soil concentration has gradually increased to over 100 mg kg^{-1} [248]. Specifically in Europe, an estimated 15% of vine lands have exceeded the 100 mg kg^{-1} threshold [249] and, consequently, can be considered Cu-polluted according to some current standards [250,251].

Copper is an essential micronutrient for all organisms because its size and the reversibility and redox potential of the Cu^{2+}/Cu^{+} pair make it irreplaceable as a cofactor of several oxidoreductases [252,253]. These Cu enzymes are mainly found in the electron transport chains of chloroplasts and mitochondria, in the detoxification of some reactive oxygen species (ROS), and, particularly in plants, in the lignification metabolism [254,255]. However, Cu in excess within cells freely catalyzes the generation of ROS [256,257] through the Fenton and Haber–Weiss reactions, which increase the cytoplasm concentration of the superoxide radical (O^{2-}) and hydrogen peroxide (H_2O_2) species [258]. Then, the O^{2-} and

H_2O_2 generate other ROS, and all together oxidize proteins, nucleic acids, and lipids, thus destroying cell structures, particularly membrane systems [259].

In addition to the oxidative stress, Cu ions directly disrupt the structure of proteins in the cell membranes and in the cytosol, as well as the structure of nucleic acids, thus explaining how excessive Cu concentrations are biocidal [257]. Although plants can protect themselves better against excessive Cu than fungi or bacteria through, e.g., vacuole sequestration and restrained root-to-shoot transfer [258], the ROS generation boosted by excess Cu has also been described occurring in plant cells [260,261]. Copper-generated ROS damage in plants shows up as morphological and physiological impairments and as nutritional imbalances in most organs, particularly in roots [262,263].

Copper is applied to vines in formulations that prevent its absorption by the plant organs, hence trying to avoid the development of phytotoxicity and berry Cu build-up [247]. Whatever the case, vine Cu-treated leaves eventually fall and decompose and/or rainfall washes Cu down to the soil where it may accumulate [264], and hence be absorbed and excessively built up in vine organs including grapes [265]. The plant-availability of Cu depends on the free Cu^{2+} activity in the soil pore water, which depends, in turn, on soil pH and Cu's resupplying ability from the soil solids [266]. In fact, since the solubility of Cu^{2+} minerals increases as pH decreases [267], the detrimental effects of excessive soil Cu concentrations upon plants develop, overall, in acidic soils [268–270].

In grapevines, toxic effects of Cu include oxidative stress in leaves [271], which may cause decreased photosynthetic and transpiration rates [272]. Moreover, excess Cu shows up as morphological changes in the roots: increased diameter, higher number of layers in the cortex, and rising vascular cylinder and total root areas [272], and less nitrate uptake [273]. Because of these metabolic disorders, impaired nutrient status, and root deformations, grapevines produce less dry matter [272,274]. In soils featuring pH values over 7, Cu^{2+} readily precipitates as solids like $Cu_2(OH)_2CO_3$ (malachite) and $Cu(OH)_2$ [275]. Additionally, Cu^{2+} activity in the soil solution steeply decreases from pH values of 6.5 upwards because of the formation of the $CuHCO_3^+$, $CuCO_3$, and $CuOH^+$ ion pairs and, accordingly, Cu toxicity severely decreases as soil pH raises [276]. Finally, soil solution Cu^{2+} concentration is remarkably decreased by rising levels of SOM since Cu^{2+} is tightly adsorbed by SOM as surface complexes [277–279].

In addition to vines, toxic effects of Cu have also been described on soil organisms, and for these, avoiding high soil Cu intake in soils with elevated Cu concentrations may be more difficult than for plants. Particularly, earthworms feed in the soil, and upon excessive Cu they decrease growth and survival rates [280]. Specifically, *Eisenia fetida* is affected by soil Cu concentrations from 28 mg kg^{-1} up [281], and it is, notwithstanding, more tolerant to Cu than other earthworms [282]. This 28 mg kg^{-1} Cu concentration could be overcome in the arable layer of vine soils (0–30 cm) in only 16 years, despite following the restrictive EC regulation 473/2002 [231], which establishes maximum rates of 8 kg Cu ha^{-1} yr^{-1} during the first 4 years of vineyard establishment and then 6 kg Cu ha^{-1} yr^{-1} onwards.

Zinc compounds are used as alternative to copper for fungal control in vineyards [283]. However, in general, viticulture cannot get by without Cu, and thus ways to adapt must be found, by means of which the Cu issue is minimized. Therefore, if soil Cu is rated to vines below 10 kg ha^{-1} yr^{-1}, then concentrations over 200 mg kg^{-1}, which severely impair the soil's biological community functionality [284], should not be overcome. In acidic soils, the observation of these constraints is more important, even though under such circumstances the harmful effects of Cu on vines and soil organisms can be counteracted by liming to raise the soil pH [274]. In this regard, higher soil pH values not only decrease the solubility of Cu^{2+} minerals and the activity of Cu^{2+} in the soil solution, and therefore decrease the plant Cu availability and phytotoxicity, but also the Ca^{2+} ions play a beneficial role through competition with Cu^{2+} ions for plant uptake [276,285]. Additionally, the contribution of organic matters, such as farmyard manure, crop residues, and rhizodepositions from cover crops, may increase SOM contents and thus reinforce the protective effect of higher soil pH values. Moreover, aging decreases Cu availability [275], and to maximize the benefits from the aging factor, minimum soil disturbance, i.e., minimum tillage, may help.

Additionally, for mitigation of the Cu issue, cover crops have some potential in the remediation of vineyard soils [153]. Particularly, in non-calcareous soils, which are the most Cu-troublesome, cover crops featuring crimson clover seem the most promising since they have been found able to extract more than 0.1 kg Cu ha^{-1} for one year from soils containing between 80 and 160 mg Cu kg^{-1} [154]. Note that this range corresponds to between 20 and 70 kg Cu ha^{-1} if one takes the layer from 0 to 20 cm and a bulk density of 1.5 g cm^{-3}. Other Cu-hyperaccumulator plants might be used as cover crops for Cu phytoremediation in vineyard soils, but more research is needed to provide practitioners with validated knowledge, and hence useful recommendations.

7.2. Organically Based Phytosanitary Products

In viticulture, herbicides and insecticides follow fungicides in this order as the most-used phytosanitary products by market share [15]. Being for the most part organically based, herbicides can be classified into phenoxy hormone compounds, triazines, amides, carbamates, dinitroanilines, urea derivates, sulfonyl ureas, bipiridils, and uracils [15], whereas insecticides can be classified into organochlorine, organophosphate, carbamate, pyrethroid, and neonicotinoid [286].

The herbicides are increasingly applied in vineyards worldwide with glyphosate, glufosinate, and flazasulfuron leading as the three most-used [287]. These three herbicides have been shown to decrease grapevine root mycorrhization and alter soil microorganisms, with consequences on the nutrient composition of grapevine roots, leaves, xylem sap, and grape juice [287]. Specifically regarding soil organisms, glyphosate has been shown to decrease the survival and reproduction rates of earthworms in vineyards [288], thus raising concerns about how it affects soil health. Even though its effects on weeds may vanish, glyphosate presence in the soil may last a long time as it is strongly adsorbed by SOM [289]. Moreover, from SOM, glyphosate may be gradually released to the underlying waters [290], thus polluting them.

A sustainable alternative for controlling weeds in vineyards is the integration of livestock such as sheep [226] and geese [291]. For obvious reasons, livestock introduction in vineyards may be particularly adequate during the vine-dormant stage, although new developments might extend it to the active stage [292] when, additionally, the shepherding activity might help prevent wild animals from feeding on the vines. A drawback is the likely Cu chronic poisoning of the livestock [291,293] because of the thorough application of Cu fungicides, as previously addressed. Due to all this, methods for the beneficial integration of grazing animals in vineyards should be widely investigated.

The organophosphate pesticides were developed as an alternative to organochlorine ones, with the advantage of being less stable, less lipophilic, and less volatile. However, they are more hydrosoluble and are, therefore, more readily dispersed in water until they are degraded, which spreads other, more chemically simple, but still xenobiotic compounds throughout the environment.

The organophosphates comprise many agrochemicals like malathion, parathion, chlorpyrifos, diazinon, fonofos, and disulfoton, which are formed by the esterification of phosphoric, phosphonic, phosphinic, or thiophosphoric acids with a variety of organic compounds such as those of the phenoxy, cyanide, and thiocyanate kinds [294]. The organophosphates act mainly by inhibiting esterases, among which the acetylcholinesterase stands out. The inhibition of acetylcholinesterase activity increases the levels of the neurotransmitter acetylcholine up to toxic levels, thus harming nerve communication in insects, but also in other organisms [295] including those in soil [296], with harmful effects [297] that negatively impact the survival and reproduction rates of earthworms [298].

In viticulture, organophosphates have been found to be effective enough against the mealybug *Planococcus ficus*, which is the most important insect pest affecting vines worldwide, including the Mediterranean basin [299]. Specifically, applications of 0.96 kg ha^{-1}

of chlorpyrifos and of 0.48 kg ha^{-1} of methidathion have shown a mean efficacy of 70% against the sap-sucking mealybug in vineyards in Spain, Portugal, and Greece [300].

Although the organophosphates present advantageous characteristics compared with other pesticides, they still may accumulate in soils where they damage the soil's biological community [298]. Particularly, chlorpyrifos has been found to be harmful to not only soil earthworms [297] but also bacteria [301,302] thus impairing soil health. Therefore, for the control of insects in grapevine growing, notably the mealybug, the combination of other more effective but at the same time more eco-friendly chemical insecticides such as the spirotetramat with other non-chemical strategies within an integrated pest management approach offers a promising way to enhance the soil health of vineyards [299].

7.3. Interrelation of Soil Health and Vine Health: Pest and Disease Pressure

Several pests and diseases pose a threat to grapevines, and boost the need for control over them, which is usually carried out with xenobiotic chemicals, as has been shown. The impacts of climate change are likely to alter the ecology of many harmful biological agents with, e.g., insects hitting crops harder in the temperate regions where most of grapevines are grown [303]. Therefore, adaptations in plant protection measures should unfold as we witness an increase in pest and disease pressure in vineyards [304]. Globally, the sustainable management of vineyard pests and diseases is considered a top research priority [238,305]. Soil-borne pests, in particular, can inflict significant losses on all crops, including vineyards, affecting both grape yield and quality [33]. These losses entail severe economic, environmental, and social implications and, therefore, to achieve soil pathogen suppressiveness is a desirable goal. However, the current focus also lies in preserving soil biological diversity, enhancing soil and plant health [306].

Therefore, a fundamental question is how to enhance the resilience of vineyards to pests and diseases while promoting sustainable productivity. A key solution lies in developing practices that promote general soil health and, consequently, enhancement of beneficial organisms and suppression of pests and diseases, with the eventual achievement of healthy vine growth and improvement of the environment. Specifically, the soil structure should be enhanced since it exerts a remarkable control on the soil Eh and pH, and the soil Eh-pH conditions are related to the susceptibility, tolerance, and resistance of plants to pests and pathogens across various stress conditions [307]. Indeed, it has been hypothesized that soil pathogen suppressiveness results from soil structure, leading to diverse Eh-pH niches that increase the variety of soil microorganisms. Soil structure can be enhanced through, among other measures, an adequate management of organic amendments. These amendments can be, additionally, customized to induce significant shifts in the soil microbiome, promoting favorable consortia that sustain a disease-suppressive viticulture system without the need for additional agrochemical inputs. However, it is worth noting that some organic amendments may, in certain cases, trigger other neutral or, even negative, effects [308], as will be in part addressed in the next section and, therefore, more research and developments are needed in this regard.

Soil-inhabiting microbes play pivotal roles in shaping grapevine-associated microbiota and are key factors in wine quality by means of their significant impact on vine health and productivity through various direct and indirect processes, including the triggering of vine immune responses [34]. Furthermore, the soil microbial community in the rhizosphere, which could serve as a new biological indicator of vineyard soil quality and health, is a part of alternative methods that show promise in protecting grapevines. These methods include biocontrol and microbe-assisted crop production to manage grapevine diseases and promote sustainability in viticulture [57].

According to the aforementioned facts, general practices to increase soil health for pest and disease suppressiveness should include the following [309]: (i) minimizing soil disturbance and reducing compaction, (ii) maximizing biodiversity in the soil and aboveground through mulching and cover crops or by applying various types of sanitized and unpolluted organic matter amendments, (iii) managing water to meet crop needs but also to facilitate timely fieldwork, (iv) maintaining soil Eh-pH values within adequate ranges, and (v) ensuring nutrient levels are sufficient for vines but never excessive.

8. Fertilizers and Soil Amendments for Soil Health in Viticulture

One of the primary challenges in the pursuit of agricultural sustainability and soil health lies in the effective management of fertilizers and soil amendments, both mineral and organic. The use of one and the other should be integrated, and vineyard soil fertilization strategies should be tailored to the diverse soil and climate situations encountered in actual practice. Particularly, the likely impact of vine nutrition choices on soil processes should be analyzed prior to their application to optimize soil health in vineyards [220]. In this regard, the use of models to simulate the nutrients' transformations and fate in the soil–plant–water–atmosphere system is recommended.

In the quest for soil health, soil management strategies in viticulture should extend beyond merely optimizing grapevine yield and quality: they should also aim at sustaining soil biodiversity and overall soil functioning [40]. The application of inorganic fertilizers in excess of vine requirements is known to alter plant vigor, with unwanted effects on grape production and quality. Furthermore, the soil's biological functionality is severely affected following excessive mineral fertilization, which shows up in lower essential soil enzymatic activities such as dehydrogenase, catalase, invertase, urease, casein protease, and arylsulphatase [58]. Long-term N fertilization can influence soil microbiome populations, thus reshaping vine and soil microbial communities [40]. Therefore, an excess of mineral fertilization, beyond crop requirements, harm plant and soil health. In addition, such mineral fertilizer overuse can result in secondary salinization issues and nutrient leakage from the soil–plant system into other environmental compartments through leaching and gaseous emissions, thereby polluting the environment and further compromising environmental health [129].

8.1. Mineral Nutrients Effects on Vines and Soil Health

Mineral nutrition plays a pivotal role in grape production, exerting influence over all aspects of vine growth, berry development and maturation, and ultimately, wine quality [310,311]. The primary macronutrients, i.e., N, P, and K, control critical grapevine metabolic processes, which influence grape quality and thus the eventual alcoholic fermentation process [46]. Therefore, soils deprived of essential nutrients lead to nutrient-deficient vines, resulting in low-quality grapes and imperfect wines [312,313]. To meet vine nutrition demands is imperative for achieving production and oenological objectives. However, while it is well-established that nutrients like N and K affect berry and wine quality, and despite recent advances on establishing N levels for maximizing berry quality [136], the precise levels of the remaining nutrients for attaining optimal grape quality must still be refined [59]. Therefore, to ensure a sufficient supply of nutrients, viticulturists often rate fertilizers in excess, thus eventually spoiling grape quality, as is known to occur with excessive N [136,314] and K [310].

Additionally, there are the environmental harms associated with over-fertilization, which show up as nutrient leakages to the atmosphere in the form of NH_3 and N_2O gas emissions, and to waters in the form of dissolved NO_3^- and colloidal organic P. Particularly, intensive viticulture can result in environmental pollution problems due to excessive nutrient inputs that surpass the soil–plant system's capacity to absorb [161]. Complicating matters further is the impact of climate change on grapevine mineral nutrition. For instance, under the increasing warming conditions caused by climate change, K increasingly builds up in berries, thereby decreasing must acidity, and eventually harming overall wine quality [315]. Moreover, the improper use of mineral fertilizers, particularly nitrates and phosphates, not only jeopardize atmosphere and water health, but also the vineyard soil health in terms of its ecosystem services as shown by chemical indicators, and moreover by physical and biological ones [316]. Furthermore, the use of mineral fertilizers, which are directly salts featuring K^+, NH_4^+, NO_3^-, and phosphate ions, in addition to, often,

Na^+, Ca^{2+}, Mg^{2+}, Cl^-, and SO_4^{2-}, or either indirectly have the ability to generate salts by reaction in the soil, can induce secondary salinization [317] resulting in a deterioration or loss of soil functions.

To mitigate the negative impact of fertilizers on vineyard soils, particularly the secondary salinization issue, excessive nutrient applications must be avoided, and high-purity, chloride-free, low-saline fertilizers should be selected [318]. Additionally, soil management practices to reduce the secondary salinization hazards due to fertilizer overuse include: (i) improving the efficiency of fertilizer applications through irrigation water (fertigation), (ii) improving the salt tolerance of the grapevine by the addition of different nutrients based on competition between ions, and (iii) ameliorating the deleterious effects of salt stress by the addition of humic substances or by the use of biofertilizers, which reduce the application of chemical fertilizers [318].

To mitigate over-fertilization issues in general, viticulturists will have to use several tools, such as the ones based on the current information and communications technologies. In this context, soil fertilization should be based on the spatial variability of the soil characteristics as well as on the vine nutritional status and requirements, as these change throughout the growing season [319]. Additionally, devising effective fertilization schedules for vineyards will benefit a lot from the use of models. Bespoke fertilization practices to enhance both macro- and micro-nutrient management, thereby supporting soil health, hinges on the integration of the processes governing nutrient availability, soil properties, and environmental conditions, along with vine nutrient requirements [320], and models offer an optimum way to perform such integration. Therefore, fertilizer recommendations in vineyards could be adapted to the varying nutrient uptake ability of the different *Vitis* genotypes and the influence of rootstocks on nutrient concentrations in the grafted scion [167]. As our understanding of the nutrient requirements of specific cultivar–rootstock combinations advances, site-specific fertilization management programs will be tailored to individual vineyards through the use of models. This will help reduce fertilization costs and environmental pollution, promoting vineyard sustainability [321]. For now, adequately validated models for the simulation of the nutrient dynamics in viticultural soil–plant systems are lacking and more investigation and innovation is needed in this regard.

8.2. Organic Approaches for Soil Health in Viticulture

Organic farming represents an alternative to conventional agriculture based on chemical supplies. Organic farming is based on holism principles and ecological interactions instead of reductionism postulates and artificial forcing. Therefore, organic farming does not fight nature; on the contrary, it takes advantage of the natural processes that already unfold in the agro-ecosystems so as to fulfil the objective of food- and raw materials production for human welfare. Such collaboration with nature fostered by organic farming logically implies that some agricultural produce is lost to it; specifically organic farming systems exhibit 20% lower yields than their conventional counterparts [322]. However, such loss is traded off by 34 to 53% less energy use, as well as 97% less pesticide use [322], hence enhancing environmental health.

Organic farming pays particular attention to soil health, and hence postulates the joint and integrated use of various nature-based strategies to enhance soil functions, among which the following may be cited, especially for vineyards: (i) the application of microbial and non-microbial biostimulants, (ii) fertilization with organic amendments, (iii) foliar fertilization with nature-based products, (iv) the use of cover crops, and (v) the integration of livestock [58,323]. Therefore, for organic grape production, success requires treating the vineyard as a balanced agro-ecosystem to maintain the trade-off between, on the one hand, grapevine productivity, provided vine health is enhanced and weed issues minimized, and on the other hand, soil ecosystem services [324]. Consequently, although there are variations in cultural practices in organic farming, they usually increase sustainability metrics such as soil species richness and abundance, overall soil fertility, and water infiltration and holding capacity [323].

Organic viticulture, by avoiding synthetic fertilizers, soil conditioners and pesticides as much as possible, can enhance the activity of beneficial soil organisms such as entomopathogenic nematodes and increase the presence of free-living nematodes [23]. Furthermore, considering that economic sustainability is also a concern for viticulturists, organic vineyard management offers a viable model to achieve economic goals while safeguarding the environment from issues like water pollution and depletion, greenhouse gas emissions, and soil degradation, ultimately promoting biodiversity [325]. Specifically, Tangolar et al. [326] observed in their study on cv. Black Magic grafted onto the Paulsen 1103 rootstock that microbial biomass C and enzyme activity, including dehydrogenase, urease, and phosphatase, were higher when compost was applied to a calcareous vineyard soil. Similarly, Di Giacinto et al. [327] found that organic viticultural soil management resulted in higher soil enzymatic activities, as well as a richer bacterial and fungal community, because of the use of cover crops and organic compost.

As well as composts, biochar, a stable C-rich material produced through biomass pyrolysis, i.e., burning under restricted oxygen access, represents another potential sustainable soil conditioner [328]. Biochar has been shown to continuously increase the saturated hydraulic conductivity and plant-available soil water content when applied up to rates of at least 80 Mg ha^{-1} [329,330], and to serve as an effective material for sequestration of potentially toxic elements [331]. In vineyards, the same effects on soil hydrological properties have been reported, entailing better water status and enhanced long-term photosynthetic activity of vines [332], which may ultimately increase vineyard production. Specifically, in their four-year study involving cv. Merlot grafted onto the Couderc 3309 rootstock, Genesio et al. [333] found that biochar augmented vine-water availability, thus increasing yield by 66% on average, with particular harvest-increasing effects during dry years, and no effects at all on berry quality. Interestingly, there are important variations in physicochemical characteristics resulting from differences in feedstock for biochar production and, in this regard, crop residues, as biochar raw materials, score higher in their soil-beneficial effects over manures, wood, and sludges [329].

To address soil nutrient limitations for vines and other physical and biological soil fertility constraints, the use of organic fertilizers is vital due to their potential for sustaining and restoring soil health. However, it is worth noting that organic fertilizers also entail hazards, heavy-metals concentrations being one of the most common [15]. The presence of heavy metals in significant concentrations in organic fertilizers, particularly sewage sludges, has been well-documented and is very variable [334]. Therefore, to avoid any soil pollution threat, the concentrations of heavy metals in organic materials should be known before use. Then, their rates can be conformed to safe standards, which usually also take account of key soil chemical properties, mainly pH, because in acid soils the mobility of many heavy metals increases [335].

Another point of caution in the use of organic fertilizers is the utilization of animal manures and the secondary soil salinization they may induce, which has been given much less attention in comparison with the heavy metal issue. Thus, successive applications of animal manures with high Na^+, K^+, Cl^-, and SO_4^{2-} concentrations, since these ions are commonly added into the animal diet, can also promote secondary salinization in vineyard soils [336], chiefly in arid and semi-arid regions.

To adequately rate organic materials in combination with mineral fertilizers, or alone, e.g., under an organic winegrowing scheme, the use of validated models of the soil–vine–water–atmosphere system that take into account the composition or the organic materials, including the different nutrients, heavy metals, and accompanying salts would be highly recommended.

8.3. Soil Acidity and Liming Effects on Soil Health in Vineyards

Soil acidity naturally develops due to various factors of soil formation, which include parent materials that are low in bases and humid climates that promote the dissolution of the alkali and alkali-earth minerals, including the aluminosilicates, followed by the leaching of their dissolution products [337]. However, productive agriculture also contributes to soil acidification, mainly through the use of acid fertilizers such as ammonium salts and hydrogen phosphates, resulting in adverse effects on soil health, which include the development of phytotoxicities due to hydrogen ions (H^+), aluminum (Al), and Mn, and the limited plant availability of essential nutrients [338]. The phytotoxic effects of H^+ and Al typically show up as damaged root systems in the first place, whereas those of Mn directly affect above-ground plant parts [338]. Moreover, in general, the mobility of heavy metals in the soil environment increases as pH drops [335], with likely consequences on vine health. However, vine damages by heavy metals, specifically decreases in plant growth and photochemical efficiencies, have only been studied for Cu and Zn [283], because of the particular importance of both in phytosanitary products for fungal control in viticulture, as previously addressed (Section 7.1). Therefore, more research in this regard is needed.

In addition to phytotoxic effects, soil acidity issues are characterized by the appearance of Ca deficiencies, and to a lesser extent, also of P, Mg, and Mo deficiencies [339]. Particularly, P deficiency in acid soils affects the vine reproduction more than the vegetative development [340], and is the result of high phosphate fixation by clay-sized Al and iron (Fe) oxides and hydroxides, as well as the formation of phosphate precipitates of Al, Fe, and Mn [341].

In addition to the effects of acidity on chemical fertility and plant nutrition, low soil pH values significantly influence the soil microbial community. Indeed, soil pH is currently considered the best predictor of microbiome diversity at the phylum level [162] and the primary soil driver for bacterial diversity [40]. The growth and activity of soil microorganisms in acidic soils are constrained by high H^+ concentrations, which negatively impact most soil microbial processes, which show up in low rates of SOM turnover and nutrient cycling [338]. Furthermore, since soil fungi thrive in low pH environments, acid soils experience an imbalance between fungal and bacterial populations, with fungi often dominating. This can create a favorable environment for fungal pathogens, increasing the risk of infections, and reducing nutrient mineralization [162], and hence harming soil health.

For the long-term sustainability of viticulture agro-ecosystems and soil health, managing soil acidity is crucial. Traditional approaches to address soil acidity issues involve increasing soil pH by liming with materials rich in $CaCO_3$, $CaMg(CO_3)_2$, and $Ca(OH)_2$, which neutralize acidity and increase soil Ca. However, these strategies usually affect the topsoil only [337,342], whereas vine roots can pierce deeper. Therefore, preventing or minimizing subsoil acidification is as important as mitigating the topsoil acidification issue, being essential to understand the processes of subsoil acidification such as downward movement of soluble Al [343]. In this regard, a viable strategy to address soil acidity issues, which may be developed alongside the previous liming efforts, is grafting crop plants onto acidity-tolerant rootstocks. In viticulture, practitioners have for strongly acid soils ($pH < 5.5$) several available soil-acidity-tolerant rootstocks, such as Couderc 3306, Fercal, Gravesac, Paulsen 1103, Richter 99, Richter 110, and Ruggeri 140 [344].

9. The Expression of Terroir and Soil Health

Beyond grapevine cultivars, it is well known that the sensory attributes of wine depend on the soil and climate conditions in which the grapevines are cultivated, and in its simplest form, this constitutes the terroir concept [46,345]. This way, the 'terroir' idea has been anchored to phrases such as "the art of expressing the soil". However, the wine industry, including researchers, professionals, and consumers, has collectively acknowledged that a specific terroir should not be an abstract notion like that: both wine producers and consumers should have clearer, more specific concepts [346]. Additionally, it is also recognized that cultural practices play an important role in contributing to terroir and, furthermore, the interactions among grapevine management, soil, climate, and grapevine cultivars and rootstocks. Consequently, terroir encapsulates all the biological, environmental, and cultural influences on grapevine cultivation and winemaking as well as their interactions, recognizing the complex multifaceted nature of wine composition, and hence its sensory attributes [347].

Therefore, the specific characteristics of one vineyard constitute its terroir, and consequently terroirs change as such characteristics vary. Then, the wine composition, and thus its organoleptic properties, expresses the terroir and consequently, the terroir expression changes following the vineyard characteristics throughout space and time. In the context of the soil-health loss unfolding in conventionally managed vineyards, as well as its regaining by means of improved management practices, there is a necessity to gain insight into how the spatial and temporal variations in grape and wine composition relate to the variability in soil and climate characteristics as well as management practices. This understanding will lead to the establishment of which spatial scales and resolutions are the most appropriate for managing the production of terroir wines within a specific region while preserving the environmental health [46]. In this context, the scientific community has been called upon to assist in identifying and defining the numerous interconnected aspects of terroir that collectively [347] contribute to shape the quality and characteristics of the wines resulting from one area [348].

Achieving a more profound comprehension of the connections between the vine and the environment where it thrives requires not only scientific expertise on winegrowing but also interdisciplinary approaches. Current technological capabilities in land zoning are instrumental in exploring the intricacies of terroir expression [349]. Therefore, terroir research has evolved from being predominantly descriptive and region-based, i.e., primarily focused on traditional soil and climate zoning, to becoming a more applied, technical field with two main development lines. These are based (i) on the use of metabolomics and metagenomics for assessing terroir characteristics, and (ii) on employing remote and proximal sensing to monitor soil and vine characteristics so as to give practitioners useful information to precisely manage the soil–vine system and attain predetermined oenological objectives [46].

9.1. The Factors That Determine the Terroir Expression

If focusing solely on one factor at a time, a multitude of physical and chemical soil and climate features can influence the terroir expression. For example, in rain-fed vineyards, the soil's water-holding capacity largely determines the water supply to the grapevines during the ripening process, and thus it controls grape and wine quality remarkably [350]. However, in terms of the balance between vegetative and reproductive grapevine growth, which influences grape quality attributes, distinguishing the effects of increased water supply from increased nutrient supply is often challenging [59]. In many regions, it has been shown that climate change makes vine growth start earlier and progress faster with consequences on ripening profiles and wine styles [351].

To comprehend how soil and climate features operate on terroir expression, it is crucial not only to consider the factors one by one, but also the interactions among them [352]. Then, gaining a comprehensive understanding of the mechanistic connections between soil and climate and grapevines, mediated through the human interventions, which modulate the effects of soil and climate on the vines, can facilitate the analysis of the effects the soil health status have on wine terroir expression, and how to counteract them if needed. In this regard, since climate change affects grape and wine quality, and hence terroir, the identification of new suitable vineyard locations, and the precise vine management to achieve specific oenological objectives can be eased [349]. Specifically, our insight on the terroir expression significantly increases when considering the interactions among soil, climate, and human factors as revealed by, e.g., topography and fruit zone microclimate [353], and ground microbiome.

Particularly relevant is topography, which is shaped by the forces of tectonics, erosion, and land mass movements [354], often intermediated by human actions. Topography serves as an important nexus between soil and climate in the terroir expression, affecting not only the wine quality but also the feasibility of cultivating vines in climates that overall fail to meet grapevine requirements [353]. Another connection of soil and climate with terroir occurs at the level of the microclimatic variations in the bunch zone, which depend on the soil characteristics, singularly the albedo, and are also regulated by the canopy management [352].

Therefore, the human factor interacts with soil and climate and thus modulates the terroir expression through winegrowing practices that can be based on inherited culture and novel scientific knowledge. Even more, the decisions made by viticulturists regarding soil and overall vineyard management, directly or indirectly influence soil health and wine terroir. In this regard, examples exist of how wine growers' practices that promote secondary soil salinization lead to outcomes in terroir expression [355], featuring soapy and salty tastes, as well as loss of fruitiness, and increases in sour, bitter, and oxidative perceptions in wine [356]. Therefore, through grape build-up of Na^{+} and Cl^{-} ions, soil salinization becomes one of the clearest drivers of changes on terroir expression because of soil-health loss.

9.2. The Soil Environment and the Terroir Expression

When it comes to grape quality attributes, the significance of the soil's physical and chemical properties cannot be overstated [319]. However, despite some exceptions, e.g., salts, the soil impact on wine terroir is often notably intricate because soil influences so many aspects of vine development, from mineral nutrition to water uptake, but also rooting depth and temperature within the root zone [349]. Therefore, despite years of research on how soil factors contribute to the terroir expression in wine, there are still fundamental questions that require more investigation. Specifically, two questions linger: (i) which are the primary soil factors that influence terroir expression, and (ii) how do these factors interact among each other and with the vine to influence grape and, ultimately, wine quality and terroir expression.

Soils differ in their capacity to retain water, and also in the quantity and composition of the mineral nutrients they contain, and the extent to which these nutrients are available for uptake by vine roots [167]. Thus, an understanding of the spatial variability of soil fertility and plant nutrition is crucial for planning and implementing site-specific vineyard management, making viable the adoption of differential management strategies [319]. The information regarding soil spatial variability is of paramount importance in the identification and classification of terroir [349].

Although the precise influences of specific soil characteristics on terroir are not yet fully understood [46], it seems that thermal properties, pH, salinity, macro- and micronutrients, and the biological activity of soil organisms all contribute to the terroir expression. Beyond the influence of specific soil components on the attributes of wine quality, which is still a topic of research, some ecophysiological mechanisms have been shown to be highly influential, particularly in relation to water supply and N nutrition during critical stages of the vegetative cycle [46].

It is widely accepted that the soil's physical characteristics, particularly structure, exert significant control over the vine water supply [353], which strongly impacts N dynamics [46], as well as the qualitative response of the berries [349]. Specifically, the soil water availability has been found to influence the hormonal balance of each grapevine variety, subsequently regulating genotype expression [46]. Additionally, as with vine water status, soil N availability critically affects grape quality [136]. For example, in vineyard soils high in N, vines present excessive vigor, resulting in shaded canopies and subsequent poor grape quality [357]. Regarding other soil chemical properties, significant associations between them and wine's organoleptic characteristics have been found [353]. For example, Tomasi et al. [358] reported the effect of high limestone content on obtaining higher aromatic wines, and Arnó et al. [319] found that high calcium carbonate and pH lead to Mn deficit and, consequently, grapes low in phenolic contents.

In addition to the physical and chemical properties, the soil microbial diversity associated with the vineyard location is a key aspect directly involved in vine cultivation and growth, as well as grape quality and winemaking [348]. However, the global and local variability of soil microbial communities and the biotic and abiotic factors that drive their differentiation are still poorly understood [359]. The richer the soil in life forms, the more capable it is of not only providing essential ecosystem services [354], but of influencing the terroir expression [360]. Emerging technologies such as metabolomics, with the ability to identify the chemical fingerprint of a particular cell, organ, organism, or tissue type, and metagenomics, with the ability to sequence the entire microbial genomes present in vineyard samples, hold promise for understanding the terroir formation with regards to this [46]. Additionally, although the debate on whether there is a microbiological component in the terroir expression is currently not very influential, researchers like Zarraonaindia and Gilbert [38] have highlighted the importance of the soil microbiome in shaping the microbiology of the vine, and possibly directly shaping the characteristics of wine.

Indeed, the recent advancements in omics techniques have already allowed for a more detailed exploration of soil functionality, microbial diversity, and vine-associated microorganisms [46]. For instance, Lazcano et al. [138] found that soil may serve as a reservoir for grape microorganisms that are carried over into fermentation, greatly influencing the organoleptic properties of wine. This microbial presence exhibits a strong biogeographical pattern that enables the differentiation of terroirs, leading several scientists to propose the concept of the "microbial terroir" [361,362]. In fact, the soil microbiome should be considered as an important variable in identifying viticultural sites for the definition of homogeneous functional zones such as basic terroir units [348]. Additionally, Vink et al. [363] demonstrated how specific rootstock–cultivar combinations influence bacterial diversity in grafted grapevines. This could affect the capacity of specific combinations to chemically attract specific bacteria associated with plant protection, performance, and productivity, through root exudates, as well as colonization in grapes, potentially affecting wine quality.

9.3. *Terroir Expression and Soil Health in Zoning Procedures*

Terroir forms the primary basis for representing regional wine quality. Even on a small vineyard, certain factors can introduce significant variations in local soil, resulting in remarkable variability in terroir characteristics [364]. Therefore, it is evident that a complex issue arises when maintaining terroirs unchanged is an objective in a scenario where soil health is changing alongside time and space, either worsening or improving [138]. Particularly, soil management, which significantly regulates overall soil-health status, is anticipated to affect terroir and therefore, wine quality attributes. This may be a drawback, but also an advantage. For example, because agrochemicals substantially influence the soil environment, alternatives like organic grape production represent an opportunity for grape growers and winemakers to develop wines with unique terroir, distinct from conventional wines [324]. However, several questions still remain: (i) how much each soil property contributes to terroir expression, (ii) how terroir expression changes as soil degrades, and (iii) how terroir expression changes as CAPs are introduced in vineyards and soil health is enhanced.

In this paper, we aim to establish the primary parameters to consider when the concept of soil health relates to vineyard soils. Given the evolving nature and scope of the terroir concept, identifying highly healthy soils may be one of the most crucial steps in vineyard zoning. However, considering that every winegrowing terroir can be defined as a spatial and temporal entity characterized by homogeneous or dominant attributes [365], and considering the legal-social framework associated with designations of origin and geographical indications aimed at ensuring wine origin for consumers, conflicts of interest and disputes seem likely. Perhaps the creation of new hierarchical scales of vineyard entities, incorporating soil-health parameters into quantitative and environmental/ecophysiological zoning criteria, could not only ensure a more comprehensive conflict resolution system between stakeholders but also promote the adoption of sustainability practices in grape production.

The wine industry has a fresh opportunity to play a vital role in promoting soil-health restoration and mitigating the impact of climate change through regenerative soil management practices. By minimizing soil disturbance and promoting biodiversity, it can create healthier and more resilient vineyards, potentially enhancing grape quality and the flavor of wine [366]. Consequently, managing for soil health becomes a strategy to enable the full expression of terroir [138].

10. Chemical, Physical, and Biological Indicators of Soil Health

As reviewed, to be considered as healthy, a soil should present the levels of its physical, chemical, and biological properties within the ranges that enable it to function within its ecological boundaries to sustain productivity, maintain environmental quality, and promote plant and animal health. However, the quantification of these ranges is not usually straightforward because soils are complex systems and hence all their properties are interrelated among each other, with the result that the evaluation of soil health depends on that interrelationship rather than on the particular values the attributes present [129,367]. Consequently, soil health and soil quality may be better assessed on the basis of indexes that make use of several selected soil properties [368–371]. However, alternatively, soil health may be assessed on the basis of indicators that make use of an integrative soil property, which is ideally a biological one [372–374], for which SOM or SOC are often advantageously used as proxies.

10.1. Soil Organic Carbon

Soil organic carbon (SOC) is recognized worldwide as having great potential to be used as an indicator of soil health because it correlates well with many attributes of soil's biological functionality [375–379], and hence with soil-health indexes [380], and because it is easily measured through, e.g., soil sampling and ensuing classical titration or gravimetric methods in the laboratory [212]. However, the adoption of SOC as a soil-health indicator is not without drawbacks. First, there is no universal consensus on which levels of SOC have to be considered as healthy [68], because SOC levels depend on climate and soil conditions. Consequently, when soil health is evaluated on the SOC basis, soil health seems to increase as temperature drops, and rainfall rates and soil clay contents increase [381,382]. One option to sort out this issue would be the calculation of normalized SOC levels (NSOC). These NSOC levels would be calculated with regards to the potential SOC (PSOC) that represents the soil class and climate type to which one soil belongs by using equations such as the Equation (1):

$$\mathrm{NSOC} = \mathrm{SOC}/\mathrm{PSOC} \quad (1)$$

where PSOC levels could be established by survey or by modelling or both.

In addition to normalization, some kind of concave score function that takes into account both the positive and the negative impacts of SOC on soil health should be adopted for use of SOC as soil-health indicator. These negative impacts mainly arise due to the effects high SOC levels have on the fate of the phytosanitary products in the soil environment. Indeed, the adsorption of many pesticides by SOM increases as the SOC level rises [383]. Consequently, as SOM increases the activity of pesticides against their target organisms decreases, and their loss to groundwater in association with the dissolved SOC fraction may increase, thus boosting the need for pesticide reapplication with on-farm as well as off-farm harmful effects on animal and human health [384].

10.2. Soil Aggregate Stability

Soil aggregate stability also has the potential to be used as indicator of soil health because it correlates well enough with many attributes of soil quality [381,385], and of susceptibility to water erosion [120,386]. Moreover, soil aggregate stability can be measured through easy-to-implement and cost-effective methods in the laboratory following field sampling [386,387]. However, since aggregate stability strongly depends on SOC [74,75], it will present the same drawbacks for soil-health assessment than SOC. Moreover, regarding SOC, soil aggregate stability presents two additional drawbacks. First, it is more variable and, second, its determination lacks a worldwide standard procedure and measurement benchmark. Therefore, the different methods in use, though correlated with one another, provide values of different magnitude that hinder comparisons [388,389].

10.3. Soil Bulk Density and Infiltration Capacity

Soil bulk density and infiltration capacity have also the potential to be used as indicators of soil health because both correlate well with many of its healthy attributes, though bulk density stands out in this regard [378]. Both soil characteristics are remarkably related to each other with soil infiltration capacity decreasing, in general, as bulk density increases [390,391]. However, bulk density is easier to measure than infiltration capacity and it is less variable, thus making bulk density preferable as an indicator of soil health over infiltration capacity. Additionally, bulk density presents the advantage of being directly traceable to the primary standards of weight and length thus making the interpretation of bulk density measurements straightforward, and even easier if its counterpart, porosity, is calculated and presented alongside. As drawbacks, bulk density is more variable than other likely indicators of soil health, e.g., SOC. Moreover, in general, bulk density is slowly responsive to soil management, which makes soil-health monitoring through bulk density unpractical in the short to medium term.

10.4. Soil Chemistry: pH, Salinity, Sodicity, Nutrients, and Pollutants

Classical soil chemistry measurements such as pH, electrical conductivity as surrogate of salinity, sodicity, and the contents of nutrients and pollutants are traditional soil fertility indicators. Therefore, they address only the soil fertility and partially the quality and plant- and animal-health aspects of soil health. They do not directly integrate ecological soil attributes and, therefore, they lack basis to be used as standalone soil-health indicators, even though they may form part of soil-health indexes [371].

10.5. Soil Biology

For soil assessment on the basis of the health paradigm, a range of promising methods are being devised featuring biological parameters. On this point, researchers like Burns et al. [392] have proposed to identify soil bacterial communities, related to soil resources and vineyard management, through 16S rDNA fingerprinting as a biological indicator to monitor soil-quality or health in vineyard soils. However, most of these new soil-health assessment methods require further validation and enhancement concerning their relevance, scientific validity, practicality, and adaptability to local conditions [268]. Particularly, a major challenge in the practical application to soil health assessment of these the new proposed biological and biochemical indicators is the absence of reference values for interpreting them, as well as the integration of these with the standard procedures to provide a comprehensive assessment of a particular soil-health status [393]. The challenge of soil-health evaluation may be addressed through innovative biological approaches; however, they should also integrate standardized physical and chemical techniques to be able to disclose the functioning of vineyard soils and be, therefore, readily interpretable.

11. Conclusions

Vine agro-ecosystems face several challenges, including (i) soil compaction and erosion, (ii) SOM decline, (iii) biological simplification, and (iv) soil and water pollution due to the use of chemicals [26,393]. All these processes constrain the soil's ability to fulfil its ecological functions, with both on-farm and off-farm detrimental effects. Consequently, these degradation processes significantly affect soil health, thus jeopardizing the long-run sustainability of vine cropping [26,394]. For example, a notable issue in vineyard soils is the accumulation of Cu due to the application of Cu-based pesticides for vine protection against fungal pathogens (Section 7). The Cu accumulation in vineyard soils harms the functionality of the soil biological community, hence disrupting soil fauna burrowing activities and specific nutrient cycling pathways [395,396]. Such effects, in turn, may make vine growing more dependent on tillage, mineral fertilizers, and herbicides. In particular, high levels of soil Cu will hamper the integration of grazing livestock in vineyards for sustainable weed control and nutrient recycling [293]. Although instances of Cu-phytotoxicity symptoms on vines have been rare, mainly due to their deep rooting systems [397], it is abundantly clear that addressing this soil-health issue is highly needed because it may prevent the development of other sustainable vineyard management practices such as cover cropping and the integration of livestock. Therefore, vineyards require more sustainable ways of applying Cu, alternative plant-protection treatments and remediation procedures for vineyard Cu-polluted soils, which must be all investigated further.

Vineyard management is influenced by various factors, including climate, water availability for irrigation, soil type, grapevine variety, rootstock, agro-environmental policies, and, most importantly, the decisions and attitudes of viticulturists [187]. Note that distinctively from other agricultural activities, viticulture often prioritizes enhancing harvest quality over yield, specifically in winemaking, with the goal of obtaining terroir wines by means of highlighting specific flavors that please consumers and earn their loyalty. However, as support for planetary health expands, contemporary wine consumers are increasingly more concerned about the environmental impacts of their wine choices. In this regard, the continued ability of vineyard soils to function as living environments that support plant, animal, and human health, as encompassed by the soil-health concept, has been shown to go through the adoption of conservative agricultural management practices [3]. However, despite the good deal of research on the application of these sustainable soil management practices to viticulture, doubts persist in practitioners about how to manage vineyards to optimize both soil health and terroir expression [138]. Therefore, the key challenge in modern viticulture of enhancing sustainability in grape production processes without endangering productivity and, importantly, harvest quality and hence profitability, is a major research field [398,399]. Complicating matters further is global warming, which challenges traditional vine-management guidelines and demands their modernization to adapt to its consequences [399–401]. The adoption of the soil health framework is therefore even more necessary to cope with the significant effects climate change is having on viticulture [399].

The concept of soil health is gradually emerging, and research and vineyard operation are adopting this new paradigm. However, conventional agricultural soil management is still dominant, and revolves around the soil-nutrient status, often overlooking the significance of many soil-biological characteristics for plant health and agricultural sustainability [402]. Consequently, viticulturists still mostly rely on analytical chemical procedures to assess both soil fertility and quality, as the latter has been traditionally understood. As the soil-health concept gains ground, standard procedures will have to be adapted or replaced, thus offering research and development opportunities in this regard.

Advances in soil health are grounded in four principal development lines: soil biology understanding, the integrated interpretation of soil data, and improved in-field measurements [403], in addition to the modelling of the soil–vine–water–atmosphere system. These advances should collectively promote sustainable practices that contribute to significantly reduce the exhaustion viticulture causes on the natural resources it depends upon, thereby mitigating the environmental impact of their activities and especially sustaining soil health [404]. Particularly, given the substantial impact of soil biodiversity in vineyards by means of its effects on several regulating services like pest control and SOM decomposition, through which it supports nutrient cycling and availability, and its eventual influence on both grape yield and quality [26], it is likely that soil-health research will increasingly focus on this area. Nonetheless, it remains clear that soil health should be approached holistically and established as a scientific field open to contributions from various disciplines.

Therefore, researchers in viticulture, as well as in other agronomic fields, should embrace soil health as a guiding principle to which they can contribute knowledge, viewing it not merely as a property to measure but as a well-established scientific field that, remarkably, can benefit from insights from various disciplines [64,82]. In the pursuit of improving the soil health of winegrowing systems so that vineyard soils fulfil their ecosystem services, as well as the harvest amount and quality aims of viticulturists, collaboration at the landscape level will be essential. This cooperation should involve winegrowers, other agriculturalists, researchers from various disciplines, policymakers, consumers, and media. Importantly, however, such collaboration should be undertaken without undermining the sovereignty and independence of local viticulturists in their quest for producing wines that meet the standards they have committed to [405].

Author Contributions: Conceptualization, F.V. and M.Á.O.; methodology, F.V. and M.Á.O.; investigation, F.V., R.L. and M.Á.O.; resources, F.V.; data curation, F.V., R.L. and M.Á.O.; writing—original draft preparation, F.V., R.L. and M.Á.O.; writing—review and editing, F.V., R.L. and M.Á.O.; visualization, F.V., R.L. and M.Á.O.; supervision, F.V.; project administration, F.V. All authors have read and agreed to the published version of the manuscript.

Funding: This research received no external funding.

Data Availability Statement: No new data were created or analyzed in this study. Data sharing is not applicable to this article.

Conflicts of Interest: The authors declare no conflict of interest.

References

1. International Organisation of Vine and Wine. *Annual Assessment of the World Vine and Wine Sector in 2022*; International Organisation of Vine and Wine: Dijon, France, 2023.
2. Poore, J.; Nemecek, T. Reducing Food's Environmental Impacts through Producers and Consumers. *Science* **2018**, *360*, 987–992. [CrossRef]
3. Cataldo, E.; Fucile, M.; Mattii, G.B. A Review: Soil Management, Sustainable Strategies and Approaches to Improve the Quality of Modern Viticulture. *Agronomy* **2021**, *11*, 2359. [CrossRef]
4. International Health Conference. Constitution of the World Health Organization. 1946. *Bull. World Health Organ.* **2002**, *80*, 983–984.
5. Minami, K. Soil Is a Living Substance. *Soil Sci. Plant Nutr.* **2021**, *67*, 26–30. [CrossRef]
6. Doran, J.W.; Zeiss, M.R. Soil Health and Sustainability: Managing the Biotic Component of Soil Quality. *Appl. Soil Ecol.* **2000**, *15*, 3–11. [CrossRef]
7. Yang, T.; Siddique, K.H.M.; Liu, K. Cropping Systems in Agriculture and Their Impact on Soil Health—A Review. *Glob. Ecol. Conserv.* **2020**, *23*, e01118. [CrossRef]
8. Doran, J.W.; Parkin, T.B. Defining and Assessing Soil Quality. In *Defining Soil Quality for a Sustainable Environment*; Doran, J.W., Coleman, D.C., Bezdicek, D.F., Stewart, B.A., Eds.; SSSA: Madison, WI, USA, 2015; pp. 3–21; ISBN 9780891189305.
9. Blum, W.E.H. Functions of Soil for Society and the Environment. *Rev. Environ. Sci. Biotechnol.* **2005**, *4*, 75–79. [CrossRef]
10. Riesenfeld, C.S.; Goodman, R.M.; Handelsman, J. Uncultured Soil Bacteria Are a Reservoir of New Antibiotic Resistance Genes. *Environ. Microbiol.* **2004**, *6*, 981–989. [CrossRef] [PubMed]
11. Van Goethem, M.W.; Pierneef, R.; Bezuidt, O.K.I.; Van De Peer, Y.; Cowan, D.A.; Makhalanyane, T.P. A Reservoir of "historical" Antibiotic Resistance Genes in Remote Pristine Antarctic Soils. *Microbiome* **2018**, *6*, 40. [CrossRef] [PubMed]
12. van de Meent, D.; de Bruijn, J.H.M. A Modeling Procedure to Evaluate the Coherence of Independently Derived Environmental Quality Objectives for Air, Water and Soil. *Environ. Toxicol. Chem.* **1995**, *14*, 177–186. [CrossRef]
13. Karlen, D.L.; Mausbach, M.J.; Doran, J.W.; Cline, R.G.; Harris, R.F.; Schuman, G.E. Soil Quality: A Concept, Definition, and Framework for Evaluation (A Guest Editorial). *Soil Sci. Soc. Am. J.* **1997**, *61*, 4–10. [CrossRef]
14. Calistri, P.; Iannetti, S.; Danzetta, M.L.; Narcisi, V.; Cito, F.; Di Sabatino, D.; Bruno, R.; Sauro, F.; Atzeni, M.; Carvelli, A.; et al. The Components of 'One World-One Health' Approach. *Transbound. Emerg. Dis.* **2013**, *60*, 4–13. [CrossRef]
15. United Nations. FAOSTAT. Available online: https://www.fao.org/faostat/en/#home (accessed on 6 December 2023).
16. Ohno, T.; Hettiarachchi, G.M. Soil Chemistry and the One Health Initiative: Introduction to the Special Section. *J. Environ. Qual.* **2018**, *47*, 1305–1309. [CrossRef] [PubMed]
17. Van Bruggen, A.H.C.; Goss, E.M.; Havelaar, A.; van Diepeningen, A.D.; Finckh, M.R.; Morris, J.G. One Health-Cycling of Diverse Microbial Communities as a Connecting Force for Soil, Plant, Animal, Human and Ecosystem Health. *Sci. Total Environ.* **2019**, *664*, 927–937. [CrossRef] [PubMed]
18. Banerjee, S.; van der Heijden, M.G.A. Soil Microbiomes and One Health. *Nat. Rev. Microbiol.* **2022**, *21*, 6–20. [CrossRef] [PubMed]

19. Bonilla-Aldana, D.K.; Dhama, K.; Rodriguez-Morales, A.J. Revisiting the One Health Approach in the Context of COVID-19: A Look into the Ecology of This Emerging Disease. *Adv. Anim. Vet. Sci.* **2020**, *8*, 234–237. [CrossRef]
20. Abbott, L.; Murphy, D. *Soil Biological Fertility: A Key to Sustainable Land Use in Agriculture*; Springer: Berlin/Heidelberg, Germany, 2007.
21. Stirling, G.; Hayden, H.; Pattison, T.; Stirling, M. *Soil Health, Soil Biology, Soilborne Diseases and Sustainable Agriculture*; CSIRO: Canberra, Australia, 2016.
22. Hatfield, J.L.; Walthall, C.L. Soil Biological Fertility: Foundation for the Next Revolution in Agriculture? *Commun. Soil Sci. Plant Anal.* **2015**, *46*, 753–762. [CrossRef]
23. Blanco-Pérez, R.; Vicente-Díez, I.; Ramos-Sáez de Ojer, J.L.; Marco-Mancebón, V.S.; Pérez-Moreno, I.; Campos-Herrera, R. Organic Viticulture Enhanced the Activity of Native Entomopathogenic Nematodes in DOCa Rioja Soils (North of Spain). *Agric. Ecosyst. Environ.* **2022**, *332*, 107931. [CrossRef]
24. Rodrigo-Comino, J. Five Decades of Soil Erosion Research in "Terroir". The State-of-the-Art. *Earth Sci. Rev.* **2018**, *179*, 436–447. [CrossRef]
25. Karimi, B.; Cahurel, J.Y.; Gontier, L.; Charlier, L.; Chovelon, M.; Mahé, H.; Ranjard, L. A Meta-Analysis of the Ecotoxicological Impact of Viticultural Practices on Soil Biodiversity. *Environ. Chem. Lett.* **2020**, *18*, 1947–1966. [CrossRef]
26. Giffard, B.; Winter, S.; Guidoni, S.; Nicolai, A.; Castaldini, M.; Cluzeau, D.; Coll, P.; Cortet, J.; Le Cadre, E.; d'Errico, G.; et al. Vineyard Management and Its Impacts on Soil Biodiversity, Functions, and Ecosystem Services. *Front. Ecol. Evol.* **2022**, *10*, 850272. [CrossRef]
27. Marques, M.J.; Bienes, R.; Cuadrado, J.; Ruiz-Colmenero, M.; Barbero-Sierra, C.; Velasco, A. Analysing Perceptions Attitudes and Responses of Winegrowers about Sustainable Land Management in Central Spain. *Land Degrad. Dev.* **2015**, *26*, 458–467. [CrossRef]
28. Khmelevtsova, L.E.; Sazykin, I.S.; Azhogina, T.N.; Sazykina, M.A. Influence of Agricultural Practices on Bacterial Community of Cultivated Soils. *Agriculture* **2022**, *12*, 371. [CrossRef]
29. Wu, T.; Milner, H.; Díaz-Pérez, J.C.; Ji, P. Effects of Soil Management Practices on Soil Microbial Communities and Development of Southern Blight in Vegetable Production. *Appl. Soil Ecol.* **2015**, *91*, 58–67. [CrossRef]
30. García-Orenes, F.; Guerrero, C.; Roldán, A.; Mataix-Solera, J.; Cerdà, A.; Campoy, M.; Zornoza, R.; Bárcenas, G.; Caravaca, F. Soil Microbial Biomass and Activity under Different Agricultural Management Systems in a Semiarid Mediterranean Agroecosystem. *Soil Tillage Res.* **2010**, *109*, 110–115. [CrossRef]
31. Pampulha, M.E.; Oliveira, A. Impact of an Herbicide Combination of Bromoxynil and Prosulfuron on Soil Microorganisms. *Curr. Microbiol.* **2006**, *53*, 238–243. [CrossRef]
32. Young, I.M.; Crawford, J.W. Interactions and Self-Organization in the Soil-Microbe Complex. *Science* **2004**, *304*, 1634–1637. [CrossRef] [PubMed]
33. Marín, D.; Armengol, J.; Carbonell-Bejerano, P.; Escalona, J.M.; Gramaje, D.; Hernández-Montes, E.; Intrigliolo, D.S.; Martínez-Zapater, J.M.; Medrano, H.; Mirás-Avalos, J.M.; et al. Challenges of Viticulture Adaptation to Global Change: Tackling the Issue from the Roots. *Aust. J. Grape Wine Res.* **2020**, *27*, 8–25. [CrossRef]
34. Darriaut, R.; Lailheugue, V.; Masneuf-Pomarède, I.; Marguerit, E.; Martins, G.; Compant, S.; Ballestra, P.; Upton, S.; Ollat, N.; Lauvergeat, V. Grapevine Rootstock and Soil Microbiome Interactions: Keys for a Resilient Viticulture. *Hortic. Res.* **2022**, *9*, uhac019. [CrossRef] [PubMed]
35. Agnelli, A.; Bol, R.; Trumbore, S.E.; Dixon, L.; Cocco, S.; Corti, G. Carbon and Nitrogen in Soil and Vine Roots in Harrowed and Grass-Covered Vineyards. *Agric. Ecosyst. Environ.* **2014**, *193*, 70–82. [CrossRef]
36. Ritz, K.; Harris, J.; Murray, P. *The Role of Soil Biota in Soil Fertility and Quality, and Approaches to Influencing Soil Communities to Enhance Delivery of These Functions*; Department for Environment, Food and Rural Affairs: London, UK, 2010.
37. Bettenfeld, P.; Cadena i Canals, J.; Jacquens, L.; Fernandez, O.; Fontaine, F.; van Schaik, E.; Courty, P.E.; Trouvelot, S. The Microbiota of the Grapevine Holobiont: A Key Component of Plant Health. *J. Adv. Res.* **2022**, *40*, 1–15. [CrossRef] [PubMed]
38. Zarraonaindia, I.; Owens, S.M.; Weisenhorn, P.; West, K.; Hampton-Marcell, J.; Lax, S.; Bokulich, N.A.; Mills, D.A.; Martin, G.; Taghavi, S.; et al. The Soil Microbiome Influences Grapevine-Associated Microbiota. *mBio* **2015**, *6*, e02527-14. [CrossRef] [PubMed]
39. Custódio, V.; Gonin, M.; Stabl, G.; Bakhoum, N.; Oliveira, M.M.; Gutjahr, C.; Castrillo, G. Sculpting the Soil Microbiota. *Plant J.* **2022**, *109*, 508–522. [CrossRef]
40. Pingel, M.; Reineke, A.; Leyer, I. A 30-Years Vineyard Trial: Plant Communities, Soil Microbial Communities and Litter Decomposition Respond More to Soil Treatment than to N Fertilization. *Agric. Ecosyst. Environ.* **2019**, *272*, 114–125. [CrossRef]
41. Trouvelot, S.; Bonneau, L.; Redecker, D.; van Tuinen, D.; Adrian, M.; Wipf, D. Arbuscular Mycorrhiza Symbiosis in Viticulture: A Review. *Agron. Sustain. Dev.* **2015**, *35*, 1449–1467. [CrossRef]
42. Lemtiri, A.; Colinet, G.; Alabi, T.; Cluzeau, D.; Zirbes, L.; Haubruge, É.; Francis, F. Impacts of Earthworms on Soil Components and Dynamics. A Review. *Biotechnol. Agron. Soc. Environ.* **2014**, *18*, 121–133.
43. Neemisha. Role of Soil Organisms in Maintaining Soil Health, Ecosystem Functioning, and Sustaining Agricultural Production. In *Soil Health*; Springer: Berlin/Heidelberg, Germany, 2020; pp. 313–335.
44. Tian, T.; Reverdy, A.; She, Q.; Sun, B.; Chai, Y. The Role of Rhizodeposits in Shaping Rhizomicrobiome. *Environ. Microbiol. Rep.* **2019**, *12*, 160–172. [CrossRef] [PubMed]

45. Raghavendra, M.P.; Santhoshkannada, A.N. Role of Rhizomicrobiome in Maintaining Soil Fertility and Crop Production. In *Soil Health*; Springer: Berlin/Heidelberg, Germany, 2020; pp. 373–401.
46. Vaudour, E.; Costantini, E.; Jones, G.V.; Mocali, S. An Overview of the Recent Approaches to Terroir Functional Modelling, Footprinting and Zoning. *Soil* **2015**, *1*, 287–312. [CrossRef]
47. Mcnear, D., Jr. The Rhizosphere-Roots, Soil and Everything in Between. *Nat. Educ. Knowl.* **2013**, *4*, 1.
48. Whalley, W.R.; Riseley, B.; Leeds-Harrison, P.B.; Bird, N.R.A.; Leech, P.K.; Adderley, W.P. Structural Differences between Bulk and Rhizosphere Soil. *Eur. J. Soil Sci.* **2005**, *56*, 353–360. [CrossRef]
49. Youssef, R.A.; Chino, M. Root-Induced Changes in the Rhizosphere of Plants. II. Distribution of Heavy Metals across the Rhizosphere in Soils. *Soil Sci. Plant Nutr.* **1989**, *35*, 609–621. [CrossRef]
50. Youssef, R.A.; Chino, M. Root-Induced Changes in the Rhizosphere of Plants. I. PH Changes in Relation to the Bulk Soil. *Soil Sci. Plant Nutr.* **1989**, *35*, 461–468. [CrossRef]
51. Wang, L.; Rengel, Z.; Zhang, K.; Jin, K.; Lyu, Y.; Zhang, L.; Cheng, L.; Zhang, F.; Shen, J. Ensuring Future Food Security and Resource Sustainability: Insights into the Rhizosphere. *iScience* **2022**, *25*, 104168. [CrossRef] [PubMed]
52. Gianfreda, L. Enzymes of Importance to Rhizosphere Processes. *J. Soil Sci. Plant Nutr.* **2015**, *15*, 283–306. [CrossRef]
53. Custos, J.-M.; Moyne, C.; Sterckeman, T. How Root Nutrient Uptake Affects Rhizosphere PH: A Modelling Study. *Geoderma* **2020**, *369*, 114314. [CrossRef]
54. Afridi, M.S.; Fakhar, A.; Kumar, A.; Ali, S.; Medeiros, F.H.V.; Muneer, M.A.; Ali, H.; Saleem, M. Harnessing Microbial Multitrophic Interactions for Rhizosphere Microbiome Engineering. *Microbiol. Res.* **2022**, *265*, 127199. [CrossRef]
55. Bao, L.; Sun, B.; Wei, Y.; Xu, N.; Zhang, S.; Gu, L.; Bai, Z. Grape Cultivar Features Differentiate the Grape Rhizosphere Microbiota. *Plants* **2022**, *11*, 1111. [CrossRef]
56. Dries, L.; Bussotti, S.; Pozzi, C.; Kunz, R.; Schnell, S.; Löhnertz, O.; Vortkamp, A. Rootstocks Shape Their Microbiome—Bacterial Communities in the Rhizosphere of Different Grapevine Rootstocks. *Microorganisms* **2021**, *9*, 822. [CrossRef]
57. Dries, L.; Hendgen, M.; Schnell, S.; Löhnertz, O.; Vortkamp, A. Rhizosphere Engineering: Leading towards a Sustainable Viticulture? *OENO One* **2021**, *55*, 353–363. [CrossRef]
58. Baweja, P.; Kumar, S.; Kumar, G. Fertilizers and Pesticides: Their Impact on Soil Health and Environment. In *Soil Health*; Springer: Berlin/Heidelberg, Germany, 2020; pp. 265–285.
59. Schreiner, R.P. Spatial and Temporal Variation of Roots, Arbuscular Mycorrhizal Fungi, and Plant and Soil Nutrients in a Mature Pinot Noir (*Vitis vinifera* L.) Vineyard in Oregon, USA. *Plant Soil* **2005**, *276*, 219–234. [CrossRef]
60. Aguilera, P.; Ortiz, N.; Becerra, N.; Turrini, A.; Gaínza-Cortés, F.; Silva-Flores, P.; Aguilar-Paredes, A.; Romero, J.K.; Jorquera-Fontena, E.; Mora, M.d.L.L.; et al. Application of Arbuscular Mycorrhizal Fungi in Vineyards: Water and Biotic Stress Under a Climate Change Scenario: New Challenge for Chilean Grapevine Crop. *Front. Microbiol.* **2022**, *13*, 826571. [CrossRef]
61. Gazioglu Sensoy, R. Effects of Arbuscular Mycorrhizal Fungus (AMF) and Whey Applications on the Grapevine (*Vitis Vinifera* L.) Cutting Exposed to Salt Stress. *J. Elem.* **2022**, *27*, 507–519. [CrossRef]
62. Cardinale, M.; Minervini, F.; De Angelis, M.; Papadia, P.; Migoni, D.; Dimaglie, M.; Dinu, D.G.; Quarta, C.; Selleri, F.; Caccioppola, A.; et al. Vineyard Establishment under Exacerbated Summer Stress: Effects of Mycorrhization on Rootstock Agronomical Parameters, Leaf Element Composition and Root-Associated Bacterial Microbiota. *Plant Soil* **2022**, *478*, 613–634. [CrossRef]
63. Lumini, E.; Orgiazzi, A.; Borriello, R.; Bonfante, P.; Bianciotto, V. Disclosing Arbuscular Mycorrhizal Fungal Biodiversity in Soil through a Land-use Gradient Using a Pyrosequencing Approach. *Environ. Microbiol.* **2010**, *12*, 2165–2179. [CrossRef]
64. Lehmann, J.; Kleber, M. The Contentious Nature of Soil Organic Matter. *Nature* **2015**, *528*, 60–68. [CrossRef] [PubMed]
65. Kögel-Knabner, I. The Macromolecular Organic Composition of Plant and Microbial Residues as Inputs to Soil Organic Matter: Fourteen Years On. *Soil Biol. Biochem.* **2017**, *105*, A3–A8. [CrossRef]
66. Stockmann, U.; Padarian, J.; McBratney, A.; Minasny, B.; de Brogniez, D.; Montanarella, L.; Hong, S.Y.; Rawlins, B.G.; Field, D.J. Global Soil Organic Carbon Assessment. *Glob. Food Secur.* **2015**, *6*, 9–16. [CrossRef]
67. Hengl, T.; Mendes de Jesus, J.; Heuvelink, G.B.M.; Ruiperez Gonzalez, M.; Kilibarda, M.; Blagotić, A.; Shangguan, W.; Wright, M.N.; Geng, X.; Bauer-Marschallinger, B.; et al. SoilGrids250m: Global Gridded Soil Information Based on Machine Learning. *PLoS ONE* **2017**, *12*, e0169748. [CrossRef] [PubMed]
68. Loveland, P.; Webb, J. Is There a Critical Level of Organic Matter in the Agricultural Soils of Temperate Regions: A Review. *Soil Tillage Res.* **2003**, *70*, 1–18. [CrossRef]
69. Murphy, B.W. Impact of Soil Organic Matter on Soil Properties—A Review with Emphasis on Australian Soils. *Soil Res.* **2015**, *53*, 605. [CrossRef]
70. Foley, J.A. The Sensitivity of the Terrestrial Biosphere to Climatic Change: A Simulation of the Middle Holocene. *Global Biogeochem. Cycles* **1994**, *8*, 505–525. [CrossRef]
71. Ruddiman, W.F. The Anthropogenic Greenhouse Era Began Thousands of Years Ago. *Clim. Change* **2003**, *61*, 261–293. [CrossRef]
72. Lal, R. Soil Carbon Sequestration Impacts on Global Climate Change and Food Security. *Science* **2004**, *304*, 1623–1627. [CrossRef] [PubMed]
73. Feller, C.; Manlay, R.J.; Swift, M.J.; Bernoux, M. Functions, Services and Value of Soil Organic Matter for Human Societies and the Environment: A Historical Perspective. *Geol. Soc. Lond. Spec. Publ.* **2006**, *266*, 9–22. [CrossRef]
74. Tisdall, J.M.; Oades, J.M. Organic Matter and Water-stable Aggregates in Soils. *J. Soil Sci.* **1982**, *33*, 141–163. [CrossRef]

75. Kong, A.Y.Y.; Six, J.; Bryant, D.C.; Denison, R.F.; van Kessel, C. The Relationship between Carbon Input, Aggregation, and Soil Organic Carbon Stabilization in Sustainable Cropping Systems. *Soil Sci. Soc. Am. J.* **2005**, *69*, 1078–1085. [CrossRef]
76. Chen, H.; Hou, R.; Gong, Y.; Li, H.; Fan, M.; Kuzyakov, Y. Effects of 11 Years of Conservation Tillage on Soil Organic Matter Fractions in Wheat Monoculture in Loess Plateau of China. *Soil Tillage Res.* **2009**, *106*, 85–94. [CrossRef]
77. Mustafa, A.; Minggang, X.; Ali Shah, S.A.; Abrar, M.M.; Nan, S.; Baoren, W.; Zejiang, C.; Saeed, Q.; Naveed, M.; Mehmood, K.; et al. Soil Aggregation and Soil Aggregate Stability Regulate Organic Carbon and Nitrogen Storage in a Red Soil of Southern China. *J. Environ. Manag.* **2020**, *270*, 110894. [CrossRef]
78. Xu, N.; Amgain, N.R.; Rabbany, A.; Capasso, J.; Korus, K.; Swanson, S.; Bhadha, J.H. Interaction of Soil Health Indicators to Different Regenerative Farming Practices on Mineral Soils. *Agrosystems Geosci. Environ.* **2022**, *5*, e20243. [CrossRef]
79. Cresswell, H.P.; Kirkegaard, J.A. Subsoil Amelioration by Plant-Roots—The Process and the Evidence. *Aust. J. Soil Res.* **1995**, *33*, 221. [CrossRef]
80. Yunusa, I.A.M.; Newton, P.J. Plants for Amelioration of Subsoil Constraints and Hydrological Control: The Primer-Plant Concept. *Plant Soil* **2003**, *257*, 261–281. [CrossRef]
81. Lavelle, P.; Spain, A.; Blouin, M.; Brown, G.; Decaëns, T.; Grimaldi, M.; Jiménez, J.J.; McKey, D.; Mathieu, J.; Velasquez, E.; et al. Ecosystem Engineers in a Self-Organized Soil: A Review of Concepts and Future Research Questions. *Soil Sci.* **2016**, *181*, 91–109. [CrossRef]
82. Lehmann, A.; Leifheit, E.F.; Rillig, M.C. Mycorrhizas and Soil Aggregation. In *Mycorrhizal Mediation of Soil*; Elsevier: Amsterdam, The Netherlands, 2017; pp. 241–262.
83. Wilpiszeski, R.L.; Aufrecht, J.A.; Retterer, S.T.; Sullivan, M.B.; Graham, D.E.; Pierce, E.M.; Zablocki, O.D.; Palumbo, A.V.; Elias, D.A. Soil Aggregate Microbial Communities: Towards Understanding Microbiome Interactions at Biologically Relevant Scales. *Appl. Environ. Microbiol.* **2019**, *85*, e00324-19. [CrossRef] [PubMed]
84. Yu, Z.; Zhang, J.; Zhang, C.; Xin, X.; Li, H. The Coupling Effects of Soil Organic Matter and Particle Interaction Forces on Soil Aggregate Stability. *Soil Tillage Res.* **2017**, *174*, 251–260. [CrossRef]
85. Li, S.; Wang, B.; Zhang, X.; Wang, H.; Yi, Y.; Huang, X.; Gao, X.; Zhu, P.; Han, W. Soil Particle Aggregation and Aggregate Stability Associated with Ion Specificity and Organic Matter Content. *Geoderma* **2023**, *429*, 116285. [CrossRef]
86. Li, Z.; Zhang, X.; Liu, Y. Pore-Scale Simulation of Gas Diffusion in Unsaturated Soil Aggregates: Accuracy of the Dusty-Gas Model and the Impact of Saturation. *Geoderma* **2017**, *303*, 196–203. [CrossRef]
87. Hu, W.; Jiang, Y.; Chen, D.; Lin, Y.; Han, Q.; Cui, Y. Impact of Pore Geometry and Water Saturation on Gas Effective Diffusion Coefficient in Soil. *Appl. Sci.* **2018**, *8*, 2097. [CrossRef]
88. Prifling, B.; Weber, M.; Ray, N.; Prechtel, A.; Phalempin, M.; Schlüter, S.; Vetterlein, D.; Schmidt, V. Quantifying the Impact of 3D Pore Space Morphology on Soil Gas Diffusion in Loam and Sand. *Transp. Porous Media* **2023**, *149*, 501–527. [CrossRef]
89. Ehlers, W. Observations on Earthworm Channels and Infiltration on Tilled and Untilled Loess Soil. *Soil Sci.* **1975**, *119*, 242–249. [CrossRef]
90. Franzluebbers, A.J. Water Infiltration and Soil Structure Related to Organic Matter and Its Stratification with Depth. *Soil Tillage Res.* **2002**, *66*, 197–205. [CrossRef]
91. Lipiec, J.; Kuś, J.; Słowińska-Jurkiewicz, A.; Nosalewicz, A. Soil Porosity and Water Infiltration as Influenced by Tillage Methods. *Soil Tillage Res.* **2006**, *89*, 210–220. [CrossRef]
92. Kirkby, C.A.; Kirkegaard, J.A.; Richardson, A.E.; Wade, L.J.; Blanchard, C.; Batten, G. Stable Soil Organic Matter: A Comparison of C:N:P:S Ratios in Australian and Other World Soils. *Geoderma* **2011**, *163*, 197–208. [CrossRef]
93. Tipping, E.; Somerville, C.J.; Luster, J. The C:N:P:S Stoichiometry of Soil Organic Matter. *Biogeochemistry* **2016**, *130*, 117–131. [CrossRef] [PubMed]
94. Coonan, E.C.; Kirkby, C.A.; Kirkegaard, J.A.; Amidy, M.R.; Strong, C.L.; Richardson, A.E. Microorganisms and Nutrient Stoichiometry as Mediators of Soil Organic Matter Dynamics. *Nutr. Cycl. Agroecosyst.* **2020**, *117*, 273–298. [CrossRef]
95. Schulten, H.-R.; Schnitzer, M. The Chemistry of Soil Organic Nitrogen: A Review. *Biol. Fertil. Soils* **1997**, *26*, 1–15. [CrossRef]
96. Schroth, A.W.; Bostick, B.C.; Graham, M.; Kaste, J.M.; Mitchell, M.J.; Friedland, A.J. Sulfur Species Behavior in Soil Organic Matter during Decomposition. *J. Geophys. Res. Biogeosci.* **2007**, *112*, G04011. [CrossRef]
97. Spohn, M. Increasing the Organic Carbon Stocks in Mineral Soils Sequesters Large Amounts of Phosphorus. *Glob. Change Biol.* **2020**, *26*, 4169–4177. [CrossRef] [PubMed]
98. Parfitt, R.L.; Giltrap, D.J.; Whitton, J.S. Contribution of Organic Matter and Clay Minerals to the Cation Exchange Capacity of Soils. *Commun. Soil Sci. Plant Anal.* **1995**, *26*, 1343–1355. [CrossRef]
99. Oorts, K.; Vanlauwe, B.; Merckx, R. Cation Exchange Capacities of Soil Organic Matter Fractions in a Ferric Lixisol with Different Organic Matter Inputs. *Agric. Ecosyst. Environ.* **2003**, *100*, 161–171. [CrossRef]
100. Fageria, N.; Baligar, V.; Clark, R. Micronutrients in Crop Production. *Adv. Agron.* **2002**, *77*, 185–268.
101. Lindsay, W.L. Iron Oxide Solubilization by Organic Matter and Its Effect on Iron Availability. *Plant Soil* **1991**, *130*, 27–34. [CrossRef]
102. Dhaliwal, S.S.; Naresh, R.K.; Mandal, A.; Singh, R.; Dhaliwal, M.K. Dynamics and Transformations of Micronutrients in Agricultural Soils as Influenced by Organic Matter Build-up: A Review. *Environ. Sustain. Indic.* **2019**, *1–2*, 100007. [CrossRef]
103. Merino, C.; Kuzyakov, Y.; Godoy, K.; Jofré, I.; Nájera, F.; Matus, F. Iron-Reducing Bacteria Decompose Lignin by Electron Transfer from Soil Organic Matter. *Sci. Total Environ.* **2021**, *761*, 143194. [CrossRef]
104. Jones, D.L. Organic Acids in the Rhizosphere-A Critical Review. *Plant Soil* **1998**, *205*, 25–44. [CrossRef]

105. Goldberg, S. Reactions of Boron with Soils. *Plant Soil* **1997**, *193*, 35–48. [CrossRef]
106. Sauvé, S.; Hendershot, W.; Allen, H.E. Solid-Solution Partitioning of Metals in Contaminated Soils: Dependence on PH, Total Metal Burden, and Organic Matter. *Environ. Sci. Technol.* **2000**, *34*, 1125–1131. [CrossRef]
107. Wichard, T.; Mishra, B.; Myneni, S.C.B.; Bellenger, J.-P.; Kraepiel, A.M.L. Storage and Bioavailability of Molybdenum in Soils Increased by Organic Matter Complexation. *Nat. Geosci.* **2009**, *2*, 625–629. [CrossRef]
108. Tolls, J. Sorption of Veterinary Pharmaceuticals in Soils: A Review. *Environ. Sci. Technol.* **2001**, *35*, 3397–3406. [CrossRef] [PubMed]
109. Nam, J.J.; Gustafsson, O.; Kurt-Karakus, P.; Breivik, K.; Steinnes, E.; Jones, K.C. Relationships between Organic Matter, Black Carbon and Persistent Organic Pollutants in European Background Soils: Implications for Sources and Environmental Fate. *Environ. Pollut.* **2008**, *156*, 809–817. [CrossRef] [PubMed]
110. Chien, S.-W.C.; Chen, S.-H.; Li, C.-J. Effect of Soil PH and Organic Matter on the Adsorption and Desorption of Pentachlorophenol. *Environ. Sci. Pollut. Res.* **2017**, *25*, 5269–5279. [CrossRef]
111. Cui, S.; Qi, Y.; Zhu, Q.; Wang, C.; Sun, H. A Review of the Influence of Soil Minerals and Organic Matter on the Migration and Transformation of Sulfonamides. *Sci. Total Environ.* **2023**, *861*, 160584. [CrossRef]
112. Lo, K.S.L.; Yang, W.F.; Lin, Y.C. Effects of Organic Matter on the Specific Adsorption of Heavy Metals by Soil. *Toxicol. Environ. Chem.* **1992**, *34*, 139–153. [CrossRef]
113. Salati, S.; Quadri, G.; Tambone, F.; Adani, F. Fresh Organic Matter of Municipal Solid Waste Enhances Phytoextraction of Heavy Metals from Contaminated Soil. *Environ. Pollut.* **2010**, *158*, 1899–1906. [CrossRef] [PubMed]
114. Li, J.; Li, Y.; Yang, X.; Zhang, J.; Lin, Z.; Zhao, B. Microbial Community Structure and Functional Metabolic Diversity Are Associated with Organic Carbon Availability in an Agricultural Soil. *J. Integr. Agric.* **2015**, *14*, 2500–2511. [CrossRef]
115. Zhang, L.; Yan, H.; Ren, T.; Li, S.; Yang, Y.; Peng, G.; Yu, J.; Liu, G. Effects of Organic Matter on Nutrient, Enzyme Activity and Functional Diversity of Microbial Community in Tobacco Planting Soil. *Acta Tabacaria Sin.* **2019**, *25*, 55–62.
116. Johnson, D.C.; Teague, R.; Apfelbaum, S.; Thompson, R.; Byck, P. Adaptive Multi-Paddock Grazing Management's Influence on Soil Food Web Community Structure for: Increasing Pasture Forage Production, Soil Organic Carbon, and Reducing Soil Respiration Rates in Southeastern USA Ranches. *PeerJ* **2022**, *10*, e13750. [CrossRef] [PubMed]
117. Sokol, N.W.; Kuebbing, S.E.; Karlsen-Ayala, E.; Bradford, M.A. Evidence for the Primacy of Living Root Inputs, Not Root or Shoot Litter, in Forming Soil Organic Carbon. *New Phytol.* **2018**, *221*, 233–246. [CrossRef] [PubMed]
118. Chaney, K.; Swift, R.S. The Influence of Organic Matter on Aggregate Stability in Some British Soils. *J. Soil Sci.* **1984**, *35*, 223–230. [CrossRef]
119. Chenu, C.; Le Bissonnais, Y.; Arrouays, D. Organic Matter Influence on Clay Wettability and Soil Aggregate Stability. *Soil Sci. Soc. Am. J.* **2000**, *64*, 1479–1486. [CrossRef]
120. Le Bissonnais, Y.; Blavet, D.; De Noni, G.; Laurent, J.-Y.; Asseline, J.; Chenu, C. Erodibility of Mediterranean Vineyard Soils: Relevant Aggregate Stability Methods and Significant Soil Variables. *Eur. J. Soil Sci.* **2007**, *58*, 188–195. [CrossRef]
121. Burras, C.; Nyasimi, M.; Butler, L. Soils, Human Health, and Wealth: A Complicated Relationship. In *Soils and Human Health;* CRC Press: Boca Raton, FL, USA, 2013; pp. 215–226.
122. Gu, B.; Chen, D.; Yang, Y.; Vitousek, P.; Zhu, Y.-G. Soil-Food-Environment-Health Nexus for Sustainable Development. *Research* **2021**, *2021*, 9804807. [CrossRef]
123. Parras-Alcántara, L.; Martín-Carrillo, M.; Lozano-García, B. Impacts of Land Use Change in Soil Carbon and Nitrogen in a Mediterranean Agricultural Area (Southern Spain). *Solid Earth* **2013**, *4*, 167–177. [CrossRef]
124. Jakšić, S.; Ninkov, J.; Milić, S.; Vasin, J.; Banjac, D.; Jakšić, D.; Živanov, M. The State of Soil Organic Carbon in Vineyards as Affected by Soil Types and Fertilization Strategies (Tri Morave Region, Serbia). *Agronomy* **2020**, *11*, 9. [CrossRef]
125. Maetens, W.; Vanmaercke, M.; Poesen, J.; Jankauskas, B.; Jankauskiene, G.; Ionita, I. Effects of Land Use on Annual Runoff and Soil Loss in Europe and the Mediterranean: A Meta-Analysis of Plot Data. *Prog. Phys. Geogr. Earth Environ.* **2012**, *36*, 599–653. [CrossRef]
126. Prosdocimi, M.; Cerdà, A.; Tarolli, P. Soil Water Erosion on Mediterranean Vineyards: A Review. *Catena* **2016**, *141*, 1–21. [CrossRef]
127. Rodrigo Comino, J.; Senciales, J.M.; Ramos, M.C.; Martínez-Casasnovas, J.A.; Lasanta, T.; Brevik, E.C.; Ries, J.B.; Ruiz Sinoga, J.D. Understanding Soil Erosion Processes in Mediterranean Sloping Vineyards (Montes de Málaga, Spain). *Geoderma* **2017**, *296*, 47–59. [CrossRef]
128. Gristina, L.; Novara, A.; Minacapilli, M. Rethinking Vineyard Ground Management to Counter Soil Tillage Erosion. *Soil Tillage Res.* **2022**, *217*, 105275. [CrossRef]
129. Kibblewhite, M.G.; Ritz, K.; Swift, M.J. Soil Health in Agricultural Systems. *Philos. Trans. R. Soc. Lond. B Biol. Sci.* **2008**, *363*, 685–701. [CrossRef] [PubMed]
130. García-Ruiz, J.M. The Effects of Land Uses on Soil Erosion in Spain: A Review. *Catena* **2010**, *81*, 1–11. [CrossRef]
131. Lozano-Parra, J.; Maldonado-Clavelle, C.F.; Caballero-Calvo, A.; Pulido-Fernández, M.; Gyasi-Agyei, Y.; Rodrigo-Comino, J. Estimation of Soil Mobilization Rates by a Rainy Period and Intense Tillage Practices in Vineyards—A Case Study in the Maule Region (Chile). *Hydrol. Process.* **2023**, *37*, e14829. [CrossRef]
132. Martínez-Casasnovas, J.A.; Ramos, M.C. Soil Alteration Due to Erosion, Ploughing and Levelling of Vineyards in North East Spain. *Soil Use Manag.* **2009**, *25*, 183–192. [CrossRef]
133. Mishra, U.; Lal, R.; Slater, B.; Calhoun, F.; Liu, D.; Van Meirvenne, M. Predicting Soil Organic Carbon Stock Using Profile Depth Distribution Functions and Ordinary Kriging. *Soil Sci. Soc. Am. J.* **2009**, *73*, 614–621. [CrossRef]

134. Bai, J.; Zhang, G.; Zhao, Q.; Lu, Q.; Jia, J.; Cui, B.; Liu, X. Depth-Distribution Patterns and Control of Soil Organic Carbon in Coastal Salt Marshes with Different Plant Covers. *Sci. Rep.* **2016**, *6*, 34835. [CrossRef]
135. Murphy, B.W.; Wilson, B.R.; Koen, T. Mathematical Functions to Model the Depth Distribution of Soil Organic Carbon in a Range of Soils from New South Wales, Australia under Different Land Uses. *Soil Syst.* **2019**, *3*, 46. [CrossRef]
136. Visconti, F.; Intrigliolo, D.S.; Mirás-Avalos, J.M. Effects of the Annual Nitrogen Fertilization Rate on Vine Performance and Grape Quality for Winemaking: Insights from a Meta-Analysis. *Aust. J. Grape Wine Res.* **2023**, *2023*, 7989254. [CrossRef]
137. Baldock, J.; Skjemstad, J. Soil Organic Carbon/Soil Organic Matter. In *Soil Analysis: An Interpretation Manual*; Peverill, K., Sparrow, L., Reuter, D., Eds.; CSIRO Publishing: Melbourne, Australia, 1999; pp. 159–170.
138. Lazcano, C.; Decock, C.; Wilson, S.G. Defining and Managing for Healthy Vineyard Soils, Intersections With the Concept of Terroir. *Front. Environ. Sci.* **2020**, *8*, 68. [CrossRef]
139. Müller-Nedebock, D.; Chaplot, V. Soil Carbon Losses by Sheet Erosion: A Potentially Critical Contribution to the Global Carbon Cycle. *Earth Surf. Process Landf.* **2015**, *40*, 1803–1813. [CrossRef]
140. Li, H.; Liu, G.; Gu, J.; Chen, H.; Shi, H.; Abd Elbasit, M.A.M.; Hu, F. Response of Soil Aggregate Disintegration to the Different Content of Organic Carbon and Its Fractions during Splash Erosion. *Hydrol. Process.* **2021**, *35*, e14060. [CrossRef]
141. Vicente-Vicente, J.L.; García-Ruiz, R.; Francaviglia, R.; Aguilera, E.; Smith, P. Soil Carbon Sequestration Rates under Mediterranean Woody Crops Using Recommended Management Practices: A Meta-Analysis. *Agric. Ecosyst. Environ.* **2016**, *235*, 204–214. [CrossRef]
142. Gómez de Barreda, D.; Bautista, I.; Castell, V.; Lidón, A. Rice Straw Mulch Installation in a Vineyard Improves Weed Control and Modifies Soil Characteristics. *Agronomy* **2023**, *13*, 3068. [CrossRef]
143. Qin, T.; Wang, L.; Zhao, J.; Zhou, G.; Li, C.; Guo, L.; Jiang, G. Effects of Straw Mulching Thickness on the Soil Health in a Temperate Organic Vineyard. *Agriculture* **2022**, *12*, 1751. [CrossRef]
144. Morlat, R.; Chaussod, R. Long-Term Additions of Organic Amendments in a Loire Valley Vineyard. I. Effects on Properties of a Calcareous Sandy Soil. *Am. J. Enol. Vitic.* **2008**, *59*, 353–363. [CrossRef]
145. Gaiotti, F.; Marcuzzo, P.; Belfiore, N.; Lovat, L.; Fornasier, F.; Tomasi, D. Influence of Compost Addition on Soil Properties, Root Growth and Vine Performances of *Vitis Vinifera* Cv Cabernet Sauvignon. *Sci. Hortic.* **2017**, *225*, 88–95. [CrossRef]
146. Torres, N.; Yu, R.; Kurtural, S.K. Inoculation with Mycorrhizal Fungi and Irrigation Management Shape the Bacterial and Fungal Communities and Networks in Vineyard Soils. *Microorganisms* **2021**, *9*, 1273. [CrossRef] [PubMed]
147. Betancur-Agudelo, M.; Meyer, E.; Lovato, P.E. Growth, Heavy Metal Uptake, and Photosynthesis in "Paulsen 1103" (Vitis Berlandieri x Rupestris) Grapevine Rootstocks Inoculated with Arbuscular Mycorrhizal Fungi from Vineyard Soils with High Copper Contents. *VITIS—J. Grapevine Res.* **2020**, *59*, 169–180. [CrossRef]
148. Zhou, J.; Zang, H.; Loeppmann, S.; Gube, M.; Kuzyakov, Y.; Pausch, J. Arbuscular Mycorrhiza Enhances Rhizodeposition and Reduces the Rhizosphere Priming Effect on the Decomposition of Soil Organic Matter. *Soil Biol. Biochem.* **2020**, *140*, 107641. [CrossRef]
149. Guzmán, G.; Cabezas, J.M.; Sánchez-Cuesta, R.; Lora, Á.; Bauer, T.; Strauss, P.; Winter, S.; Zaller, J.G.; Gómez, J.A. A Field Evaluation of the Impact of Temporary Cover Crops on Soil Properties and Vegetation Communities in Southern Spain Vineyards. *Agric. Ecosyst. Environ.* **2019**, *272*, 135–145. [CrossRef]
150. Steenwerth, K.; Belina, K.M. Cover Crops Enhance Soil Organic Matter, Carbon Dynamics and Microbiological Function in a Vineyard Agroecosystem. *Appl. Soil Ecol.* **2008**, *40*, 359–369. [CrossRef]
151. Ruiz-Colmenero, M.; Bienes, R.; Marques, M.J. Soil and Water Conservation Dilemmas Associated with the Use of Green Cover in Steep Vineyards. *Soil Tillage Res.* **2011**, *117*, 211–223. [CrossRef]
152. Fleishman, S.M.; Bock, H.W.; Eissenstat, D.M.; Centinari, M. Undervine Groundcover Substantially Increases Shallow but Not Deep Soil Carbon in a Temperate Vineyard. *Agric. Ecosyst. Environ.* **2021**, *313*, 107362. [CrossRef]
153. Ortega, P.; Sánchez, E.; Gil, E.; Matamoros, V. Use of Cover Crops in Vineyards to Prevent Groundwater Pollution by Copper and Organic Fungicides. Soil Column Studies. *Chemosphere* **2022**, *303*, 134975. [CrossRef]
154. Eon, P.; Robert, T.; Goutouly, J.-P.; Aurelle, V.; Cornu, J.-Y. Cover Crop Response to Increased Concentrations of Copper in Vineyard Soils: Implications for Copper Phytoextraction. *Chemosphere* **2023**, *329*, 138604. [CrossRef] [PubMed]
155. Austin, E.E.; Wickings, K.; McDaniel, M.D.; Robertson, G.P.; Grandy, A.S. Cover Crop Root Contributions to Soil Carbon in a No-till Corn Bioenergy Cropping System. *GCB Bioenergy* **2017**, *9*, 1252–1263. [CrossRef]
156. Mortensen, E.Ø.; De Notaris, C.; Peixoto, L.; Olesen, J.E.; Rasmussen, J. Short-Term Cover Crop Carbon Inputs to Soil as Affected by Long-Term Cropping System Management and Soil Fertility. *Agric. Ecosyst. Environ.* **2021**, *311*, 107339. [CrossRef]
157. Novara, A.; Cerda, A.; Barone, E.; Gristina, L. Cover Crop Management and Water Conservation in Vineyard and Olive Orchards. *Soil Tillage Res.* **2021**, *208*, 104896. [CrossRef]
158. Buesa, I.; Mirás-Avalos, J.M.; De Paz, J.M.; Visconti, F.; Sanz, F.; Yeves, A.; Guerra, D.; Intrigliolo, D.S. Soil Management in Semi-Arid Vineyards: Combined Effects of Organic Mulching and No-Tillage under Different Water Regimes. *Eur. J. Agron.* **2021**, *123*, 126198. [CrossRef]
159. Beslic, Z.; Pantelic, M.; Dabic, D.; Todic, S.; Natic, M.; Tesic, Z. Effect of Vineyard Floor Management on Water Regime, Growth Response, Yield and Fruit Quality in Cabernet Sauvignon. *Sci. Hortic.* **2015**, *197*, 650–656. [CrossRef]
160. Reynolds, W.D.; Bowman, B.T.; Drury, C.F.; Tan, C.S.; Lu, X. Indicators of Good Soil Physical Quality: Density and Storage Parameters. *Geoderma* **2002**, *110*, 131–146. [CrossRef]

161. Kumar, K.S.A.; Karthika, K.S. Abiotic and Biotic Factors Influencing Soil Health and/or Soil Degradation. In *Soil Health*; Springer: Berlin/Heidelberg, Germany, 2020; pp. 145–161.
162. Manda, R.R.; Addanki, V.A.; Giabardo, A.; Benjamin, J.; Hossain, M.J.; Khanna, S.; Gaddam, M.; Kumar, R.; Srivastava, S. Soil Health Management and Microorganisms: Recent Development. In *Detection, Diagnosis and Management of Soil-Borne Phytopathogens*; Springer Nature Singapore: Singapore, 2023; pp. 437–493.
163. Oliver, D.P.; Bramley, R.G.V.; Riches, D.; Porter, I.; Edwards, J. Review: Soil Physical and Chemical Properties as Indicators of Soil Quality in Australian Viticulture. *Aust. J. Grape Wine Res.* **2013**, *19*, 129–139. [CrossRef]
164. Saljnikov, E.; Mueller, L.; Lavrishchev, A.; Eulenstein, F. *Advances in Understanding Soil Degradation Innovations in Landscape Research*; Springer: Müncheberg, Germany, 2022.
165. Schjønning, P. *Subsoil Compaction—A Threat to Sustainable Food Production and Soil Ecosystem Services*; RECARE Policy Brief; RECARE: Berlin, Germany, 2018.
166. Batey, T. Soil Compaction and Soil Management—A Review. *Soil Use Manag.* **2009**, *25*, 335–345. [CrossRef]
167. Keller, M. *The Science of Grapevines*, 3rd ed.; Maragioglio, N., Ed.; Elsevier: Nueva Delhi, India, 2020.
168. Kirschbaum, M.U.F. The Temperature Dependence of Soil Organic Matter Decomposition, and the Effect of Global Warming on Soil Organic C Storage. *Soil Biol. Biochem.* **1995**, *27*, 753–760. [CrossRef]
169. Keller, T.; Sandin, M.; Colombi, T.; Horn, R.; Or, D. Historical Increase in Agricultural Machinery Weights Enhanced Soil Stress Levels and Adversely Affected Soil Functioning. *Soil Tillage Res.* **2019**, *194*, 104293. [CrossRef]
170. Horn, R.; Schrader, S.; Mordhorst, A.; Fleige, H.; Schroeder, R. Soil Health and Biodiversity: Interactions with Physical Processes and Functions. In *Sustainable Soil Management as a Key to Preserve Soil Biodiversity and Stop Its Degradation*; Reyes-Sánchez, L., Horn, R., Costantini, E., Eds.; International Union of Soil Sciences (IUSS): Vienna, Austria, 2022; pp. 314–330.
171. Bogunovic, I.; Bilandzija, D.; Andabaka, Z.; Stupic, D.; Rodrigo Comino, J.; Cacic, M.; Brezinscak, L.; Maletic, E.; Pereira, P. Soil Compaction under Different Management Practices in a Croatian Vineyard. *Arab. J. Geosci.* **2017**, *10*, 340. [CrossRef]
172. Inglett, P.; Reddy, K.; Corstanje, R. Anaerobic Soils. In *Encyclopedia of Soils in the Environment*; Elsevier: Amsterdam, The Netherlands, 1998; pp. 72–78.
173. Tan, X.; Chang, S.X. Soil Compaction and Forest Litter Amendment Affect Carbon and Net Nitrogen Mineralization in a Boreal Forest Soil. *Soil Tillage Res.* **2007**, *93*, 77–86. [CrossRef]
174. Tropeano, D. Rate of Soil Erosion Processes on Vineyards in Central Piedmont (NW Italy). *Earth Surf. Process Landf.* **1984**, *9*, 253–266. [CrossRef]
175. Vršič, S. Soil Erosion and Earthworm Population Responses to Soil Management Systems in Steep-Slope Vineyards. *Plant Soil Environ.* **2011**, *57*, 258–263. [CrossRef]
176. Prosdocimi, M.; Jordán, A.; Tarolli, P.; Keesstra, S.; Novara, A.; Cerdà, A. The Immediate Effectiveness of Barley Straw Mulch in Reducing Soil Erodibility and Surface Runoff Generation in Mediterranean Vineyards. *Sci. Total Environ.* **2016**, *547*, 323–330. [CrossRef]
177. Cerdà, A.; Keesstra, S.D.; Rodrigo-Comino, J.; Novara, A.; Pereira, P.; Brevik, E.; Giménez-Morera, A.; Fernández-Raga, M.; Pulido, M.; di Prima, S.; et al. Runoff Initiation, Soil Detachment and Connectivity Are Enhanced as a Consequence of Vineyards Plantations. *J. Environ. Manag.* **2017**, *202*, 268–275. [CrossRef]
178. Pappalardo, S.E.; Gislimberti, L.; Ferrarese, F.; De Marchi, M.; Mozzi, P. Estimation of Potential Soil Erosion in the Prosecco DOCG Area (NE Italy), toward a Soil Footprint of Bottled Sparkling Wine Production in Different Land-Management Scenarios. *PLoS ONE* **2019**, *14*, e0210922. [CrossRef]
179. Stanchi, S.; Zecca, O.; Hudek, C.; Pintaldi, E.; Viglietti, D.; D'Amico, M.E.; Colombo, N.; Goslino, D.; Letey, M.; Freppaz, M. Effect of Soil Management on Erosion in Mountain Vineyards (N-W Italy). *Sustainability* **2021**, *13*, 1991. [CrossRef]
180. Alewell, C.; Ringeval, B.; Ballabio, C.; Robinson, D.A.; Panagos, P.; Borrelli, P. Global Phosphorus Shortage Will Be Aggravated by Soil Erosion. *Nat. Commun.* **2020**, *11*, 4546. [CrossRef]
181. Novara, A.; Pisciotta, A.; Minacapilli, M.; Maltese, A.; Capodici, F.; Cerdà, A.; Gristina, L. The Impact of Soil Erosion on Soil Fertility and Vine Vigor. A Multidisciplinary Approach Based on Field, Laboratory and Remote Sensing Approaches. *Sci. Total Environ.* **2018**, *622–623*, 474–480. [CrossRef]
182. Costantini, E.A.C.; Castaldini, M.; Diago, M.P.; Giffard, B.; Lagomarsino, A.; Schroers, H.-J.; Priori, S.; Valboa, G.; Agnelli, A.E.; Akça, E.; et al. Effects of Soil Erosion on Agro-Ecosystem Services and Soil Functions: A Multidisciplinary Study in Nineteen Organically Farmed European and Turkish Vineyards. *J. Environ. Manag.* **2018**, *223*, 614–624. [CrossRef]
183. Pijl, A.; Wang, W.; Straffelini, E.; Tarolli, P. Soil and Water Conservation in Terraced and Non-Terraced Cultivations: An Extensive Comparison of 50 Vineyards. *Land Degrad. Dev.* **2022**, *33*, 596–610. [CrossRef]
184. Biddoccu, M.; Ferraris, S.; Opsi, F.; Cavallo, E. Long-Term Monitoring of Soil Management Effects on Runoff and Soil Erosion in Sloping Vineyards in Alto Monferrato (North–West Italy). *Soil Tillage Res.* **2016**, *155*, 176–189. [CrossRef]
185. Dezső, J.; Lóczy, D.; Rezsek, M.; Hüppi, R.; Werner, J.; Horváth, L. Crop Growth, Carbon Sequestration and Soil Erosion in an Organic Vineyard of the Villány Wine District, Southwest Hungary. *Hung. Geogr. Bull.* **2020**, *69*, 281–298. [CrossRef]
186. Ruiz-Colmenero, M.; Bienes, R.; Eldridge, D.J.; Marques, M.J. Vegetation Cover Reduces Erosion and Enhances Soil Organic Carbon in a Vineyard in the Central Spain. *Catena* **2013**, *104*, 153–160. [CrossRef]

187. Winter, S.; Bauer, T.; Strauss, P.; Kratschmer, S.; Paredes, D.; Popescu, D.; Landa, B.; Guzmán, G.; Gómez, J.A.; Guernion, M.; et al. Effects of Vegetation Management Intensity on Biodiversity and Ecosystem Services in Vineyards: A Meta-Analysis. *J. Appl. Ecol.* **2018**, *55*, 2484–2495. [CrossRef] [PubMed]
188. Niu, Y.H.; Wang, L.; Wan, X.G.; Peng, Q.Z.; Huang, Q.; Shi, Z.H. A Systematic Review of Soil Erosion in Citrus Orchards Worldwide. *Catena* **2021**, *206*, 105558. [CrossRef]
189. Pou, A.; Gulías, J.; Moreno, M.; Tomàs, M.; Medrano, H.; Cifre, J. Cover Cropping in *Vitis vinifera* L. Cv. Manto Negro Vineyards under Mediterranean Conditions: Effects on Plant Vigour, Yield and Grape Quality. *OENO One* **2011**, *45*, 223. [CrossRef]
190. Mirás-Avalos, J.M.; Ramírez-Cuesta, J.M.; Fandiño, M.; Cancela, J.J.; Intrigliolo, D.S. Agronomic Practices for Reducing Soil Erosion in Hillside Vineyards under Atlantic Climatic Conditions (Galicia, Spain). *Soil Syst.* **2020**, *4*, 19. [CrossRef]
191. Triplett, G.B.; Dick, W.A. No-Tillage Crop Production: A Revolution in Agriculture! *Agron. J.* **2008**, *100*, S-153. [CrossRef]
192. Osterholz, W.R.; Culman, S.W.; Herms, C.; Joaquim de Oliveira, F.; Robinson, A.; Doohan, D. Knowledge Gaps in Organic Research: Understanding Interactions of Cover Crops and Tillage for Weed Control and Soil Health. *Org. Agric.* **2021**, *11*, 13–25. [CrossRef]
193. Laudicina, V.A.; Palazzolo, E.; Catania, P.; Vallone, M.; García, A.D.; Badalucco, L. Soil Quality Indicators as Affected by Shallow Tillage in a Vineyard Grown in a Semiarid Mediterranean Environment. *Land Degrad. Dev.* **2017**, *28*, 1038–1046. [CrossRef]
194. Shaheb, M.R.; Venkatesh, R.; Shearer, S.A. A Review on the Effect of Soil Compaction and Its Management for Sustainable Crop Production. *J. Biosyst. Eng.* **2021**, *46*, 417–439. [CrossRef]
195. Schwartz, R.C.; Baumhardt, R.L.; Evett, S.R. Tillage Effects on Soil Water Redistribution and Bare Soil Evaporation throughout a Season. *Soil Tillage Res.* **2010**, *110*, 221–229. [CrossRef]
196. Nunes, M.R.; Karlen, D.L.; Moorman, T.B.; Cambardella, C.A. How Does Tillage Intensity Affect Chemical Soil Health Indicators? A United States Meta-Analysis. *Agrosyst. Geosci. Environ.* **2020**, *3*, e20083. [CrossRef]
197. Linares, R.; de la Fuente, M.; Junquera, P.; Lissarrague, J.R.; Baeza, P. Effects of Soil Management in Vineyard on Soil Physical and Chemical Characteristics. *BIO Web Conf.* **2014**, *3*, 01008. [CrossRef]
198. Nunes, M.R.; Karlen, D.L.; Veum, K.S.; Moorman, T.B.; Cambardella, C.A. Biological Soil Health Indicators Respond to Tillage Intensity: A US Meta-Analysis. *Geoderma* **2020**, *369*, 114335. [CrossRef]
199. Badalíková, B. Influence of Soil Tillage on Soil Compaction. In *Soil Biology*; Dedousis, A., Bartzanas, T., Eds.; Springer: Volos, Greece, 2010; pp. 19–30.
200. Saffih-Hdadi, K.; Défossez, P.; Richard, G.; Cui, Y.-J.; Tang, A.-M.; Chaplain, V. A Method for Predicting Soil Susceptibility to the Compaction of Surface Layers as a Function of Water Content and Bulk Density. *Soil Tillage Res.* **2009**, *105*, 96–103. [CrossRef]
201. Ahmadi, I.; Ghaur, H. Effects of Soil Moisture Content and Tractor Wheeling Intensity on Traffic-Induced Soil Compaction. *J. Cent. Eur. Agric.* **2015**, *16*, 489–502. [CrossRef]
202. Schjønning, P.; Lamandé, M.; Tøgersen, F.A.; Arvidsson, J.; Keller, T. Modelling Effects of Tyre Inflation Pressure on the Stress Distribution near the Soil–Tyre Interface. *Biosyst. Eng.* **2008**, *99*, 119–133. [CrossRef]
203. Hamza, M.A.; Anderson, W.K. Soil Compaction in Cropping Systems: A Review of the Nature, Causes and Possible Solutions. *Soil Tillage Res.* **2005**, *82*, 121–145. [CrossRef]
204. Van Dijck, S.J.E.; van Asch, T.W.J. Compaction of Loamy Soils Due to Tractor Traffic in Vineyards and Orchards and Its Effect on Infiltration in Southern France. *Soil Tillage Res.* **2002**, *63*, 141–153. [CrossRef]
205. Pessina, D.; Galli, L.E.; Santoro, S.; Facchinetti, D. Sustainability of Machinery Traffic in Vineyard. *Sustainability* **2021**, *13*, 24745. [CrossRef]
206. Elaoud, A.; Chehaibi, S. Soil Compaction Due to Tractor Traffic. *J. Fail. Anal. Prev.* **2011**, *11*, 539–545. [CrossRef]
207. Lagacherie, P.; Coulouma, G.; Ariagno, P.; Virat, P.; Boizard, H.; Richard, G. Spatial Variability of Soil Compaction over a Vineyard Region in Relation with Soils and Cultivation Operations. *Geoderma* **2006**, *134*, 207–216. [CrossRef]
208. Hunter, J.J.; Archer, E.; Van Schalkwyk, D.; Strever, A.E.; Volschenk, C.G. Grapevine Roots: Interaction with Natural Factors and Agronomic Practices. In *Acta Horticulturae, Proceedings of the I International Symposium on Grapevine Roots, Chania, Greece, 30 July 2016*; International Society for Horticultural Science: Leuven, Belgium, 2016; Volume 1136, pp. 63–80.
209. Marinello, F.; Pezzuolo, A.; Cillis, D.; Chiumenti, A.; Sartori, L. Traffic Effects on Soil Compaction and Sugar Beet (*Beta Vulgaris* L.) Taproot Quality Parameters. *Span. J. Agric. Res.* **2017**, *15*, e0201. [CrossRef]
210. Döring, J.; Friedel, M.; Hendgen, M.; Stoll, M.; Kauer, R. Soil Management in Sustainable Viticultural Systems. In *Improving Sustainable Viticulture and Winemaking Practices*; Elsevier: Amsterdam, The Netherlands, 2022; pp. 85–103.
211. Burg, P.; Masan, V., Zemanek, P. Evaluation of Effect of Vegetation Cover on Soil Compaction. In Proceedings of the Engineering for Rural Development, Jelgava, Latvia, 23–25 May 2017; Latvia University of Agriculture: Jelgava, Latvia, 2017; Volume 16, pp. 1346–1353.
212. Visconti, F.; Jiménez, M.G.; de Paz, J.M. How Do the Chemical Characteristics of Organic Matter Explain Differences among Its Determinations in Calcareous Soils? *Geoderma* **2022**, *406*, 115454. [CrossRef]
213. Hartwig, N.L.; Ammon, H.U. Cover Crops and Living Mulches. *Weed Sci.* **2002**, *50*, 688–699. [CrossRef]
214. Chrysargyris, A.; Xylia, P.; Litskas, V.; Stavrinides, M.; Heyman, L.; Demeestere, K.; Höfte, M.; Tzortzakis, N. Assessing the Impact of Drought Stress and Soil Cultivation in Chardonnay and Xynisteri Grape Cultivars. *Agronomy* **2020**, *10*, 670. [CrossRef]
215. Capri, C.; Gatti, M.; Fiorini, A.; Ardenti, F.; Tabaglio, V.; Poni, S. A Comparative Study of Fifteen Cover Crop Species for Orchard Soil Management: Water Uptake, Root Density Traits and Soil Aggregate Stability. *Sci. Rep.* **2023**, *13*, 721. [CrossRef]

216. Zengin, H.; Sabir, A. Physiological and Growth Responses of Grapevine Rootstocks (Vitis Spp.) to Organic and Synthetic Mulch Application in Arid Ecology under the Effect of Climate Change. *J. Cent. Eur. Agric.* **2022**, *23*, 655–664. [CrossRef]
217. Cabrera-Pérez, C.; Llorens, J.; Escolà, A.; Royo-Esnal, A.; Recasens, J. Organic Mulches as an Alternative for Under-Vine Weed Management in Mediterranean Irrigated Vineyards: Impact on Agronomic Performance. *Eur. J. Agron.* **2023**, *145*, 126798. [CrossRef]
218. Marks, J.N.J.; Lines, T.E.P.; Penfold, C.; Cavagnaro, T.R. Cover Crops and Carbon Stocks: How under-Vine Management Influences SOC Inputs and Turnover in Two Vineyards. *Sci. Total Environ.* **2022**, *831*, 154800. [CrossRef] [PubMed]
219. Gattullo, C.E.; Mezzapesa, G.N.; Stellacci, A.M.; Ferrara, G.; Occhiogrosso, G.; Petrelli, G.; Castellini, M.; Spagnuolo, M. Cover Crop for a Sustainable Viticulture: Effects on Soil Properties and Table Grape Production. *Agronomy* **2020**, *10*, 1334. [CrossRef]
220. Salomé, C.; Coll, P.; Lardo, E.; Metay, A.; Villenave, C.; Marsden, C.; Blanchart, E.; Hinsinger, P.; Le Cadre, E. The Soil Quality Concept as a Framework to Assess Management Practices in Vulnerable Agroecosystems: A Case Study in Mediterranean Vineyards. *Ecol. Indic.* **2016**, *61*, 456–465. [CrossRef]
221. Ferrara, G.; Nigro, D.; Torres, R.; Gadaleta, A.; Fidelibus, M.W.; Mazzeo, A. Cover Crops in the Inter-Row of a Table Grape Vineyard Managed with Irrigation Sensors: Effects on Yield, Quality and Glutamine Synthetase Activity in Leaves. *Sci. Hortic.* **2021**, *281*, 109963. [CrossRef]
222. DeVetter, L.W.; Dilley, C.A.; Nonnecke, G.R. Mulches Reduce Weeds, Maintain Yield, and Promote Soil Quality in a Continental-Climate Vineyard. *Am. J. Enol. Vitic.* **2015**, *66*, 54–64. [CrossRef]
223. Mania, E.; Isocrono, D.; Pedullà, M.L.; Guidoni, S. Plant Diversity in an Intensively Cultivated Vineyard Agro-Ecosystem (Langhe, North-West Italy). *S. Afr. J. Enol. Vitic.* **2015**, *36*, 378–388. [CrossRef]
224. Eisenhauer, N. Plant Diversity Effects on Soil Microorganisms: Spatial and Temporal Heterogeneity of Plant Inputs Increase Soil Biodiversity. *Pedobiologia* **2016**, *59*, 175–177. [CrossRef]
225. Chen, C.; Chen, H.Y.H.; Chen, X.; Huang, Z. Meta-Analysis Shows Positive Effects of Plant Diversity on Microbial Biomass and Respiration. *Nat. Commun.* **2019**, *10*, 1332. [CrossRef] [PubMed]
226. Niles, M.T.; Garrett, R.D.; Walsh, D. Ecological and Economic Benefits of Integrating Sheep into Viticulture Production. *Agron. Sustain. Dev.* **2018**, *38*, 1. [CrossRef]
227. Fermaud, M.; Smits, N.; Merot, A.; Roudet, J.; Thiéry, D.; Wery, J.; Delbac, L. New Multipest Damage Indicator to Assess Protection Strategies in Grapevine Cropping Systems. *Aust. J. Grape Wine Res.* **2016**, *22*, 450–461. [CrossRef]
228. Daane, K.M.; Cooper, M.L.; Triapitsyn, S.V.; Walton, V.M.; Yokota, G.Y.; Haviland, D.R.; Bentley, W.J.; Godfrey, K.E.; Wunderlich, L.R. Vineyard Managers and Researchers Seek Sustainable Solutions for Mealybugs, a Changing Pest Complex. *Calif. Agric.* **2008**, *62*, 167–176. [CrossRef]
229. Reineke, A.; Thiéry, D. Grapevine Insect Pests and Their Natural Enemies in the Age of Global Warming. *J. Pest Sci.* **2016**, *89*, 313–328. [CrossRef]
230. Imperatore, G.; Ghirardelli, A.; Strinna, L.; Baldoin, C.; Pozzebon, A.; Zanin, G.; Otto, S. Evaluation of a Fixed Spraying System for Phytosanitary Treatments in Heroic Viticulture in North-Eastern Italy. *Agriculture* **2021**, *11*, 833. [CrossRef]
231. European Commission Commission Regulation (EC). No 473/2002 of 15 March 2002 Amending Annexes I, II and VI to Council Regulation (EEC) No 2092/91 on Organic Production of Agricultural Products and Indications Referring Thereto on Agricultural Products and Foodstuffs, and Laying Down Detailed Rules as Regards the Transmission of Information on the Use of Copper Compounds. Available online: https://eur-lex.europa.eu/legal-content/EN/ALL/?uri=CELEX:32002R0473 (accessed on 6 December 2023).
232. Unsworth, J. *History of Pesticide Use*; International Union of Pure and Applied Chemistry: Durham, NC, USA, 2010.
233. Vocke, G. *The Green Revolution for Wheat in Developing Countries*; US Department of Agriculture: New York, NY, USA, 1986.
234. Kogan, M. Integrated Pest Management: Historical Perspectives and Contemporary Developments. *Annu. Rev. Entomol.* **1998**, *43*, 243–270. [CrossRef]
235. Barzman, M.; Bàrberi, P.; Birch, A.N.E.; Boonekamp, P.; Dachbrodt-Saaydeh, S.; Graf, B.; Hommel, B.; Jensen, J.E.; Kiss, J.; Kudsk, P.; et al. Eight Principles of Integrated Pest Management. *Agron. Sustain. Dev.* **2015**, *35*, 1199–1215. [CrossRef]
236. Gregoire, C.; Payraudeau, S.; Domange, N. Use and Fate of 17 Pesticides Applied on a Vineyard Catchment. *Int. J. Environ. Anal. Chem.* **2010**, *90*, 406–420. [CrossRef]
237. Cabras, P.; Meloni, M.; Pirisi, F.M. Pesticide Fate from Vine to Wine. *Rev. Environ. Contam. Toxicol.* **1987**, *99*, 83–117. [CrossRef] [PubMed]
238. Pertot, I.; Caffi, T.; Rossi, V.; Mugnai, L.; Hoffmann, C.; Grando, M.S.; Gary, C.; Lafond, D.; Duso, C.; Thiery, D.; et al. A Critical Review of Plant Protection Tools for Reducing Pesticide Use on Grapevine and New Perspectives for the Implementation of IPM in Viticulture. *Crop Prot.* **2017**, *97*, 70–84. [CrossRef]
239. Kullaj, E.; Shahini, S.; Varaku, S.; Çakalli, M. Evaluation of the Efficacy for Reducing Copper Use against Downy Mildew Control in Organic Mediterranean Viticulture. *Int. J. Pest. Manag.* **2016**, *63*, 3–9. [CrossRef]
240. Kraus, C.; Abou-Ammar, R.; Schubert, A.; Fischer, M. Warburgia Ugandensis Leaf and Bark Extracts: An Alternative to Copper as Fungicide against Downy Mildew in Organic Viticulture? *Plants* **2021**, *10*, 2765. [CrossRef] [PubMed]
241. Vargas-Bernal, R.; Rodríguez-Miranda, E.; Herrera-Pérez, G.; Vargas-Bernal, R.; Rodríguez-Miranda, E.; Herrera-Pérez, G. Evolution and Expectations of Enzymatic Biosensors for Pesticides. In *Pesticides-Advances in Chemical and Botanical Pesticides*; IntechOpen: London, UK, 2012; ISBN 978-953-51-0680-7.

242. Urruty, N.; Deveaud, T.; Guyomard, H.; Boiffin, J. Impacts of Agricultural Land Use Changes on Pesticide Use in French Agriculture. *Eur. J. Agron.* **2016**, *80*, 113–123. [CrossRef]
243. Droz, B.; Payraudeau, S.; Rodríguez Martín, J.A.; Tóth, G.; Panagos, P.; Montanarella, L.; Borrelli, P.; Imfeld, G. Copper Content and Export in European Vineyard Soils Influenced by Climate and Soil Properties. *Environ. Sci. Technol.* **2021**, *55*, 7327–7334. [CrossRef] [PubMed]
244. Dagostin, S.; Schärer, H.-J.; Pertot, I.; Tamm, L. Are There Alternatives to Copper for Controlling Grapevine Downy Mildew in Organic Viticulture? *Crop Prot.* **2011**, *30*, 776–788. [CrossRef]
245. Lo Scalzo, R.; Fibiani, M.; Pietromarchi, P.; Mandalà, C.; La Torre, A. Effects of Different Fungicide Treatments on Grape, Must and Wine Quality. *Commun. Agric. Appl. Biol. Sci.* **2012**, *77*, 151–161.
246. Thuerig, B.; James, E.E.; Schärer, H.; Langat, M.K.; Mulholland, D.A.; Treutwein, J.; Kleeberg, I.; Ludwig, M.; Jayarajah, P.; Giovannini, O.; et al. Reducing Copper Use in the Environment: The Use of Larixol and Larixyl Acetate to Treat Downy Mildew Caused by *Plasmopara Viticola* in Viticulture. *Pest Manag. Sci.* **2017**, *74*, 477–488. [CrossRef]
247. Lamichhane, J.R.; Osdaghi, E.; Behlau, F.; Köhl, J.; Jones, J.B.; Aubertot, J.-N. Thirteen Decades of Antimicrobial Copper Compounds Applied in Agriculture. A Review. *Agron. Sustain. Dev.* **2018**, *38*, 28. [CrossRef]
248. Widmer, J.; Norgrove, L. Identifying Candidates for the Phytoremediation of Copper in Viticultural Soils: A Systematic Review. *Environ. Res.* **2023**, *216*, 114518. [CrossRef]
249. Ballabio, C.; Panagos, P.; Lugato, E.; Huang, J.-H.; Orgiazzi, A.; Jones, A.; Fernández-Ugalde, O.; Borrelli, P.; Montanarella, L. Copper Distribution in European Topsoils: An Assessment Based on LUCAS Soil Survey. *Sci. Total Environ.* **2018**, *636*, 282–298. [CrossRef] [PubMed]
250. New South Wales Environment Protection Agency. *Environmental Management Guidelines—Use and Disposal of Biosolids Products*; New South Wales Environment Protection Agency: Parramatta, Australia, 1997.
251. Voluntary Risk Assessment Reports-Copper and Copper Compounds-ECHA. Available online: https://echa.europa.eu/es/copper-voluntary-risk-assessment-reports (accessed on 6 December 2023).
252. Sandmann, G.; Böger, P. The Enzymatological Function of Heavy Metals and Their Role in Electron Transfer Processes of Plants. In *Encyclopedia of Plant Physiology*; Lauchli, A., Bieleski, R.L., Eds.; Springer: Berlin/Heidelberg, Germany, 1983; pp. 563–596.
253. Solomon, E.I.; Heppner, D.E.; Johnston, E.M.; Ginsbach, J.W.; Cirera, J.; Qayyum, M.; Kieber-Emmons, M.T.; Kjaergaard, C.H.; Hadt, R.G.; Tian, L. Copper Active Sites in Biology. *Chem. Rev.* **2014**, *114*, 3659–3853. [CrossRef]
254. Yruela, I. Copper in Plants. *Braz. J. Plant Physiol.* **2005**, *17*, 145–156. [CrossRef]
255. Broadley, M.; Brown, P.; Cakmak, I.; Rengel, Z.; Zhao, F. Function of Nutrients: Micronutrients. In *Marschner's Nutrition of Higher Plants*, 3rd ed.; Academic Press: Cambridge, MA, USA, 2012; pp. 191–248. [CrossRef]
256. Gaetke, L. Copper Toxicity, Oxidative Stress, and Antioxidant Nutrients. *Toxicology* **2003**, *189*, 147–163. [CrossRef] [PubMed]
257. Borkow, G.; Gabbay, J. Copper, An Ancient Remedy Returning to Fight Microbial, Fungal and Viral Infections. *Curr. Chem. Biol.* **2009**, *3*, 272–278. [CrossRef]
258. Wairich, A.; De Conti, L.; Lamb, T.I.; Keil, R.; Neves, L.O.; Brunetto, G.; Sperotto, R.A.; Ricachenevsky, F.K. Throwing Copper Around: How Plants Control Uptake, Distribution, and Accumulation of Copper. *Agronomy* **2022**, *12*, 994. [CrossRef]
259. Yang, X.; Lu, M.; Wang, Y.; Wang, Y.; Liu, Z.; Chen, S. Response Mechanism of Plants to Drought Stress. *Horticulturae* **2021**, *7*, 50. [CrossRef]
260. Fernandes, J.C.; Henriques, F.S. Biochemical, Physiological, and Structural Effects of Excess Copper in Plants. *Bot. Rev.* **1991**, *57*, 246–273. [CrossRef]
261. Babu, T.S.; Akhtar, T.A.; Lampi, M.A.; Tripuranthakam, S.; Dixon, D.G.; Greenberg, B.M. Similar Stress Responses Are Elicited by Copper and Ultraviolet Radiation in the Aquatic Plant Lemna Gibba: Implication of Reactive Oxygen Species as Common Signals. *Plant Cell Physiol.* **2003**, *44*, 1320–1329. [CrossRef]
262. Sheldon, A.R.; Menzies, N.W. The Effect of Copper Toxicity on the Growth and Root Morphology of Rhodes Grass (*Chloris gayana* Knuth.) in Resin Buffered Solution Culture. *Plant Soil* **2005**, *278*, 341–349. [CrossRef]
263. Lequeux, H.; Hermans, C.; Lutts, S.; Verbruggen, N. Response to Copper Excess in *Arabidopsis thaliana*: Impact on the Root System Architecture, Hormone Distribution, Lignin Accumulation and Mineral Profile. *Plant Physiol. Biochem.* **2010**, *48*, 673–682. [CrossRef]
264. Pérez-Rodríguez, P.; Soto-Gómez, D.; De La Calle, I.; López-Periago, J.E.; Paradelo, M. Rainfall-Induced Removal of Copper-Based Spray Residues from Vines. *Ecotoxicol. Environ. Saf.* **2016**, *132*, 304–310. [CrossRef]
265. Hummes, A.P.; Bortoluzzi, E.C.; Tonini, V.; da Silva, L.P.; Petry, C. Transfer of Copper and Zinc from Soil to Grapevine-Derived Products in Young and Centenarian Vineyards. *Water Air Soil Pollut.* **2019**, *230*, 150. [CrossRef]
266. Zhao, F.; Rooney, C.P.; Zhang, H.; McGrath, S.P. Comparison of Soil Solution Speciation and Diffusive Gradients in Thin-films Measurement as an Indicator of Copper Bioavailability to Plants. *Environ. Toxicol. Chem.* **2006**, *25*, 733–742. [CrossRef]
267. Li, L.; Wu, H.; van Gestel, C.A.M.; Peijnenburg, W.J.G.M.; Allen, H.E. Soil Acidification Increases Metal Extractability and Bioavailability in Old Orchard Soils of Northeast Jiaodong Peninsula in China. *Environ. Pollut.* **2014**, *188*, 144–152. [CrossRef] [PubMed]
268. Guo, X.Y.; Zuo, Y.B.; Wang, B.R.; Li, J.M.; Ma, Y.B. Toxicity and Accumulation of Copper and Nickel in Maize Plants Cropped on Calcareous and Acidic Field Soils. *Plant Soil* **2010**, *333*, 365–373. [CrossRef]

269. Huang, J.-S.; Wei, D.-P.; Guo, X.-Y.; Ma, Y.-B. Toxicity and Accumulation of Copper and Nickel in Wheat Plants Cropped on Alkaline and Acidic Field Soils. *Huan Jing Ke Xue/Environ. Sci.* **2012**, *33*, 1369–1375.
270. Trentin, E.; Cesco, S.; Pii, Y.; Valentinuzzi, F.; Celletti, S.; Feil, S.B.; Zuluaga, M.Y.A.; Ferreira, P.A.A.; Ricachenevsky, F.K.; Stefanello, L.O.; et al. Plant Species and PH Dependent Responses to Copper Toxicity. *Environ. Exp. Bot.* **2022**, *196*, 104791. [CrossRef]
271. Miotto, A.; Ceretta, C.A.; Brunetto, G.; Nicoloso, F.T.; Girotto, E.; Farias, J.G.; Tiecher, T.L.; De Conti, L.; Trentin, G. Copper Uptake, Accumulation and Physiological Changes in Adult Grapevines in Response to Excess Copper in Soil. *Plant Soil* **2013**, *374*, 593–610. [CrossRef]
272. Ambrosini, V.G.; Rosa, D.J.; Bastos de Melo, G.W.; Zalamena, J.; Cella, C.; Simão, D.G.; Souza da Silva, L.; Pessoa dos Santos, H.; Toselli, M.; Tiecher, T.L.; et al. High Copper Content in Vineyard Soils Promotes Modifications in Photosynthetic Parameters and Morphological Changes in the Root System of 'Red Niagara' Plantlets. *Plant Physiol. Biochem.* **2018**, *128*, 89–98. [CrossRef] [PubMed]
273. Feil, S.B.; Zuluaga, M.Y.A.; Cesco, S.; Pii, Y. Copper Toxicity Compromises Root Acquisition of Nitrate in the High Affinity Range. *Front. Plant Sci.* **2023**, *13*, 1034425. [CrossRef] [PubMed]
274. Rosa, D.J.; Ambrosini, V.G.; Kokkoris, V.; Brunetto, G.; Hart, M.; Ricachenevsky, F.; Pescador, R. Lime Protection for Young Vines Exposed to Copper Toxicity. *Water Air Soil Pollut.* **2020**, *231*, 296. [CrossRef]
275. Ma, Y.; Lombi, E.; Oliver, I.W.; Nolan, A.L.; McLaughlin, M.J. Long-Term Aging of Copper Added to Soils. *Environ. Sci. Technol.* **2006**, *40*, 6310–6317. [CrossRef] [PubMed]
276. Wang, X.; Hua, L.; Ma, Y. A Biotic Ligand Model Predicting Acute Copper Toxicity for Barley (Hordeum Vulgare): Influence of Calcium, Magnesium, Sodium, Potassium and PH. *Chemosphere* **2012**, *89*, 89–95. [CrossRef] [PubMed]
277. Stevenson, F.J.; Fitch, A. Reactions with Organic Matter. In *Copper in Soils and Plants*; Loneragan, J.F., Robson, A.D., Graham, R.D., Eds.; Academic Press: Sydney, Australia, 1981; pp. 265–285.
278. Fernández-Calviño, D.; Nóvoa-Muñoz, J.C.; López-Periago, E.; Arias-Estévez, M. Changes in Copper Content and Distribution in Young, Old and Abandoned Vineyard Acid Soils Due to Land Use Changes. *Land. Degrad. Dev.* **2007**, *19*, 165–177. [CrossRef]
279. Lagomarsino, A.; Marabottini, S.; Grego, S.; Stazi, S.R. Copper Distribution among Physical and Chemical Fractions in a Former Vineyard Soil. *Agrochimica* **2010**, *LIV*, 1–12.
280. Eijsackers, H.; Beneke, P.; Maboeta, M.; Louw, J.P.E.; Reinecke, A.J. The Implications of Copper Fungicide Usage in Vineyards for Earthworm Activity and Resulting Sustainable Soil Quality. *Ecotoxicol. Environ. Saf.* **2005**, *62*, 99–111. [CrossRef]
281. Duan, X.; Xu, M.; Zhou, Y.; Yan, Z.; Du, Y.; Zhang, L.; Zhang, C.; Bai, L.; Nie, J.; Chen, G.; et al. Effects of Soil Properties on Copper Toxicity to Earthworm Eisenia Fetida in 15 Chinese Soils. *Chemosphere* **2016**, *145*, 185–192. [CrossRef]
282. Qiu, H.; Vijver, M.G.; He, E.; Peijnenburg, W.J.G.M. Predicting Copper Toxicity to Different Earthworm Species Using a Multicomponent Freundlich Model. *Environ. Sci. Technol.* **2013**, *47*, 4796–4803. [CrossRef] [PubMed]
283. Tiecher, T.L.; Tiecher, T.; Ceretta, C.A.; Ferreira, P.A.A.; Nicoloso, F.T.; Soriani, H.H.; De Conti, L.; Kulmann, M.S.S.; Schneider, R.O.; Brunetto, G. Tolerance and Translocation of Heavy Metals in Young Grapevine (*Vitis Vinifera*) Grown in Sandy Acidic Soil with Interaction of High Doses of Copper and Zinc. *Sci. Hortic.* **2017**, *222*, 203–212. [CrossRef]
284. Shaw, J.L.A.; Ernakovich, J.G.; Judy, J.D.; Farrell, M.; Whatmuff, M.; Kirby, J. Long-Term Effects of Copper Exposure to Agricultural Soil Function and Microbial Community Structure at a Controlled and Experimental Field Site. *Environ. Pollut.* **2020**, *263*, 114411. [CrossRef] [PubMed]
285. Ben Massoud, M.; Kharbech, O.; Sakouhi, L.; Ben Hassine, S.; Zhu, Y.; Chaoui, A.; Sheehan, D.; Djebali, W. Calcium and Citrate Protect Pisum Sativum Roots against Copper Toxicity by Regulating the Cellular Redox Status. *J. Soil Sci. Plant Nutr.* **2021**, *22*, 345–358. [CrossRef]
286. Tomizawa, M.; Casida, J.E. Neonicotinoid Insecticides: Highlights of a Symposium on Strategic Molecular Designs. *J. Agric. Food Chem.* **2010**, *59*, 2883–2886. [CrossRef]
287. Zaller, J.G.; Cantelmo, C.; Dos Santos, G.; Muther, S.; Gruber, E.; Pallua, P.; Mandl, K.; Friedrich, B.; Hofstetter, I.; Schmuckenschlager, B.; et al. Herbicides in Vineyards Reduce Grapevine Root Mycorrhization and Alter Soil Microorganisms and the Nutrient Composition in Grapevine Roots, Leaves, Xylem Sap and Grape Juice. *Environ. Sci. Pollut. Res. Int.* **2018**, *25*, 23215–23226. [CrossRef] [PubMed]
288. Stellin, F.; Gavinelli, F.; Stevanato, P.; Concheri, G.; Squartini, A.; Paoletti, M.G. Effects of Different Concentrations of Glyphosate (Roundup 360®) on Earthworms (*Octodrilus Complanatus*, *Lumbricus Terrestris* and *Aporrectodea Caliginosa*) in Vineyards in the North-East of Italy. *Appl. Soil Ecol.* **2018**, *123*, 802–808. [CrossRef]
289. Albers, C.N.; Banta, G.T.; Hansen, P.E.; Jacobsen, O.S. The Influence of Organic Matter on Sorption and Fate of Glyphosate in Soil–Comparing Different Soils and Humic Substances. *Environ. Pollut.* **2009**, *157*, 2865–2870. [CrossRef]
290. Gros, P.; Ahmed, A.; Kühn, O.; Leinweber, P. Glyphosate Binding in Soil as Revealed by Sorption Experiments and Quantum-Chemical Modeling. *Sci. Total Environ.* **2017**, *586*, 527–535. [CrossRef] [PubMed]
291. Massaccesi, L.; Cartoni Mancinelli, A.; Mattioli, S.; De Feudis, M.; Castellini, C.; Dal Bosco, A.; Marongiu, M.L.; Agnelli, A. Geese Reared in Vineyard: Soil, Grass and Animals Interaction. *Animals* **2019**, *9*, 179. [CrossRef]
292. Gonçalves, P.; Nóbrega, L.; Monteiro, A.; Pedreiras, P.; Rodrigues, P.; Esteves, F. SheepIT, an E-Shepherd System for Weed Control in Vineyards: Experimental Results and Lessons Learned. *Animals* **2021**, *11*, 2625. [CrossRef]
293. Trouillard, M.; Lèbre, A.; Heckendorn, F. Grazing Sheep in Organic Vineyards: An On-Farm Study about Risk of Chronic Copper Poisoning. *Sustainability* **2021**, *13*, 12860. [CrossRef]

294. O'Brien, R.D. *Toxic Phosphorous Esters: Chemistry, Metabolism, and Biological Effects*; Elsevier: New York, NY, USA, 2016.
295. Kushwaha, M.; Verma, S.; Chatterjee, S. Profenofos, an Acetylcholinesterase-Inhibiting Organophosphorus Pesticide: A Short Review of Its Usage, Toxicity, and Biodegradation. *J. Environ. Qual.* **2016**, *45*, 1478–1489. [CrossRef]
296. Booth, L.H.; Hodge, S.; O'Halloran, K. Use of Biomarkers in Earthworms to Detect Use and Abuse of Field Applications of a Model Organophosphate Pesticide. *Bull. Environ. Contam. Toxicol.* **2001**, *67*, 633–640. [CrossRef]
297. Reinecke, S.A.; Reinecke, A.J. The Impact of Organophosphate Pesticides in Orchards on Earthworms in the Western Cape, South Africa. *Ecotoxicol. Environ. Saf.* **2007**, *66*, 244–251. [CrossRef] [PubMed]
298. Pelosi, C.; Barot, S.; Capowiez, Y.; Hedde, M.; Vandenbulcke, F. Pesticides and Earthworms. A Review. *Agron. Sustain. Dev.* **2013**, *34*, 199–228. [CrossRef]
299. Mansour, R.; Belzunces, L.P.; Suma, P.; Zappalà, L.; Mazzeo, G.; Grissa-Lebdi, K.; Russo, A.; Biondi, A. Vine and Citrus Mealybug Pest Control Based on Synthetic Chemicals. A Review. *Agron. Sustain. Dev.* **2018**, *38*, 37. [CrossRef]
300. Brück, E.; Elbert, A.; Fischer, R.; Krueger, S.; Kühnhold, J.; Klueken, A.M.; Nauen, R.; Niebes, J.-F.; Reckmann, U.; Schnorbach, H.-J.; et al. Movento®, an Innovative Ambimobile Insecticide for Sucking Insect Pest Control in Agriculture: Biological Profile and Field Performance. *Crop Prot.* **2009**, *28*, 838–844. [CrossRef]
301. Pandey, S.; Singh, D.K. Total Bacterial and Fungal Population after Chlorpyrifos and Quinalphos Treatments in Groundnut (*Arachis Hypogaea* L.) Soil. *Chemosphere* **2004**, *55*, 197–205. [CrossRef]
302. Chu, X.; Fang, H.; Pan, X.; Wang, X.; Shan, M.; Feng, B.; Yu, Y. Degradation of Chlorpyrifos Alone and in Combination with Chlorothalonil and Their Effects on Soil Microbial Populations. *J. Environ. Sci.* **2008**, *20*, 464–469. [CrossRef]
303. Deutsch, C.A.; Tewksbury, J.J.; Tigchelaar, M.; Battisti, D.S.; Merrill, S.C.; Huey, R.B.; Naylor, R.L. Increase in Crop Losses to Insect Pests in a Warming Climate. *Science* **2018**, *361*, 916–919. [CrossRef]
304. Bois, B.; Zito, S.; Calonnec, A.; Ollat, N. Climate vs Grapevine Pests and Diseases Worldwide: The First Results of a Global Survey. *J. Int. Sci. Vigne Vin* **2017**, *51*, 133–139.
305. Daane, K.M.; Vincent, C.; Isaacs, R.; Ioriatti, C. Entomological Opportunities and Challenges for Sustainable Viticulture in a Global Market. *Annu. Rev. Entomol.* **2018**, *63*, 193–214. [CrossRef]
306. Lodovica Gullino, M.; Garibaldi, A.; Gamliel, A.; Katan, J. Soil Disinfestation: From Soil Treatment to Soil and Plant Health. *Plant Dis.* **2022**, *106*, 1541–1554. [CrossRef]
307. Husson, O.; Sarthou, J.-P.; Bousset, L.; Ratnadass, A.; Schmidt, H.-P.; Kempf, J.; Husson, B.; Tingry, S.; Aubertot, J.-N.; Deguine, J.-P.; et al. Soil and Plant Health in Relation to Dynamic Sustainment of Eh and PH Homeostasis: A Review. *Plant Soil* **2021**, *466*, 391–447. [CrossRef]
308. Akanmu, A.O.; Babalola, O.O.; Venturi, V.; Ayilara, M.S.; Adeleke, B.S.; Amoo, A.E.; Sobowale, A.A.; Fadiji, A.E.; Glick, B.R. Plant Disease Management: Leveraging on the Plant-Microbe-Soil Interface in the Biorational Use of Organic Amendments. *Front. Plant Sci.* **2021**, *12*, 700507. [CrossRef] [PubMed]
309. Davis, L. Soil Health, Plant Health and Pests. In *Building Soils for Better Crops*; Magdoff, F., Es, H., Eds.; Sustainable Agriculture Research and Education (SARE): New York, NY, USA, 2021; pp. 103–116.
310. Mpelasoka, B.; Schachtman, D.; Treeby, M.; Thomas, M. A Review of Potassium Nutrition in Grapevines with Special Emphasis on Berry Accumulation. *Aust. J. Grape Wine Res.* **2003**, *9*, 154–168. [CrossRef]
311. James, A.; Mahinda, A.; Mwamahonje, A.; Rweyemamu, E.W.; Mrema, E.; Aloys, K.; Swai, E.; Mpore, F.J.; Massawe, C. A Review on the Influence of Fertilizers Application on Grape Yield and Quality in the Tropics. *J. Plant Nutr.* **2022**, *46*, 2936–2957. [CrossRef]
312. Skinner, P.W.; Ishii, R.; O'Mahony, M.; Matthews, M. Sensory Attributes of Wines Made from Vines of Differing Phosphorus Status. *OENO One* **2019**, *53*, 347–361. [CrossRef]
313. Spayd, S.E.; Stevens, R.G.; Wample, R.L.; Evans, R.G.; Edwards, C.G.; Webster, D. Impact of Nitrogen Fertilization on Vine Performance and Juice and Wine Composition of "Riesling" Grapes (*Vitis Vinifera* L.) in Washington State. *Acta Hortic.* **2000**, *512*, 65–76. [CrossRef]
314. Bell, S.-J.; Henschke, P.A. Implications of Nitrogen Nutrition for Grapes, Fermentation and Wine. *Aust. J. Grape Wine Res.* **2005**, *11*, 242–295. [CrossRef]
315. Monder, H.; Maillard, M.; Chérel, I.; Zimmermann, S.D.; Paris, N.; Cuéllar, T.; Gaillard, I. Adjustment of K(+) Fluxes and Grapevine Defense in the Face of Climate Change. *Int. J. Mol. Sci.* **2021**, *22*, 10398. [CrossRef] [PubMed]
316. Krasilnikov, P.; Taboada, M.A. Amanullah Fertilizer Use, Soil Health and Agricultural Sustainability. *Agriculture* **2022**, *12*, 462. [CrossRef]
317. Montoroi, J.-P. Soil Degradation by Salinization. In *Ecology and Environmental Science: Reduction of Water Stress and Adaptation to Aridity: Proceedings of the Third Ecoscience Workshop*; Gawronski, K., Ziernicka-Wojtaszek, A., Ksiazek, L., Nizinski, G., Eds.; Wydawnictwo Uniwersytetu Rolniczego im. H. Kołłątaja: Cracovie, Poland, 2017; pp. 104–106.
318. Machado, R.M.A.; Serralheiro, R.P. Soil Salinity: Effect on Vegetable Crop Growth. Management Practices to Prevent and Mitigate Soil Salinization. *Horticulturae* **2017**, *3*, 30. [CrossRef]
319. Arnó, J.; Rosell, J.R.; Blanco, R.; Ramos, M.C.; Martínez-Casasnovas, J.A. Spatial Variability in Grape Yield and Quality Influenced by Soil and Crop Nutrition Characteristics. *Precis. Agric.* **2011**, *13*, 393–410. [CrossRef]
320. Lacombe, B. Increasing Our Knowledge on Grapevines Physiology to Increase Yield, Quality and Sustainably. *Physiol. Plant* **2022**, *174*, e13664. [CrossRef] [PubMed]

321. Lambert, J.-J.; Anderson, M.M.; Wolpert, J.A. Vineyard Nutrient Needs Vary with Rootstocks and Soils. *Calif. Agric.* **2008**, *62*, 202–207. [CrossRef]
322. Maeder, P.; Fliessbach, A.; Dubois, D.; Gunst, L.; Fried, P.; Niggli, U. Soil Fertility and Biodiversity in Organic Farming. *Science* **2002**, *296*, 1694–1697. [CrossRef]
323. Tully, K.L.; McAskill, C. Promoting Soil Health in Organically Managed Systems: A Review. *Org. Agric.* **2020**, *10*, 339–358. [CrossRef]
324. Provost, C.; Pedneault, K. The Organic Vineyard as a Balanced Ecosystem: Improved Organic Grape Management and Impacts on Wine Quality. *Sci. Hortic.* **2016**, *208*, 43–56. [CrossRef]
325. Borsato, E.; Zucchinelli, M.; D'Ammaro, D.; Giubilato, E.; Zabeo, A.; Criscione, P.; Pizzol, L.; Cohen, Y.; Tarolli, P.; Lamastra, L.; et al. Use of Multiple Indicators to Compare Sustainability Performance of Organic vs Conventional Vineyard Management. *Sci. Total Environ.* **2020**, *711*, 135081. [CrossRef]
326. Tangolar, S.; Tangolar, S.; Torun, A.A.; Ada, M.; Göçmez, S. Influence of Supplementation of Vineyard Soil with Organic Substances on Nutritional Status, Yield and Quality of "Black Magic" Grape (*Vitis Vinifera* L.) and Soil Microbiological and Biochemical Characteristics. *OENO One* **2020**, *54*, 1143–1157. [CrossRef]
327. Di Giacinto, S.; Friedel, M.; Poll, C.; Döring, J.; Kunz, R.; Kauer, R. Vineyard Management System Affects Soil Microbiological Properties. *OENO One* **2020**, *54*, 131–143. [CrossRef]
328. Cernansky, R. Agriculture: State-of-the-Art Soil. *Nature* **2015**, *517*, 258–260. [CrossRef]
329. Omondi, M.O.; Xia, X.; Nahayo, A.; Liu, X.; Korai, P.K.; Pan, G. Quantification of Biochar Effects on Soil Hydrological Properties Using Meta-Analysis of Literature Data. *Geoderma* **2016**, *274*, 28–34. [CrossRef]
330. Razzaghi, F.; Obour, P.B.; Arthur, E. Does Biochar Improve Soil Water Retention? A Systematic Review and Meta-Analysis. *Geoderma* **2020**, *361*, 114055. [CrossRef]
331. Shaheen, S.M.; Mosa, A.; Natasha; Arockiam Jeyasundar, P.G.S.; Hassan, N.E.E.; Yang, X.; Antoniadis, V.; Li, R.; Wang, J.; Zhang, T.; et al. Pros and Cons of Biochar to Soil Potentially Toxic Element Mobilization and Phytoavailability: Environmental Implications. *Earth Syst. Environ.* **2023**, *7*, 321–345. [CrossRef]
332. Baronti, S.; Magno, R.; Maienza, A.; Montagnoli, A.; Ungaro, F.; Vaccari, F.P. Long Term Effect of Biochar on Soil Plant Water Relation and Fine Roots: Results after 10 Years of Vineyard Experiment. *Sci. Total Environ.* **2022**, *851*, 158225. [CrossRef]
333. Genesio, L.; Miglietta, F.; Baronti, S.; Vaccari, F.P. Biochar Increases Vineyard Productivity without Affecting Grape Quality: Results from a Four Years Field Experiment in Tuscany. *Agric. Ecosyst. Environ.* **2015**, *201*, 20–25. [CrossRef]
334. Mortvedt, J.J. *Heavy Metal Contaminants in Inorganic and Organic Fertilizers*; Springer: Cham, Switzerland, 1996; Volume 43.
335. Roberts, D.; Nachtegaal, M.; Sparks, D.L. Speciation of Metals in Soils. In *Chemical Processes in Soils*; Tabatabai, M.A., Sparks, D.L., Eds.; Soil Science Society of America: Madison, WI, USA, 2005; pp. 619–654.
336. Yao, L.-X.; Li, G.-L.; Tu, S.-H.; Gavin, S.; He, Z.-H. Salinity of Animal Manure and Potential Risk of Secondary Soil Salinization through Successive Manure Application. *Sci. Total Environ.* **2007**, *383*, 106–114. [CrossRef]
337. Olego, M.A.; Visconti, F.; Quiroga, M.J.; de Paz, J.M.; Garzón-Jimeno, E. Assessing the Effects of Soil Liming with Dolomitic Limestone and Sugar Foam on Soil Acidity, Leaf Nutrient Contents, Grape Yield and Must Quality in a Mediterranean Vineyard. *Span. J. Agric. Res.* **2016**, *14*, e1102. [CrossRef]
338. Foy, C.D. Physiological Effects of Hydrogen, Aluminum, and Manganese Toxicities in Acid Soil. *Soil Acidity Liming* **2015**, *12*, 57–97. [CrossRef]
339. Hue, N. Soil Acidity: Development, Impacts, and Management. In *Structure and Functions of Pedosphere*; Springer Nature: Cham, Switzerland, 2022; pp. 103–131B14 9789811687709.
340. Skinner, P.W.; Matthews, M.A. Reproductive Development in Grape (*Vitis Vinifera* L.) under Phosphorus-Limited Conditions. *Sci. Hortic.* **1989**, *38*, 49–60. [CrossRef]
341. Čoga, L.; Slunjski, S.; Herak Ćustić, M.; Maslać, J.; Petek, M.; Ćosić, T.; Pavlović, I. Influence of Soil Reaction on Phosphorus, Potassium, Calcium and Magnesium Dynamics in Grapevine (*Vitis Vinifera* L.). *Agric. Conspec. Sci.* **2009**, *74*, 39–43.
342. Quiroga, M.J.; Olego, M.Á.; Sánchez-García, M.; Medina, J.E.; Visconti, F.; Coque, J.J.R.; Garzón Jimeno, J.E. Effects of Liming on Soil Properties, Leaf Tissue Cation Composition and Grape Yield in a Moderately Acid Vineyard Soil. Influence on Must and Wine Quality. *OENO One* **2017**, *51*, 343. [CrossRef]
343. Tang, C. Causes and Management of Subsoil Acidity. In Proceedings of the SuperSoil 2004: 3rd Australian New Zealand Soils Conference, Camperdown, NSW, Australia, 5 December 2004; University of Sydney: Camperdown, Sydney, NSW, Australia, 2004; pp. 1–7.
344. Grapevine Rootstock Selector Tool—Rootstock Selector. Available online: https://grapevinerootstock.com/ (accessed on 31 January 2024).
345. Van Leeuwen, C.; de Rességuier, L. Major Soil-Related Factors in Terroir Expression and Vineyard Siting. *Elements* **2018**, *14*, 159–165. [CrossRef]
346. Ferretti, C.G. Relationship between the Geology, Soil Assessment, and Terroir of Gewürztraminer Vineyards: A Case Study in the Dolomites of Northern Italy. *Catena* **2019**, *179*, 74–84. [CrossRef]

347. Jones, G.V. Grapevines in a Changing Environment. In *Grapevine in a Changing Environment*; Gerós, H., Chaves, M., Medrano Gil, H., Delrot, S., Eds.; Wiley: Hoboken, NJ, USA, 2015; pp. 1–17.
348. Gobbi, A.; Acedo, A.; Imam, N.; Santini, R.G.; Ortiz-Álvarez, R.; Ellegaard-Jensen, L.; Belda, I.; Hansen, L.H. A Global Microbiome Survey of Vineyard Soils Highlights the Microbial Dimension of Viticultural Terroirs. *Commun. Biol.* **2022**, *5*, 241. [CrossRef]
349. Bonfante, A.; Brillante, L. Terroir Analysis and Its Complexity. *OENO One* **2022**, *56*, 375–388. [CrossRef]
350. Reynard, J.-S.; Zufferey, V.; Nicol, G.-C.; Murisier, F. Vine Water Status as a Parameter of the « terroir » Effect under the Non-Irrigated Conditions of the Vaud Viticultural Area (Switzerland). *OENO One* **2011**, *45*, 139. [CrossRef]
351. Jones, G.V. The Climate Component of Terroir. *Elements* **2018**, *14*, 167–172. [CrossRef]
352. Van Leeuwen, C. Soils and Terroir Expression in Wines. In *Soil and Culture*; Springer Netherlands: Dordrecht, The Netherlands, 2010; pp. 453–465.
353. Bohmrich, R. Terroir: Competing Perspectives on the Roles of Soil, Climate and People. *J. Wine Res.* **1996**, *7*, 33–46. [CrossRef]
354. Certini, G.; Scalenghe, R. The Crucial Interactions between Climate and Soil. *Sci. Total Environ.* **2023**, *856*, 159169. [CrossRef]
355. Coli, M.S.; Rangel, A.G.P.; Souza, E.S.; Oliveira, M.F.; Chiaradia, A.C.N. Chloride Concentration in Red Wines: Influence of Terroir and Grape Type. *Food Sci. Technol.* **2015**, *35*, 95–99. [CrossRef]
356. Zhou-Tsang, A.; Wu, Y.; Henderson, S.W.; Walker, A.R.; Borneman, A.R.; Walker, R.R.; Gilliham, M. Grapevine Salt Tolerance. *Aust. J. Grape Wine Res.* **2021**, *27*, 149–168. [CrossRef]
357. Wheeler, S.; Pickering, G. Effects of Soil Management Techniques on Grape and Wine Quality. In *Fruits: Growth, Nutrition and Quality*; Dris, R., Ed.; Wfl: Helsinki, Finland, 2005; pp. 195–208.
358. Tomasi, D.; Marcuzzo, P.; Nardi, T.; Lonardi, A.; Lovat, L.; Flamini, R.; Mian, G. Influence of Soil Chemical Features on Aromatic Profile of *V. vinifera* Cv. Corvina Grapes and Wines: A Study-Case in Valpolicella Area (Italy) in a Calcareous and Non-Calcareous Soil. *Agriculture* **2022**, *12*, 1980. [CrossRef]
359. Coller, E.; Cestaro, A.; Zanzotti, R.; Bertoldi, D.; Pindo, M.; Larger, S.; Albanese, D.; Mescalchin, E.; Donati, C. Microbiome of Vineyard Soils Is Shaped by Geography and Management. *Microbiome* **2019**, *7*, 140. [CrossRef] [PubMed]
360. Nanetti, E.; Palladino, G.; Scicchitano, D.; Trapella, G.; Cinti, N.; Fabbrini, M.; Cozzi, A.; Accetta, G.; Tassini, C.; Iannaccone, L.; et al. Composition and Biodiversity of Soil and Root-Associated Microbiome in *Vitis vinifera* Cultivar Lambrusco Distinguish the Microbial Terroir of the Lambrusco DOC Protected Designation of Origin Area on a Local Scale. *Front. Microbiol.* **2023**, *14*, 1108036. [CrossRef]
361. Gilbert, J.A.; van der Lelie, D.; Zarraonaindia, I. Microbial Terroir for Wine Grapes. *Proc. Natl. Acad. Sci. USA* **2014**, *111*, 5–6. [CrossRef]
362. Capozzi, V.; Garofalo, C.; Chiriatti, M.A.; Grieco, F.; Spano, G. Microbial Terroir and Food Innovation: The Case of Yeast Biodiversity in Wine. *Microbiol. Res.* **2015**, *181*, 75–83. [CrossRef]
363. Vink, S.N.; Dini-Andreote, F.; Höfle, R.; Kicherer, A.; Salles, J.F. Interactive Effects of Scion and Rootstock Genotypes on the Root Microbiome of Grapevines (*Vitis* Spp. L.). *Appl. Sci.* **2021**, *11*, 1615. [CrossRef]
364. Qi, Y.; Wang, R.; Qin, Q.; Sun, Q. Soil Affected the Variations in Grape and Wine Properties along the Eastern Foot of Helan Mountain, China. *Acta Agric. Scand. B Soil Plant Sci.* **2019**, *69*, 494–502. [CrossRef]
365. Vaudour, E.; Shaw, A. A Worldwide Perspective on Viticultural Zoning. *S. Afr. J. Enol. Vitic.* **2005**, *26*, 106–116. [CrossRef]
366. Villat, J. The Roots of a Better Future: Regenerative Viticulture. Available online: https://portoprotocol.com/the-roots-of-a-better-future-regenerative-viticulture/ (accessed on 11 December 2023).
367. Bezuidenhout, C.N.; van Antwerpen, R.; Berry, S.D. An Application of Principal Component Analyses and Correlation Graphs to Assess Multivariate Soil Health Properties. *Soil Sci.* **2012**, *177*, 498–505. [CrossRef]
368. Smith, J.L.; Halvorson, J.J.; Papendick, R.I. Using Multiple-Variable Indicator Kriging for Evaluating Soil Quality. *Soil Sci. Soc. Am. J.* **1993**, *57*, 743–749. [CrossRef]
369. Andrews, S.S.; Karlen, D.L.; Cambardella, C.A. The Soil Management Assessment Framework. *Soil Sci. Soc. Am. J.* **2004**, *68*, 1945–1962. [CrossRef]
370. Zornoza, R.; Acosta, J.A.; Bastida, F.; Domínguez, S.G.; Toledo, D.M.; Faz, A. Identification of Sensitive Indicators to Assess the Interrelationship between Soil Quality, Management Practices and Human Health. *Soil* **2015**, *1*, 173–185. [CrossRef]
371. Moebius-Clune, B.N.; Moebius-Clune, D.J.; Gugino, B.K.; Idowu, O.J.; Schindelbeck, R.R.; Ristow, A.J.; Van Es, H.M.; Thies, J.E.; Shayler, H.A.; Mcbride, M.B.; et al. *Comprehensive Assessment of Soil Health—The Cornell Framework*; 3.2; Cornell University: Geneva, NY, USA, 2016; ISBN 0967650763.
372. Alkorta, I.; Aizpurua, A.; Riga, P.; Albizu, I.; Amézaga, I.; Garbisu, C. Soil Enzyme Activities as Biological Indicators of Soil Health. *Rev. Environ. Health* **2003**, *18*, 65–73. [CrossRef] [PubMed]
373. Rochfort, S.J.; Ezernieks, V.; Yen, A.L. NMR-Based Metabolomics Using Earthworms as Potential Indicators for Soil Health. *Metabolomics* **2008**, *5*, 95–107. [CrossRef]
374. Chae, Y.; Cui, R.; Woong Kim, S.; An, G.; Jeong, S.-W.; An, Y.-J. Exoenzyme Activity in Contaminated Soils before and after Soil Washing: SS-Glucosidase Activity as a Biological Indicator of Soil Health. *Ecotoxicol. Environ. Saf.* **2017**, *135*, 368–374. [CrossRef]
375. Madejon, E.; Moreno, F.; Murillo, J.; Pelegrin, F. Soil Biochemical Response to Long-Term Conservation Tillage under Semi-Arid Mediterranean Conditions. *Soil Tillage Res.* **2007**, *94*, 346–352. [CrossRef]

376. Gong, W.; Yan, X.; Wang, J.; Hu, T.; Gong, Y. Long-Term Manure and Fertilizer Effects on Soil Organic Matter Fractions and Microbes under a Wheat–Maize Cropping System in Northern China. *Geoderma* **2009**, *149*, 318–324. [CrossRef]
377. Niemeyer, J.C.; Lolata, G.B.; de Carvalho, G.M.; Da Silva, E.M.; Sousa, J.P.; Nogueira, M.A. Microbial Indicators of Soil Health as Tools for Ecological Risk Assessment of a Metal Contaminated Site in Brazil. *Appl. Soil Ecol.* **2012**, *59*, 96–105. [CrossRef]
378. Norkaew, S.; Miles, R.J.; Brandt, D.K.; Anderson, S.H. Effects of 130 Years of Selected Cropping Management Systems on Soil Health Properties for Sanborn Field. *Soil Sci. Soc. Am. J.* **2019**, *83*, 1479–1490. [CrossRef]
379. Caudle, C.; Osmond, D.; Heitman, J.; Ricker, M.; Miller, G.; Wills, S. Comparison of Soil Health Metrics for a Cecil Soil in the North Carolina Piedmont. *Soil Sci. Soc. Am. J.* **2020**, *84*, 978–993. [CrossRef]
380. Ye, R.; Parajuli, B.; Szogi, A.A.; Sigua, G.C.; Ducey, T.F. Soil Health Assessment after 40 Years of Conservation and Conventional Tillage Management in Southeastern Coastal Plain Soils. *Soil Sci. Soc. Am. J.* **2021**, *85*, 1214–1225. [CrossRef]
381. Fine, A.K.; van Es, H.M.; Schindelbeck, R.R. Statistics, Scoring Functions, and Regional Analysis of a Comprehensive Soil Health Database. *Soil Sci. Soc. Am. J.* **2017**, *81*, 589–601. [CrossRef]
382. Liptzin, D.; Norris, C.E.; Cappellazzi, S.B.; Bean, G.M.; Cope, M.; Greub, K.L.H.; Rieke, E.L.; Tracy, P.W.; Aberle, E.; Ashworth, A.; et al. An Evaluation of Carbon Indicators of Soil Health in Long-Term Agricultural Experiments. *Soil Biol. Biochem.* **2022**, *172*, 108708. [CrossRef]
383. Katayama, A.; Bhula, R.; Burns, G.R.; Carazo, E.; Felsot, A.; Hamilton, D.; Harris, C.; Kim, Y.-H.; Kleter, G.; Koedel, W.; et al. Bioavailability of Xenobiotics in the Soil Environment. In *Reviews of Environmental Contamination and Toxicology*; Springer: Berlin/Heidelberg, Germany, 2010; pp. 1–86.
384. Sojka, R.E.; Upchurch, D.R. Reservations Regarding the Soil Quality Concept. *Soil Sci. Soc. Am. J.* **1999**, *63*, 1039–1054. [CrossRef]
385. Chamberlain, L.A.; Aguayo, T.; Zerega, N.J.C.; Dybzinski, R.; Egerton-Warburton, L.M. Rapid Improvement in Soil Health Following the Conversion of Abandoned Farm Fields to Annual or Perennial Agroecosystems. *Front. Sustain. Food Syst.* **2022**, *6*, 1010298. [CrossRef]
386. Barthès, B.; Roose, E. Aggregate Stability as an Indicator of Soil Susceptibility to Runoff and Erosion; Validation at Several Levels. *Catena* **2002**, *47*, 133–149. [CrossRef]
387. Kemper, W.D.; Rosenau, R.C. Aggregate Stability and Size Distribution. Methods of Soil Analysis, Part 1. In *Physical and Mineralogical Methods*; Klute, A., Ed.; SSSA: Madison, WI, USA, 2018; pp. 425–442.
388. Amézketa, E. Soil Aggregate Stability: A Review. *J. Sustain. Agric.* **1999**, *14*, 83–151. [CrossRef]
389. Rieke, E.L.; Bagnall, D.K.; Morgan, C.L.S.; Flynn, K.D.; Howe, J.A.; Greub, K.L.H.; Mac Bean, G.; Cappellazzi, S.B.; Cope, M.; Liptzin, D.; et al. Evaluation of Aggregate Stability Methods for Soil Health. *Geoderma* **2022**, *428*, 116156. [CrossRef]
390. Suleiman, A.A.; Ritchie, J.T. Estimating Saturated Hydraulic Conductivity from Soil Porosity. *Trans. ASAE* **2001**, *44*, 235. [CrossRef]
391. Li, Z.; Wu, P.; Feng, H.; Zhao, X.; Huang, J.; Zhuang, W. Simulated Experiment on Effect of Soil Bulk Density on Soil Infiltration Capacity. *Nongye Gongcheng Xuebao/Trans. Chin. Soc. Agric. Eng.* **2009**, *25*, 40–45.
392. Burns, K.N.; Bokulich, N.A.; Cantu, D.; Greenhut, R.F.; Kluepfel, D.A.; O'Geen, A.T.; Strauss, S.L.; Steenwerth, K.L. Vineyard Soil Bacterial Diversity and Composition Revealed by 16S RRNA Genes: Differentiation by Vineyard Management. *Soil Biol. Biochem.* **2016**, *103*, 337–348. [CrossRef]
393. Viers, J.H.; Williams, J.N.; Nicholas, K.A.; Barbosa, O.; Kotzé, I.; Spence, L.; Webb, L.B.; Merenlender, A.; Reynolds, M. Vinecology: Pairing Wine with Nature. *Conserv. Lett.* **2013**, *6*, 287–299. [CrossRef]
394. Stoleru, V.; Liliana, R.; Mihai, M. Fertilization with Green Manure on Chasselas Dore Grape Vine as an Alternative for Sustainable Agriculture. *J. Food Agric. Environ.* **2011**, *9*, 8.
395. Cesco, S.; Pii, Y.; Borruso, L.; Orzes, G.; Lugli, P.; Mazzetto, F.; Genova, G.; Signorini, M.; Brunetto, G.; Terzano, R.; et al. A Smart and Sustainable Future for Viticulture Is Rooted in Soil: How to Face Cu Toxicity. *Appl. Sci.* **2021**, *11*, 907. [CrossRef]
396. MacKie, K.A.; Müller, T.; Kandeler, E. Remediation of Copper in Vineyards—A Mini Review. *Environ. Pollut.* **2012**, *167*, 16–26. [CrossRef]
397. Komárek, M.; Čadková, E.; Chrastný, V.; Bordas, F.; Bollinger, J.C. Contamination of Vineyard Soils with Fungicides: A Review of Environmental and Toxicological Aspects. *Environ. Int.* **2010**, *36*, 138–151. [CrossRef]
398. Casson, A.; Ortuani, B.; Giovenzana, V.; Brancadoro, L.; Corsi, S.; Gharsallah, O.; Guidetti, R.; Facchi, A. A Multidisciplinary Approach to Assess Environmental and Economic Impact of Conventional and Innovative Vineyards Management Systems in Northern Italy. *Sci. Total Environ.* **2022**, *838*, 156181. [CrossRef]
399. Jones, G.V.; White, M.A.; Cooper, O.R.; Storchmann, K. Climate Change and Global Wine Quality. *Clim. Change* **2005**, *73*, 319–343. [CrossRef]
400. Webb, L.B.; Whetton, P.H.; Barlow, E.W.R. Climate Change and Winegrape Quality in Australia. *Clim. Res.* **2008**, *36*, 99–111. [CrossRef]
401. Hall, A.; Mathews, A.J.; Holzapfel, B.P. Potential Effect of Atmospheric Warming on Grapevine Phenology and Post-Harvest Heat Accumulation across a Range of Climates. *Int. J. Biometeorol.* **2016**, *60*, 1405–1422. [CrossRef]
402. Lehmann, J.; Bossio, D.A.; Kögel-Knabner, I.; Rillig, M.C. The Concept and Future Prospects of Soil Health. *Nat. Rev. Earth Environ.* **2020**, *1*, 544–553. [CrossRef]

403. Karlen, D.L.; Veum, K.S.; Sudduth, K.A.; Obrycki, J.F.; Nunes, M.R. Soil Health Assessment: Past Accomplishments, Current Activities, and Future Opportunities. *Soil Tillage Res.* **2019**, *195*, 104365. [CrossRef]
404. Montalvo-Falcón, J.V.; Sánchez-García, E.; Marco-Lajara, B.; Martínez-Falcó, J. Sustainability Research in the Wine Industry: A Bibliometric Approach. *Agronomy* **2023**, *13*, 871. [CrossRef]
405. Villat, J. Down to Earth: Identifying and Promoting Regenerative Viticulture Practices for Soil and Human Health. Ph.D. Thesis, Harvard University, Cambridge, MA, USA, 2021.

MDPI AG
Grosspeteranlage 5
4052 Basel
Switzerland
Tel.: +41 61 683 77 34

Horticulturae Editorial Office
E-mail: horticulturae@mdpi.com
www.mdpi.com/journal/horticulturae

www.ingramcontent.com/pod-product-compliance
Lightning Source LLC
Chambersburg PA
CBHW061458210926
51850CB00008B/1757